Statistical Techniques for Project Control

Industrial Innovation Series

Series Editor

Adedeji B. Badiru

Department of Systems and Engineering Management
Air Force Institute of Technology (AFIT) – Dayton, Ohio

PUBLISHED TITLES

Computational Economic Analysis for Engineering and Industry,
Adedeji B. Badiru & Olufemi A. Omitaomu

Conveyors: Applications, Selection, and Integration, *Patrick M. McGuire*

Global Engineering: Design, Decision Making, and Communication,
Carlos Acosta, V. Jorge Leon, Charles Conrad, and Cesar O. Malave

Handbook of Industrial Engineering Equations, Formulas, and Calculations,
Adedeji B. Badiru & Olufemi A. Omitaomu

Handbook of Industrial and Systems Engineering, *Adedeji B. Badiru*

Handbook of Military Industrial Engineering, *Adedeji B.Badiru & Marlin U. Thomas*

Industrial Control Systems: Mathematical and Statistical Models and Techniques
Adedeji B. Badiru, Oye Ibidapo-Obe, & Babatunde J. Ayeni

Industrial Project Management: Concepts, Tools, and Techniques
Adedeji B. Badiru, Abidemi Badiru, & Adetokunboh Badiru

Inventory Management: Non-Classical Views, *Mohamad Y. Jaber*

Kansei Engineering - 2 volume set

- Innovations of Kansei Engineering, *Mitsuo Nagamachi & Anitawati Mohd Lokman*
- Kansei/Affective Engineering, *Mitsuo Nagamachi*

Knowledge Discovery from Sensor Data,
Auroop R. Ganguly, João Gama, Olufemi A. Omitaomu, Mohamed Medhat Gaber, &s Ranga Raju Vatsavai

Learning Curves: Theory, Models, and Applications, *Mohamad Y. Jaber*

Modern Construction: Lean Project Delivery and Integrated Practices, *Lincoln Harding Forbes & Syed M. Ahmed*

Moving from Project Management to Project Leadership: A Practical Guide to Leading Groups,
R. Camper Bull

Project Management: Systems, Principles, and Applications, *Adedeji B. Badiru*

Quality Management in Construction Projects, *Abdul Razzak Rumane*

Social Responsibility: Failure Mode Effects and Analysis, *Holly Alison Duckworth & Rosemond Ann Moore*

Statistical Techniques for Project Control, *Adedeji B. Badiru & Tina Agustiady*

STEP Project Management: Guide for Science, Technology, and Engineering Projects,
Adedeji B. Badiru

Systems Thinking: Coping with 21st Century Problems, *John Turner Boardman & Brian J. Sauser*

Techonomics: The Theory of Industrial Evolution, *H. Lee Martin*

Triple C Model of Project Management: Communication, Cooperation, Coordination,
Adedeji B. Badiru

FORTHCOMING TITLES

Essentials of Engineering Leadership and Innovation, *Pamela McCauley-Bush & Lesia L. Crumpton-Young*

Project Management: Systems, Principles, and Applications, *Adedeji B. Badiru*

Sustainability: Utilizing Lean Six Sigma Techniques, *Tina Agustiady & Adedeji Badiru*

Technology Transfer and Commercialization of Environmental Remediation Technology,
Mark N. Goltz

Statistical Techniques for Project Control

Adedeji B. Badiru
Tina Agustiady

CRC Press
Taylor & Francis Group
Boca Raton London New York

CRC Press is an imprint of the
Taylor & Francis Group, an **informa** business

CRC Press
Taylor & Francis Group
6000 Broken Sound Parkway NW, Suite 300
Boca Raton, FL 33487-2742

First issued in paperback 2020

ISBN-13: 978-1-4200-8317-0 (hbk)
ISBN-13: 978-0-367-78376-1 (pbk)

Visit the Taylor & Francis Web site at
http://www.taylorandfrancis.com

and the CRC Press Web site at
http://www.crcpress.com

Contents

Preface

Project management is the process of managing, allocating, and timing resources to achieve a given goal in an efficient and expeditious manner. The objectives that constitute the specified goal may be in terms of time, cost, or technical results. A project can be simple or complex. In each case, proven project management processes must be followed. In all cases of project management implementation, control must be exercised to assure that project objectives are achieved.

This book fills the void that exists in the application of statistical techniques to project control. Although the book is designed as a practical guide for project management professionals, it will also appeal to students, researchers, and instructors. It is applicable to several programs including industrial engineering, systems engineering, construction engineering, operations research, engineering management, business management, general management, business administration, mechanical engineering, civil engineering, production management, industrial management, and operations management.

Adedeji B. Badiru
Tina Agustiady

Authors

Adedeji Badiru is the head of Systems and Engineering Management at the Air Force Institute of Technology. He was previously department head of Industrial and Information Engineering at the University of Tennessee in Knoxville and formerly professor of industrial engineering and dean of University College at the University of Oklahoma. He is a registered professional engineer. Dr. Badiru is a fellow of the Institute of Industrial Engineers and a fellow of the Nigerian Academy of Engineering. He holds a BS in industrial engineering, an MS in mathematics, an MS in industrial engineering from Tennessee Technological University, and a PhD in industrial engineering from the University of Central Florida. His areas of interest include mathematical modeling, project modeling and analysis, economic analysis, systems engineering, and productivity analysis and improvement. He is the author of several books and technical journal articles. He is the editor of the *Handbook of Industrial and Systems Engineering* and co-editor of the *Handbook of Military Industrial Engineering*. He is a member of several professional associations including the Institute of Industrial Engineers (IIE), the Institute of Electrical and Electronics Engineers (IEEE), the Society of Manufacturing Engineers (SME), the Institute for Operations Research and Management Science (INFORMS), the American Society for Engineering Education (ASEE), the American Society for Engineering Management (ASEM), the New York Academy of Science (NYAS), and the Project Management Institute (PMI). He has served as a consultant to several organizations around the world including Russia, Mexico, Taiwan, Nigeria, and Ghana. Dr. Badiru has conducted customized training workshops for numerous organizations including Sony, AT&T, Seagate Technology, U.S. Air Force, Oklahoma Gas and Electric, Oklahoma Asphalt Pavement Association, Hitachi, Nigeria National Petroleum Corporation, and ExxonMobil. He has won awards for his teaching, research, publications, administration, and professional accomplishments.

Tina Agustiady is a certified Six Sigma Master Black Belt. Her responsibilities include conducting training and improvement programs using Six Sigma for the baking industry. Prior to joining Dawn Foods, she worked at Nestlé Prepared Foods as a Six Sigma product and process design specialist responsible for driving optimum fit of product design and current manufacturing process capability and reducing total manufacturing cost and consumer complaints. Agustiady received a BS in industrial and manufacturing systems engineering from Ohio University. She earned her Black Belt and Master Black Belt certifications at Clemson University. Agustiady is also a member of the Institute of Industrial Engineers (IIE) Lean Division board of directors and an editor for the *International Journal of Six Sigma and Competitive Advantage*. She is also an instructor for Lean and Six Sigma for IIE and Six Sigma Digest.

1

Fundamentals of Project Management

"No man can efficiently direct work about which he knows nothing."

Col. Thurman H. Bane
Dayton, Ohio, 1919

As the opening quote above professes, we must know all aspects of our project before we can expect to direct and manage it. Knowledge of a project can come through direct observation, experiential knowledge base, or statistical inference. The basic philosophy of project control is to not leave anything to chance. Preemptive and proactive actions provide the best bets for project control. It is instructive to adopt a relevant Latin phrase when approaching project control: *Spera optimum para pessimum,* which literally translates to "hope for the best, prepare for the worst," and serves as a direction to a single person such as a project team member. To direct the phrase to a group such as a project team, the proper wording is *Sperate optimum parate pessimum.*

Project management, which relies on knowledge thus acquired, is the pursuit of organizational goals within the constraints of time, cost, and quality expectations. This can be summarized by a few basic questions:

- What needs to be done?
- What can be done?
- What will be done?
- Who will do it?
- When will it be done?
- Where will it be done?
- How will it be done?

The factors of time, cost, and quality are synchronized to answer the above questions and must be managed and controlled within the constraints of the iron triangle depicted in Figure 1.1. In this case, quality represents the composite collection of project requirements. In a situation where precise optimization is not possible, trade-offs among the three factors of success are required. A rigid (iron) triangle of constraints encases the project. Everything must be accomplished within the boundaries of time, cost, and quality. If better quality is expected, a compromise along the axes of time and cost must be executed, thereby altering the shape of the triangle.

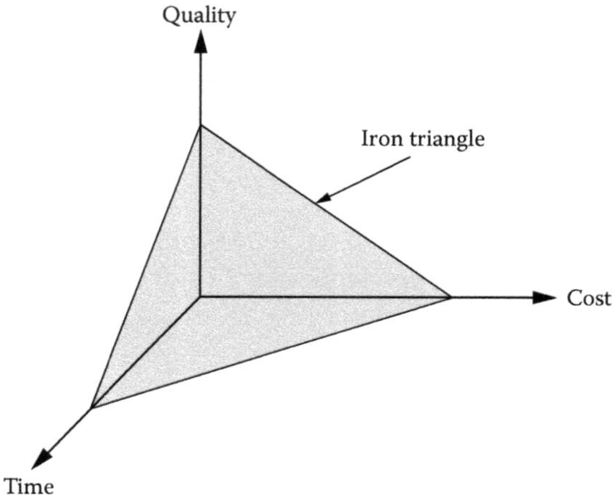

FIGURE 1.1
Project constraints of cost, time, and quality.

The trade-off relationships are not linear and must be visualized in a multidimensional context. This is better articulated by a three-dimensional (3-D) view of the systems constraints as shown in Figure 1.2. Scope requirements determine the project boundary and trade-offs must be achieved within that boundary. If we label the eight corners of the box as (a), (b), (c), …, (h), we can iteratively assess the best operating point for a project. For example, we can address the following two operational questions:

1. From the point of view of the project sponsor, which corner is the most desired operating point in terms of combination of requirements, time, and cost?
2. From the point of view of the project executor, which corner is the most desired operating point in terms of combination of requirements, time, and cost?

Note that all the corners represent extreme operating points. We notice that point (e) is the do-nothing state, where there are no requirements, no time allocations, and no cost incurrence. This cannot be the desired operating state of any organization that seeks to remain productive. Point (a) represents an extreme case of meeting all requirements with no investment of time or cost allocation. This is an unrealistic extreme in any practical environment. It represents a case of getting something for nothing. Yet, it is the most desired operating point for the project sponsor.

By comparison, point (c) provides the maximum possible for requirements, cost, and time. In other words, the highest levels of requirements can be met if the maximum possible time is allowed and the highest possible budget is

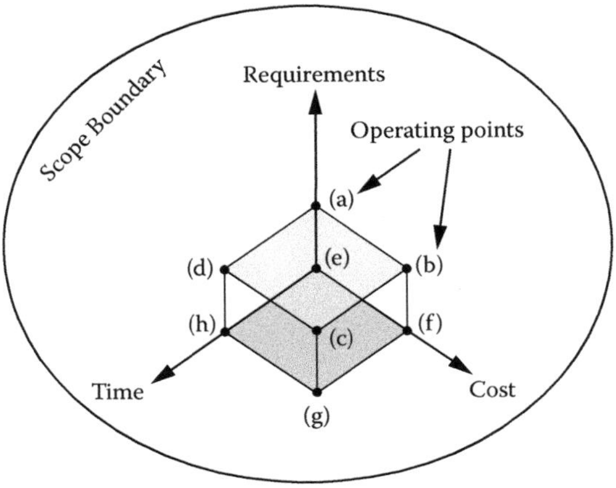

FIGURE 1.2
Systems constraints of cost, time, and quality within iron triangle.

allocated. This is an unrealistic expectation in any resource-conscious organization. You cannot get everything you ask for to execute a project. Yet, it is the most desired operating point for the project executor.

Considering the two extreme points of (a) and (c), it is obvious that a project must be executed within some compromise region inside the scope boundary. Figure 1.3 shows a possible compromise surface with peaks and valleys representing give-and-take trade-off points within the constrained box. With proper control strategies, the project team can guide the project in the appropriate directions. The challenge is to devise an analytical modeling technique to guide decision-making over the compromise region. If we could collect sets of data over several repetitions of identical projects, we could model a decision surface to guide future executions of similar projects.

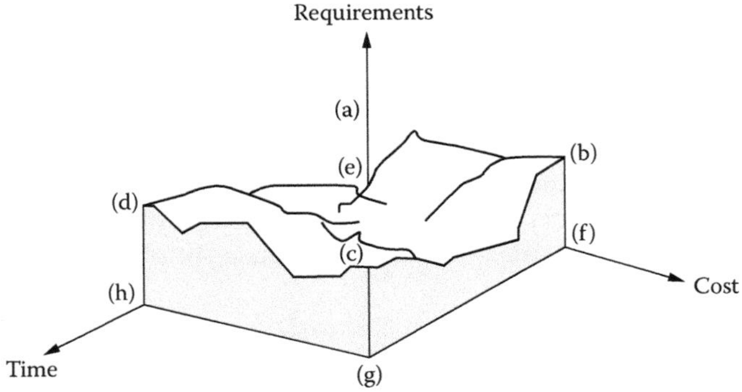

FIGURE 1.3
Compromise surface for cost, time, and requirements trade-off.

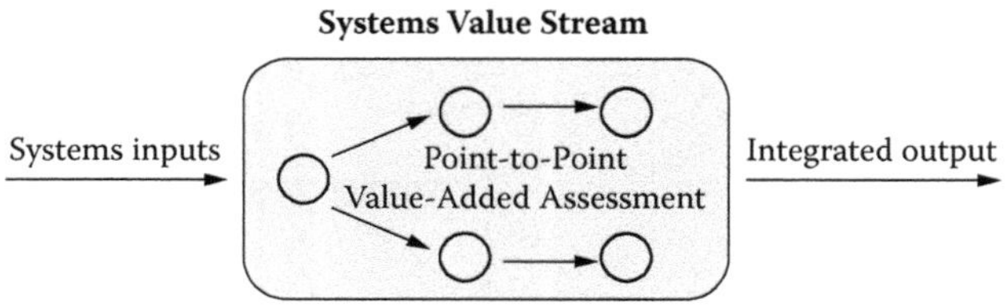

FIGURE 1.4
Systems value stream structure.

Such typical repetitions of an identical project are most readily apparent in construction projects, for example, residential home developments.

Organizations often inadvertently fall into unstructured management "blobs" because they are simple, involve low costs, and consume little time until a problem develops. A desired alternative is to model a project system using a systems value-stream structure as shown in Figure 1.4. It follows a proactive and problem-preempting approach to execute projects and presents the following advantages:

- Problem diagnosis is easier.
- Accountability is higher.
- Operating waste is minimized.
- Conflict resolution is faster.
- Value points are traceable.

Why Projects Fail

In spite of concerted efforts to maintain control of a project, many projects still fail. If no project ever fails, it would be a perfect world. But we all know that there is no perfection in human existence. In order to maintain a better control of a project, albeit no perfect control, we must understand the common reasons that projects fail. With this knowledge, we can better preempt project problems. Below are some common causes of project failure:

- Lack of communication
- Lack of cooperation
- Lack of coordination
- Diminished interest
- Diminished resources

- Change of objectives
  - Change of stakeholders
  - Change of priority
  - Change of perspective
- Change of ownership
  - Change of scope
  - Budget cut
  - Shift in milestone
- New technology
- New personnel
- Lack of training
- New unaligned capability
- Market shift
- Change of management philosophy
- Manager moves on
- Depletion of project goodwill
- Lack of user involvement
- Three strikes followed by an out (too many mistakes)

Management by Project

Project management continues to grow as an effective means of managing functions in any organization. Project management should be an enterprise-wide systems-based endeavor. Enterprise-wide project management is the application of project management techniques and practices across the full scope of an enterprise. This concept is also referred to as management by project (MBP), an approach that employs project management techniques in various functions in an organization. MBP recommends pursuing endeavors as project-oriented activities. It is an effective way to conduct any business activity. It represents a disciplined approach that defines each work assignment as a project. Under MBP, every undertaking is viewed as a project that must be managed just like a traditional project. The characteristics required of each project so defined are:

- Identified scope and goal
- Desired completion time
- Availability of resources
- Defined performance measure
- Measurement scale for review of work

An MBP approach to operations helps identify unique entities within functional requirements. This identification helps determine where functions overlap and how they are interrelated, thus paving the way for better planning, scheduling, and control. Enterprise-wide project management facilitates a unified view of organizational goals and provides a way for project teams to use information generated by other departments to carry out their functions.

The use of project management continues to grow rapidly. The need to develop effective management tools increases with the increasing complexity of new technologies and processes.

The life cycle of a new product to be introduced into a competitive market is a good example of a complex process that must be managed with integrative project management approaches. The product will encounter management functions as it goes from one stage to the next. Project management will be needed throughout the design and production stages and will be needed in developing marketing, transportation, and delivery strategies for the product. When the product finally reaches customers, project management will be needed to integrate its use with those of other products within customer organizations.

The need for a project management approach is established by the fact that a project will always tend to increase in size even if its scope is narrowing. The following four literary laws are applicable to any project environment:

Parkinson's law—Work expands to fill the available time or space.

Peter's principle—People rise to the level of their incompetence.

Murphy's law—Whatever can go wrong will.

Badiru's rule—The grass is always greener where you most need it to be dead.

An integrated systems project management approach can help diminish the adverse impacts of these laws through good project planning, organizing, scheduling, and control.

Integrated Project Implementation

Project management tools can be classified into three major categories:

1. Qualitative tools are the managerial aids that facilitate the interpersonal and organizational processes required for project management.
2. Quantitative tools are analytical techniques that aid in the computational aspects of project management.

3. Computer tools consist of software and hardware that simplify the processes of planning, organizing, scheduling, and controlling a project. Software tools can help in both the qualitative and quantitative analyses needed for project management.

While many books deal with management principles, optimization models, and computer tools, few guidelines exist for the integration of these three areas for project management purposes. In this book, we integrate the three areas as a comprehensive guide to project management. We introduce the *Triad Approach* to improve the effectiveness of project management with respect to schedule, cost, and performance constraints within the context of systems modeling. Figure 1.5 illustrates the concept that considers both the management of the project and also the management of all the functions that support the project. It is helpful to have a quantitative model, but it is far more effective to be able to apply the model to real-world problems in a practical form.

A system approach helps increase the intersection of the three categories of project management tools and hence improve overall management effectiveness. Crisis should not be the instigator for the use of project management techniques. Project management approaches should be used up front to prevent avoidable problems rather than fight them when they develop. What is worth doing is worth doing well, right from the beginning.

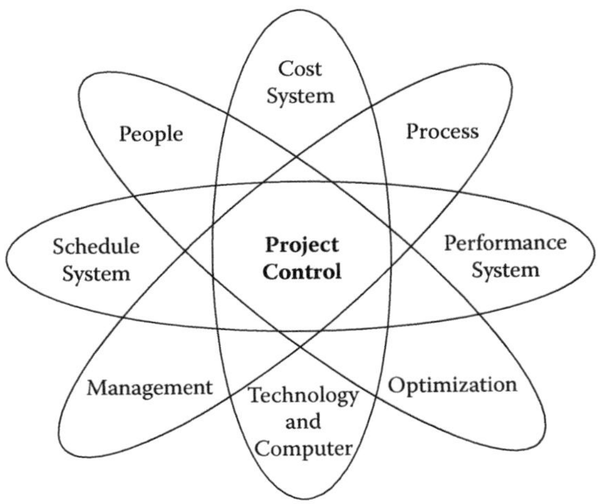

FIGURE 1.5
Systems view encompassing all aspects of project.

Critical Factors for Project Success

The premise of this book is that the critical factors for systems success revolve around people and the personal commitment and dedication of each person. No matter how good a technology is or how enhanced a process may be, the people involved ultimately determine success. This makes it imperative to resolve people issues first in the overall systems approach to project management. Many organizations recognize this, but only a few have been able to actualize the ideals of managing people productively. Execution of operational strategies requires forthrightness, openness, and commitment to get things done. Lip service and arm waving are not sufficient. Tangible programs that cater to the needs of people must be implemented. It is essential to provide incentives, encouragement, and empowerment for people to be self-actuating in determining how best to accomplish their job functions.

The first critical factor for systems success is total system management (hardware, software, and people) based on:

- Operational effectiveness
- Operational efficiency
- System suitability
- System resilience
- System affordability
- System supportability
- System life cycle cost
- System performance
- System schedule
- System cost

Systems engineering tools, techniques, and processes are essential for project life cycle management to make goals possible in the context of SMART (specific, measurable, achievable, realistic, timed) principles, which are represented as follows:

- Specific: Pursue specific and explicit outputs.
- Measurable: Design of outputs that can be tracked, measured, and assessed.
- Achievable: Make outputs achievable and align them with organizational goals.
- Realistic: Pursue only goals that are realistic and result-oriented.
- Timed: Make outputs timed to facilitate accountability.

Early Systems Engineering

Systems engineering provides the technical foundation for executing a project successfully. A systems approach is particularly essential in the early stages of the project in order to avoid having to re-engineer a project at the end of its life cycle. Early systems engineering makes it possible to proactively assess the feasibility of meeting user needs, adaptability of new technology, and integration of solutions into regular operations.

DODAF Systems Architecture for Project Management

The US military has been a major force in establishing systems engineering platforms for executing projects. Through the DODAF (Department of Defense Architecture Framework), US military projects are executed under a consistent platform of an "architecture" perspective borrowed from the conventional physical infrastructure architectural design processes. DODAF is used to standardize the format for architecture descriptions. It seeks to provide a mechanism for operating more efficiently while attending to multiple requirements spread over multiple and diverse geographic locations. One approach of DODAF adapts traditional architecture to *capability architecture* as a result of the widespread belief that scores of defense systems are either redundant or do not meet operational needs. As a result, many recent acquisition reform efforts have been aimed at pursuing interoperable and cost-effective *joint* military capabilities.

Traditional architects integrate structure and function with environment. Their end products, the blueprints, merge various stakeholders' visions and requirements into an acceptable product. They provide sheets or *views* that are relevant to the homeowner, the plumber, the electrician, the framer, the painter, the residents, and perhaps even the neighbors. By contract, the application of systems architecting in the military is not centered on a single location such as a house; the focus is on interoperable weapon systems and diverse spectra of warfare, thus requiring a lot of inter-component coordination. Only a systems view can provide this level of comprehensive appreciation of capability, interdependency, and symbiosis.

Systems architecture supports logical interface of capabilities, operations planning, resource requirements, tool development, portfolio management, goal formulation, acquisition, information management, and project phase-out. Some specific requirements for applying systems architecting to program management within the military include:

- The Joint Capabilities Integration and Development System (JCIDS) requires that each *capabilities document* contain an annex with a standard DOD-formatted architecture. Users and program offices partner to provide the architecture descriptions.
- The Defense Acquisition System (DAS) requires architecture to develop systems and manage interoperability of components.
- Systems that communicate must have information support plans (ISPs); each plan must be accompanied by a complete integrated architecture.
- DOD and Congress require systems architecture for defense business information systems that cost $1 million or more.

Just as a traditional architect provides specific *views* to different subcontractors involved in the construction of a house, DODAF prescribes views for various stakeholders involved in specific capabilities or requirements. DODAF organized twenty-six total views into three categories:

- Operational views (OVs)
- Systems and services views (SVs)
- Technical standards views (TVs)

The views are combinations of pictures, diagrams, and spreadsheets maintained in an electronic database. The OVs communicate mission-level information and document operational requirements from a user view. The SVs communicate design-level information for use by designers and maintainers. Finally, the TVs document the information technology standards (construction codes) that have been developed for networking compatibility (net-centricity).

DODAF architecture descriptions are the blueprints for linking key inputs and capabilities for planners, designers, and acquirers. For everyone involved in a large and complex project, a consistent architecture framework can guide the systems-of-systems engineering process. DODAF-integrated architectures provide insight into complex operational relationships, interoperability requirements, and systems-related structures.

Project Management Body of Knowledge (PMBOK®)

The general body of knowledge (PMBOK®) for project management is published and disseminated by the project management institute (PMI). The body of knowledge comprises specific knowledge areas, which are organized into the following broad areas:

1. Project **Integration** Management
2. Project **Scope** Management
3. Project **Time** Management
4. Project **Cost** Management
5. Project **Quality** Management
6. Project **Human** Resource Management
7. Project **Communications** Management
8. Project **Risk** Management
9. Project **Procurement** Management

These segments of the body of knowledge of project management cover the range of functions associated with all projects, particularly complex ones. Multinational projects particularly pose unique challenges pertaining to reliable power supply, efficient communication systems, credible government support, dependable procurement processes, consistent availability of technology, progressive industrial climate, trustworthy risk mitigation infrastructure, regular supply of skilled labor, uniform focus on quality of work, global consciousness, hassle-free bureaucratic processes, coherent safety and security system, steady law and order, unflinching focus on customer satisfaction, and fair labor relations. Assessing and resolving concerns about these issues in a step-by-step fashion will create a foundation of success for a large project. While no system can be perfect and satisfactory in all aspects, a tolerable trade-off of the factors is essential for project success.

Components of Knowledge Areas

The key components of each element of the body of knowledge are summarized below:

- Integration
    - Integrative project charter
    - Project scope statement
    - Project management plan
    - Project execution management
    - Change control
- Scope management
    - Focused scope statements
    - Cost and benefits analysis
    - Project constraints
    - Work breakdown

    - Responsibility breakdown
    - Change control
- Time management
    - Schedule planning and control
    - PERT and Gantt charts
    - Critical path method
    - Network models
    - Resource loading
    - Reporting
- Cost management
    - Financial analysis
    - Cost estimating
    - Forecasting
    - Cost control
    - Cost reporting
- Quality management
    - Total quality management
    - Quality assurance
    - Quality control
    - Cost of quality
    - Quality conformance
- Human resources management
    - Leadership skill development
    - Team building
    - Motivation
    - Conflict management
    - Compensation
    - Organizational structures
- Communications
    - Communication matrix
    - Communication vehicles
    - Listening and presenting skills
    - Communication barriers and facilitators
- Risk management
    - Risk identification
    - Risk analysis

    - Risk mitigation
    - Contingency planning
- Procurement and subcontracts
    - Material selection
    - Vendor pre-qualification
    - Contract types
    - Contract risk assessment
    - Contract negotiation
    - Contract change orders

Step-by-Step and Component-by-Component Implementation

The efficacy of the systems approach is based on step-by-step and component-by-component implementation of the project management process. The major knowledge areas of project management are administered in a structured outline covering six basic clusters:

1. Initiating
2. Planning
3. Executing
4. Monitoring
5. Controlling
6. Closing

The implementation clusters represent five process groups that are followed throughout a project life cycle. Each cluster consists of several functions and operational steps. When the clusters are overlaid on the nine knowledge areas, we obtain a two-dimensional matrix that spans forty-four major process steps. Table 1.1 shows an overlay of the project management knowledge areas and implementation clusters. The monitoring and controlling clusters are usually administered as a single process group (monitoring and controlling). In some cases, it may be helpful to separate them to highlight the essential attributes of each cluster of functions over the project lifecycle. In practice, the processes and clusters do overlap. Thus, there is no crisp demarcation of when and where one process ends and where another one begins over the project lifecycle. In general, project lifecycle defines the following: (1) resources that will be needed in each phase of the project life cycle; and (2) specific work to be accomplished in each phase of the life cycle

Figure 1.6 shows the major phases of a project life cycle from the conceptual phase through close-out. It should be noted that a project life cycle is

TABLE 1.1

Overlay of Project Management Areas and Implementation Clusters

Knowledge Areas	Initiating	Planning	Executing	Monitoring and Controlling	Closing
Project integration	Develop project charter Develop preliminary project scope	Develop project management plan	Direct and manage project execution	Monitor and control project work Integrated change control	
Scope		Scope planning Scope definition Create WBS		Scope verification Scope control	
Time		Activity definition Activity sequencing Activity resource estimating Activity duration estimating Schedule development		Schedule control	
Cost		Cost estimating Cost budgeting		Cost control	
Quality		Quality planning	Perform quality assurance	Perform quality control	
Human resources		Human resource planning	Acquire project team Develop project team	Manage project team	
Communication		Communication planning	Information distribution	Performance reporting Manage stakeholders	
Risk		Risk management planning Risk identification Qualitative risk analysis Quantitative risk analysis Risk response planning		Risk monitoring and control	
Procurement		Plan purchases and acquisitions Plan contracting	Request seller responses Select sellers	Contract administration	Contract closure

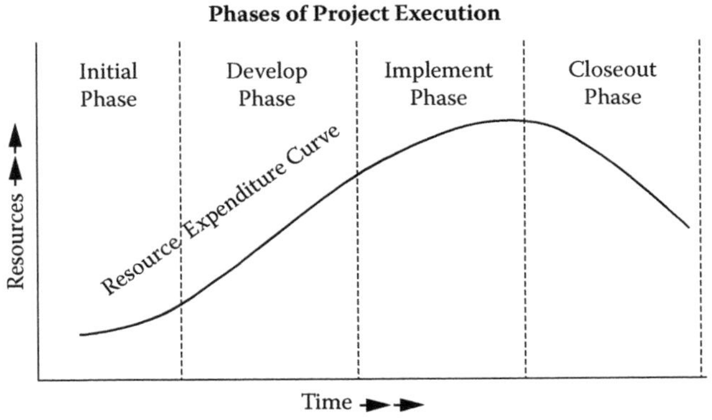

FIGURE 1.6
Project execution phases for systems implementation.

distinguished from a product life cycle. A project life cycle does not explicitly address operational issues; a product life cycle is concerned about operational issues starting from product delivery to the end of its useful life. Note that for science, technology, and engineering procurement (STEP) projects, the shape of the life cycle curve may be expedited due to the rapid developments that often occur in science, technology, and engineering (STE) activities. For example, for a high technology project, the entire life cycle may be shortened by a very rapid initial phase, even though the conceptualization stage may be very long. In a typical project life cycle, include the following:

1. Cost and staffing requirements are lowest at the beginning of the project and ramp up during the initial development stages.
2. The probability of successfully completing a project is lowest at the beginning and highest at the end because many unknowns (risks and uncertainties) exist at the beginning. As a project nears its end, the possibilities of risks and uncertainties diminish.
3. The risks to the project organization (project owner) are lowest at the beginning and highest at the end. This is because not much investment has gone into the project at the beginning, whereas much has been committed by the end of the project. There is a higher sunk cost manifested at the end of the project.
4. The ability of the stakeholders to influence the final project outcome (cost, quality, and schedule) is highest at the beginning and decreases progressively toward the end. This is intuitive because influence is best exerted at the beginning of an endeavor.
5. The value of scope changes decreases over time during a project life cycle while the cost of scope changes increases over time. The suggestion is to decide and finalize scope as early as possible and if changes are necessary, implement them as early as possible.

Project Systems Structure

Problem Identification

Problem identification is the stage at which a need for a proposed project is identified, defined, and justified. A project may involve the development of new products, implementation of new processes, or improvement of existing facilities.

Project Definition

Project definition is the phase at which the purpose of the project is clarified. A *mission statement* is the major output of this stage. For example, a prevailing low level of productivity may indicate a need for a new manufacturing technology. In general, the definition should specify how project management may be used to avoid missed deadlines, poor scheduling, inadequate resource allocation, lack of coordination, poor quality, and conflicting priorities.

Project Planning

A plan outlines the series of actions needed to accomplish a goal. Project planning determines how to initiate a project and execute its objectives. Planning may require a simple statement of a project goal or may be a detailed account of procedures to be followed to complete a project. Project planning is discussed in detail in Chapter 2. Planning consists of:

- Objectives
- Project definition
- Team organization
- Performance criteria (time, cost, quality)

Project Organization

Project organization specifies how to integrate the functions of the personnel involved in a project. Organizing is usually done concurrently with project planning. Directing (guiding and supervising project personnel) is an important aspect of project organization and is a crucial aspect of the management function. Directing requires skillful managers who can interact with subordinates effectively via good communication and motivational techniques. A good project manager will facilitate project success by directing his or her staff through proper task assignments toward the project goal.

Workers perform better when they have clearly defined expectations. They need to know how their job functions contribute to the overall goals of the project. Workers should be given some flexibility for self-direction in performing their functions. Individual worker needs and limitations should be recognized by the manager. Directing a project requires motivating, supervising, and delegating skills.

Resource Allocation

Project goals and objectives are accomplished by allocating resources to functional requirements. Resources consist of money, people, equipment, tools, facilities, information, and skills, and are usually in short supply. The people needed for a particular task may be committed to other ongoing projects. A crucial piece of equipment may be under the control of another team. Chapter 5 addresses resource allocation in detail.

Project Scheduling

Timeliness is the essence of project management. Scheduling is often the major focus in project management. The main purpose of scheduling is to allocate resources so that the overall project objectives are achieved within a reasonable time span. Project objectives are generally conflicting in nature. For example, minimization of the completion time and minimization of cost are conflicting objectives. That is, one objective is improved at the expense of worsening the other objective. Therefore, project scheduling is a multiple-objective decision-making problem.

In general, scheduling involves the assignment of time periods to specific tasks on the work schedule. Resource availability, time limitations, urgency level, required performance level, precedence requirements, work priorities, technical constraints, and other factors complicate the scheduling process. Thus, the assignment of a time slot to a task does not necessarily ensure that the task will be performed satisfactorily in accordance with the schedule. Consequently, careful control must be developed and maintained throughout project scheduling.

Project Tracking and Reporting

This phase involves checking whether results conform to project plans and performance specifications. Tracking and reporting are prerequisites for project control. A properly organized report of status will help identify deficiencies in the progress of the project and help pinpoint corrective actions.

Project Control

Project control requires that appropriate actions be taken to correct unacceptable deviations from expected performance. Control is actuated through measurement, evaluation, and corrective action. Measurement is the process of measuring the relationship between planned and actual performance of project objectives. The variables to be measured, the measurement scales, and the measuring approaches should be clearly specified during the planning stage. Corrective actions may involve rescheduling, reallocation of resources, or expedition of task performance. Project control is discussed in detail in Chapter 6 and involves:

- Tracking and reporting
- Measurement and evaluation
- Corrective action (plan revision, rescheduling, updating)

Project Termination

Termination is the final stage. The phase-out of a project is as important as its initiation and should be implemented expeditiously. A project should not be allowed to drag on after the expected completion time. A terminal activity should be defined during the planning phase. An example of a terminal activity may be the submission of a final report, the power-on of new equipment, or the signing of a release order. The conclusion of such an activity should be viewed as the completion of the project. Arrangements may be made for follow-up activities to improve or extend the outcome. These follow-up or spinoff projects should be managed as new projects but with proper input and output relationships within the sequence of the main project.

Project Systems Implementation Outline

While this book is aligned with the main tenets of PMI's PMBOK, it follows the traditional project management textbook framework encompassing the broad sequence of categories below:

Planning → Organizing → Scheduling → Control → Termination

An outline of the functions to be carried out during a project should be made during the planning stage. A model for such an outline is presented below. It may be necessary to rearrange the components of the outline to fit the specific needs of a project.

Planning

I. Specify project background
  A. Define current situation and process
    i. Understand process
    ii. Identify important variables
    iii. Quantify variables
  B. Identify areas for improvement
    i. List and discuss the areas
  C. Study potential strategies for solution
II. Define unique terminologies relevant to project
  A. Industry-specific terminologies
  B. Company-specific terminologies
  C. Project-specific terminologies
III. Define project goal and objectives
  A. Write mission statement
  B. Solicit inputs and ideas from personnel
IV. Establish performance standards
  A. Schedule
  B. Performance
  C. Cost
V. Conduct formal project feasibility study
  A. Determine impact on cost
  B. Determine impact on organization
  C. Determine project deliverables
VI. Secure management support

Organizing

I. Identify project management team
  A. Specify project organization structure
    i. Matrix structure
    ii. Formal and informal structures
    iii. Justify structure
  B. Specify departments involved and key personnel
    i. Purchasing
    ii. Materials management
    iii. Engineering, design, manufacturing, and so on

C. Define project management responsibilities
   i. Select project manager
   ii. Write project charter
   iii. Establish project policies and procedures

II. Implement triple C model
   A. Communication
      i. Determine communication interfaces
      ii. Develop communication matrix
   B. Cooperation
      i. Outline cooperation requirements, policies, and procedures
   C. Coordination
      i. Develop work breakdown structure
      ii. Assign task responsibilities
      iii. Develop responsibility chart

Scheduling (Resource Allocation)

I. Develop master schedule
   A. Estimate task duration
   B. Identify task precedence requirements
      i. Technical precedence
      ii. Resource-imposed precedence
      iii. Procedural precedence
   C. Use analytical models
      i. CPM (critical path method)
      ii. PERT (program evaluation and review technique)
      iii. Gantt chart
      iv. Optimization models

Control (Tracking, Reporting, and Correction)

I. Establish guidelines for tracking, reporting, and control
   A. Define data requirements
      i. Data categories
      ii. Data characterization
      iii. Measurement scales
   B. Develop data documentation

i. Data update requirements
ii. Data quality control
iii. Establish data security measures

II. Categorize control points
A. Schedule audit
i. Activity network and Gantt charts
ii. Milestones
iii. Delivery schedule
B. Performance audit
i. Employee performance
ii. Product quality
C. Cost audit
i. Cost containment measures
ii. Percent completion versus budget depletion

III. Identify implementation process
A. Comparison with targeted schedules
B. Corrective course of action
i. Rescheduling
ii. Reallocation of resources

Termination (Close- or Phase-Out)

I. Conduct performance review
II. Develop strategy for follow-up projects
III. Arrange for personnel retention, release, and reassignment

Documentation

I. Document project outcome
II. Submit final report
III. Archive report for future reference

Value of Lean Times

In biblical times, the "seven lean years" were followed by a period of plenty, affluence, and prosperity. Facing a lean period in project management creates value by determining how to eliminate or reduce the operational waste inherent in many human-governed processes. It is a fact that having to "make do" with limited resources creates opportunities for resourcefulness and innovation, requiring an integrated systems view of what is available and what can be leveraged. The lean principles now embraced by business, industry, and government have been around for a long time. It is just now that we are being forced to implement lean practices due to the escalating shortages of resources. It is unrealistic to expect that problems enrooted in different parts of an organization can be solved by a single-point attack. Rather, a systematic probing of all the nooks and corners of the problem must be assessed and tackled in an integrated manner. Like the biblical Joseph whose life was on the line as he interpreted dreams for the Egyptian pharaoh, decision makers cannot afford to misinterpret systems' warning signs when managing large and complex projects.

Contrary to the contention in some technocratic circles that budget cuts stifle innovation, reduction of resources often forces more creativity in identifying wastes and leveraging opportunities that lie fallow in the nooks and crannies of an organization. This is not an issue of *wanting* more for less; rather it means *doing* more with less. Only through a systems viewpoint can new opportunities to innovate be spotted. Necessity does indeed spur invention.

Project Decision Analysis

Systems decision analysis facilitates a proper consideration of the essential elements of decisions in a project systems environment. These essential elements include the problem statement, information, performance measure, decision model, and implementation of the decision. The recommended steps are enumerated below.

Step 1: Problem statement—A problem involves choosing between competing and probably conflicting alternatives. The components of problem solving in project management include:

- Describing the problem (goals, performance measures).
- Defining a model to represent the problem.
- Solving the model.
- Testing the solution.
- Implementing and maintaining the solution.

Problem definition is crucial. In many cases, *symptoms* of a problem are recognized more readily than its *cause* and *location*. Even after the problem is accurately identified and defined, a benefit and cost analysis may be needed to determine whether the cost of solving it is justified.

Step 2: Data and information requirements—Information is the driving force for making project decisions. It clarifies the relative states of past, present, and future events. The collection, storage, retrieval, organization, and processing of raw data are important steps for generating information. Without data, there can be no information. Without good information, it is impossible to make a valid decision. To generate information, it is essential to:

- Ensure that an effective data collection procedure is followed.
- Determine the type and appropriate amount of data to collect.
- Evaluate data collected for its information potential.
- Evaluate the cost of collecting required data.

For example, suppose a manager is presented with a fact that *"sales for the last quarter are 10,000 units."* This constitutes ordinary data that can be used in many ways to make decisions, depending on the manager's value system. An analyst, however, can ensure the proper use of the data by transforming it into information, such as, *"sales of 10,000 units for last quarter are within x percent of the targeted value."* This type of information is more useful for decision making.

Step 3: Performance measure—A performance measure for assessing the competing alternatives should be specified. The decision maker assigns perceived values to the available alternatives. Setting measures of performance is crucial for defining and selecting alternatives. Some performance measures commonly used in project management are project cost, completion time, resource usage, and stability of the workforce.

Step 4: Decision model—This model serves as a basis for the analysis and synthesis of information and is the mechanism used to compare competing alternatives. To be effective, a decision model must be based on a systematic and logical framework for guiding project decisions. A decision model can be a verbal, graphical, or mathematical representation of the ideas in the decision-making process. A project decision model should encompass:

- A simplified representation of the actual situation
- An explanation and prediction of the actual situation
- Validity and appropriateness
- Applicability to similar problems

The formulation of a decision model involves three essential components:

- Abstraction: Determining relevant factors
- Construction: Combining the factors into a logical model
- Validation: Assuring that the model adequately represents the problem

The five basic types of decision models for project management are described next.

Descriptive models are directed at describing a decision scenario and identifying the associated problem. For example, a project analyst may use a critical path method (CPM) network model to identify bottleneck tasks in a project.

Prescriptive models furnish procedural guidelines for implementing actions. The triple C approach (Badiru 2008) prescribes the procedures for achieving communication, cooperation, and coordination in a project environment.

Predictive models are used to predict future events in a problem environment. They are typically based on historical data. For example, a regression model based on past data may be used to predict future productivity gains associated with expected levels of resource allocation. *Simulation models* can be used when uncertainties exist in the task durations or resource requirements.

"Satisficing" models provide trade-off strategies for achieving a satisfactory solution to a problem within certain givenconstraints. Satisficing, a concatenation of the words satisfy and suffice, is a decision-making strategy that attempts to meet criteria for adequacy rather than identify an optimal solution. It is used where achieving an optimum is not practicable. Goal programming and other multi-criteria techniques provide good satisficing solutions. These models are helpful where time limitations, resource shortages, and performance requirements constrain project implementation.

Optimization models are designed to find the best available solution to a problem subject to certain constraints. For example, a linear programming model can be used to determine the optimal product mix in a production environment.

In many situations, two or more of the models may be involved in the solution of a problem. For example, a descriptive model might provide insights into the nature of the problem; an optimization model might provide the optimal set of actions to take in solving the problem; a satisficing model might temper the optimal solution with reality; a prescriptive model might suggest the procedures for implementing the selected solution; and a predictive model may foresee an outcome.

Step 5: Making decision—Using the available data, information, and the decision model, the decision maker will determine the real-world actions needed to solve the stated problem. A sensitivity analysis may be useful for determining what changes in parameter values might cause a change in the decision.

Step 6: Implementing decision—A decision represents the selection of an alternative that satisfies the objective cited in the problem statement. A good decision is useless until implemented. An important aspect of a decision is specifying how it is to be implemented. Selling a project and a decision to management requires a well-organized, persuasive presentation. The way a decision is presented can directly influence whether it is adopted. The presentation of a decision should cover at minimum an executive summary, technical and managerial aspects of the decision, resources required to implement the decision, costs, time frame for implementation of the decision, the time frame for implementing the decision, and the risks associated with the decision.

Systems Group Decision-Making Models

Systems decisions are often complex, diffuse, distributed, and poorly understood. No one person has all the information to make all decisions accurately. As a result, crucial decisions are made by groups. Some organizations use outside consultants with appropriate expertise to make recommendations for important decisions. Other organizations set up internal consulting groups without outside experts. Decisions can be made through linear responsibility, in which case one person makes the final decision based on inputs from others. Decisions can also be made through shared responsibility, in which case a group of people share the responsibility for making joint decisions. The major advantages of group decision making are listed below:

1. Facilitation of a systems view of the problem environment.
2. Ability to share experience, knowledge, and resources. Many heads are better than one. A group has better collective ability to solve a given decision problem.
3. Increased credibility. Decisions made by a group often carry more weight in an organization.
4. Improved morale. Personnel can be positively influenced when many people participate in the decision-making process.
5. Better rationalization. The opportunity to observe views of others may improve an individual's reasoning processes.
6. Ability to accumulate more knowledge and facts from diverse sources.
7. Access to broader perspectives spanning different problem scenarios.
8. Ability to generate and consider alternatives from different perspectives.
9. Broader based involvement, leading to a higher likelihood of support.
10. Possibility for group leverage for networking, communication, and political clout.

In spite of the desirable advantages, group decision making also presents flaws, for example:

1. Difficulty in arriving at a decision.
2. Slow operating timeframe.
3. Possibility for conflicting views and objectives.
4. Reluctance of some individuals to implement a decision.
5. Potential for power struggle and conflicts.
6. Loss of productive employee time.

7. Too much compromise may lead to less-than-optimal group output.
8. Risk of domination of group by one individual.
9. Over-reliance on group process may impede agility of management to make decisions quickly.
10. Risk of delay due to repeated and iterative group meetings.

Brainstorming

Brainstorming is a way to generate new ideas. A decision group comes together to discuss alternate ways of solving a problem. The members may be from different departments, may have different backgrounds and training, and may not even know one another. The diversity of the participants helps create a stimulating environment for generating many ideas from different viewpoints. The technique encourages free outward expression of new ideas, no matter how far-fetched they may appear. No criticism of any new idea is permitted during brainstorming. A major concern is that extroverts may take control of the discussions. For this reason, an experienced and respected individual should manage the discussions. The group leader establishes the procedure for proposing ideas, keeps the discussions in line with the group mission, discourages disruptive statements, and encourages the participation of all members.

After the group runs out of ideas, open discussions are held to weed out the unsuitable ones. Even the rejected ideas may stimulate the generation of other ideas that may eventually lead to other new ideas. Guidelines for improving brainstorming sessions are presented as follows:

- Focusing on a specific decision problem
- Keeping ideas relevant to the intended decision
- Being receptive to all new ideas
- Evaluating the ideas on a comparative basis after exhausting new ideas
- Maintaining an atmosphere conducive to cooperative discussions
- Keeping records of ideas generated

Delphi Method

The traditional approach to group decision making is to obtain the opinions of experienced participants through open discussions. An attempt is made to reach a consensus among the participants. However, open group discussions are often biased because of subtle intimidation from dominant individuals. Even when the threat of a dominant individual is not present, opinions may still be swayed by group pressure. This is called the "bandwagon effect."

The Delphi method attempts to overcome these difficulties by requiring individuals to present their opinions anonymously through an intermediary.

The method differs from the other interactive group methods because it eliminates face-to-face confrontations. It was originally developed for forecasting applications, but has been modified in various ways for application to different types of decision-making methods. Delphi can be useful for project management decisions. It is particularly effective when decisions must be based on a broad set of factors. The Delphi method is normally implemented as follows:

1. Problem definition. A decision problem that is considered significant is identified and clearly described.
2. Group selection. An appropriate group of experts or experienced individuals is formed to address the specific decision problem. Internal and external experts may be involved in the Delphi process. A leading individual is appointed to serve as the administrator. The group may operate via email or meet in a room. In both cases, all opinions are expressed anonymously on paper. If the group meets in the same room, care should be taken to provide enough privacy so that members are assured that no one accidentally or deliberately observed their responses.
3. Initial opinion poll. The technique is initiated by describing the problem to be addressed in unambiguous terms. The group members are requested to submit a list of major areas of concern in their specialty areas as they relate to the decision problem.
4. Questionnaire design and distribution. Questionnaires are prepared to address the areas of concern related to the decision problem. The written responses are collected and organized by the administrator who then aggregates the responses in a statistical format. For example, the average, mode, and median of the responses may be computed and the analysis is distributed to the decision group. Each member can then see how his or her responses compare with the anonymous views of the other members.
5. Iterative balloting. Additional questionnaires based on the previous responses are passed to the members. The members again submit their responses; they may choose to alter or not alter their previous responses.
6. Silent discussions and consensus. The iterative balloting may involve anonymous written discussions of why some responses are correct or incorrect. The process continues until a consensus is reached. A consensus may be declared after five or six iterations of the balloting or when a specified percentage (e.g., 80%) of the group agrees on the ballots. If a consensus on a particular point cannot be declared, it may be displayed to the whole group with a note that it does not represent a consensus.

In addition to its use in technological forecasting, the Delphi method has been widely used in other general decision making. Its major characteristics of anonymity of responses, statistical summary of responses, and controlled procedure make it a reliable mechanism for obtaining numeric data from subjective opinions. Its major limitations are:

1. Its effectiveness may be limited in cultures where strict hierarchy, seniority, and age influence decision-making processes.
2. Some experts may not readily accept the contributions of non-experts.
3. Since opinions are expressed anonymously, some members may take the liberty of making ludicrous statements. However, if the group composition is carefully reviewed, this problem may be avoided.

Nominal Group Technique

This technique is a silent version of brainstorming and is useful for reaching consensus. Rather than asking people to state their ideas aloud, the team leader asks each member to write a specific number of ideas. The ideas are then compiled as a single list on a chalkboard for the whole group to see. The group then discusses the ideas and eliminates some until a final decision choice is made. The nominal group technique is easier to control. Unlike brainstorming that may lead to shouting matches, the nominal group technique permits members to present their views silently. It also allows introverted members to contribute to the decision without the pressure of having to speak to the group.

In all of the group decision-making techniques, an important aspect that can enhance and expedite the process is to require that members review all pertinent data before going to a group meeting. This will ensure that the decision process is not impeded by trivial preliminary discussions. Some disadvantages of group decision making are

1. Peer pressure in a group situation may influence a member's opinion or discussions.
2. In a large group, some members may not participate effectively in the discussions.
3. A member's relative reputation in the group may influence how well his or her opinion is rated.
4. A member with a dominant personality may overwhelm other members in the discussions.
5. The limited time available to the group may create a time pressure that forces some members to present their opinions without fully evaluating the ramifications of the available data.
6. It is often difficult to get all members of a decision group together at the same time.

Despite these disadvantages, group decision making presents many advantages that may nullify the shortcomings. The advantages as presented earlier will exert varying levels of effects in different organizations. The triple C principle presented in Chapter 2 may also be used to improve the success of decision teams. Teamwork in making decisions can be enhanced by adhering to the following guidelines:

1. Get a willing group of people together.
2. Set an achievable goal for the group.
3. Determine the limitations of the group.
4. Develop a set of guiding rules.
5. Create an atmosphere conducive to group synergism.
6. Identify the questions to be addressed in advance.
7. Plan to address only one topic per meeting.

For major decisions and long-term group activities, arrange team training that allows the group to learn the decision rules and responsibilities together. The steps for the technique are:

1. Generate ideas in writing—silently.
2. Record ideas without discussion.
3. Conduct group discussion for clarification of meaning, not argument.
4. Vote to establish the priority or rank of each item.
5. Discuss vote.
6. Cast final vote.

Interviews, Surveys, and Questionnaires

Interviews, surveys, and questionnaires are important information gathering techniques. They also foster cooperative working relationships and encourage direct participation and inputs into project decision-making processes. These tools provide an opportunity for employees at the lower levels of an organization to contribute ideas and inputs for decision making. The greater the number of people involved in the interviews, surveys, and questionnaires, the more valid the final decision. The following guidelines are useful for using these participatory techniques to collect information for project decisions:

1. Collect and organize background information and supporting documents about the subject of the interview, survey, or questionnaire.
2. Outline the items to be covered and list the major questions to be asked.
3. Use a suitable medium of interaction and communication: telephone, fax, electronic mail, personal contact, observation, meeting, poster, or memo.

4. Inform respondents about the purpose of the interview, survey, or questionnaire, and indicate how long participation will take.
5. Use open-ended questions that stimulate ideas from the respondents.
6. Minimize the use of yes or no questions.
7. Encourage expressive statements that indicate respondents' views.
8. Use the who, what, where, when, why, and how approach to elicit specific information.
9. Thank the respondents for their participation.
10. Let the respondents know the outcome of the exercise.

Multivoting

This is a sequence of votes that leads to a group decision. It can be used to assign priorities to items on a list. After brainstorming generates a long list of items, multivoting can reduce a long list to a few items. The steps are listed below:

1. Take a first vote. Each person is allowed only one vote per item.
2. Circle the items receiving higher numbers of votes than other items.
3. Take a second vote. Each person votes for a number of items equal to one-half the total number of items circled in step 2. Only one vote per item is permitted.
4. Repeat steps 2 and 3 until the list is reduced to three to five items, depending on the needs of the group. Multivoting down to a single item is not recommended.
5. Perform further analysis of the items selected in step 4, if needed.

Hierarchy of Project Control

The traditional concepts of systems analysis are applicable to the project process. The definitions of a project system and its components are presented next from a point-to-point control perspective.

System—A group of interrelated elements organized for the purpose of achieving a common goal. The elements work synergistically to generate a unified output that is greater than the sum of the individual outputs.

Program—A very large and prolonged undertaking. Programs often span several years and are usually associated with specific systems, for example, a space exploration program within a national defense system.

Project—A time-phased effort smaller in scope and duration than a program. Programs are sometimes viewed as sets of projects. Government projects are often called *programs* because of their broad and comprehensive nature. Industry tends to use the *project* term because of the short-term and focused nature of most efforts.

Task—A functional element of a project. A project consists of a sequence of tasks that contribute to an overall goal.

Activity—A single element of a project. Activities are generally smaller in scope than tasks. In a detailed analysis of a project, an activity may be viewed as the smallest, almost indivisible work element. For example, we can regard a manufacturing plant as a system. A plant-wide endeavor to improve productivity can be viewed as a program. The installation of a flexible manufacturing system is a project within a productivity improvement program. The process of identifying and selecting equipment vendors is a task; the placement of an order with a preferred vendor is an activity. The systems structure of a project is illustrated in Figure 1.7.

The emergence of systems development has greatly affected project management in recent years. A system can be defined as a collection of interrelated elements brought together to achieve a specified objective. In a management context, a system is intended to develop and manage operational procedures and facilitate effective decision making. Some common characteristics of a system include:

1. Interaction with the environment
2. Objective
3. Self-regulation
4. Self-adjustment

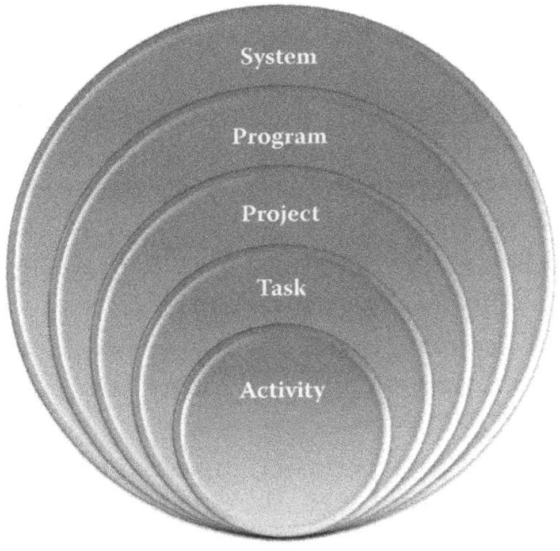

FIGURE 1.7
Hierarchy of project system.

Representative components of a project system are the organizational subsystem, planning subsystem, scheduling subsystem, information management subsystem, control subsystem, and project delivery subsystem. The primary responsibilities of project analysts involve ensuring the proper flow of information throughout the project system. The classical approach to the decision process follows rigid lines of organizational charts. By contrast, the systems approach considers all the interactions necessary among the various elements of an organization in the decision process.

The various elements (or subsystems) of the organization act simultaneously in a separate but interrelated fashion to achieve a common goal. This synergism helps to expedite the decision process and enhance the effectiveness of decisions. The supporting commitments from other subsystems of the organization serve to counterbalance weaknesses. Thus, the overall effectiveness of a system is greater than the sum of the individual results from the subsystems.

The increasing complexity of organizations and projects makes the systems approach essential in today's management environment. As the number of complex projects increases, the need for project management professionals who can function as systems integrators also increases. Project management techniques can be applied to implement a system as shown in these guidelines:

1. *Systems definition*. Define the system and associated problems using keywords that signify the importance of the problem to the organization. Locate experts who are willing to contribute to the effort. Prepare and announce the development plan.
2. *Personnel assignment*. The project group and the respective tasks should be announced, a qualified project manager should be appointed, and a solid line of command should be established and enforced.
3. *Project initiation*. Arrange an organizational meeting during which a general approach to the problem should be discussed. Prepare a specific development plan and arrange for the installation of needed hardware and tools.
4. *System prototype*. Develop a prototype system, test it, and learn more about the problem from the test results.
5. *Full system development*. Expand the prototype to a full system, evaluate the user interface structure, and incorporate user training facilities and documentation.
6. *System verification*. Get experts and potential users involved, ensure that the system performs as designed, and debug the system as needed.
7. *System validation*. Ensure that the system yields expected outputs. Validate it by evaluating performance level, such as percentages of

successes in a number of trials, measuring the level of deviation from expected outputs, and measuring the effectiveness of the system output in solving the problem.

8. *System integration.* Implement the full system as planned, ensure that it can coexist with systems already in operation, and arrange for technology transfer to other projects.
9. *System maintenance.* Arrange for continuing maintenance of the system. Update solution procedures as new information becomes available. Retain responsibility for system performance or delegate to well-trained and authorized personnel.
10. *Documentation.* Prepare full documentation of the system, prepare a user's guide, and appoint a user consultant. Include pictures and graphics in user's guides.

Systems integration permits sharing of resources. Physical equipment, concepts, information, and skills may be shared. Systems integration is now a major concern of many organizations. Even organizations that traditionally compete and typically shun cooperative efforts are beginning to appreciate the value of integrating their operations. For these reasons, systems integration has emerged as a major business focus. It may involve the physical integration of technical components, objective integration of operations, conceptual integration of management processes, or a combination.

Systems integration involves the linking of components to form subsystems and the linking of subsystems to form composite systems within and/or across departments. It facilitates the coordination of technical and managerial efforts to enhance organizational functions, reduce cost, save energy, improve productivity, and increase resource utilization. Systems integration emphasizes the identification and coordination of the interface requirements among the components of an integrated system. The components and subsystems operate synergistically to optimize total system performance. Systems integration ensures that all performance goals are satisfied with minimum expenditures of time and resources. Integration can be in several forms:

1. *Dual-use integration.* This is the use of a single component by separate subsystems to reduce the initial and operating costs during a project life cycle.
2. *Dynamic resource integration.* This involves integrating resources that flow between two normally separate subsystems of a project so that they flow from one subsystem to or through the other subsystem to minimize total resource requirements.
2. *Restructuring of functions.* This requires restructuring of functions and reintegration of subsystems to optimize costs when a new subsystem is introduced into a project environment.

Systems integration is particularly important when introducing new technology into an existing system. It involves coordinating new operations to coexist with existing operations. It may require the adjustment of functions to permit the sharing of resources, development of new policies to accommodate product integration, or realignment of managerial responsibilities. It can affect both hardware and software components. The following questions are relevant for systems integration.

What are the unique characteristics of each component in the integrated system?

How do the characteristics complement one another?

What physical interfaces exist among the components?

What data and information interfaces exist among the components?

What ideological differences exist among the components?

What are the data flow requirements for the components?

Do similar integrated systems operate elsewhere?

What are the reporting requirements for an integrated system?

Do any hierarchical restrictions affect the operations of the components of the integrated system?

What internal and external factors are expected to influence the integrated system?

How can the performance of the integrated system be measured?

What benefit and cost documentations are required?

What is the cost of designing and implementing the integrated system?

What are the relative priorities assigned to each component of the integrated system?

What are the strengths of the integrated system?

What are the weaknesses of the integrated system?

What resources are needed to keep the integrated system operating satisfactorily?

Which section of the organization will have primary responsibility for the operation of the integrated system?

What are the quality specifications and requirements for the integrated system?

The integrated approach to project management recommended in this book is represented by the flowchart in Figure 1.8. Figure 1.9 illustrates the application of matrix organization structure in the pursuit of a function goal, where horizontal and vertical lines of responsibilities share knowledge, resources, and personnel.

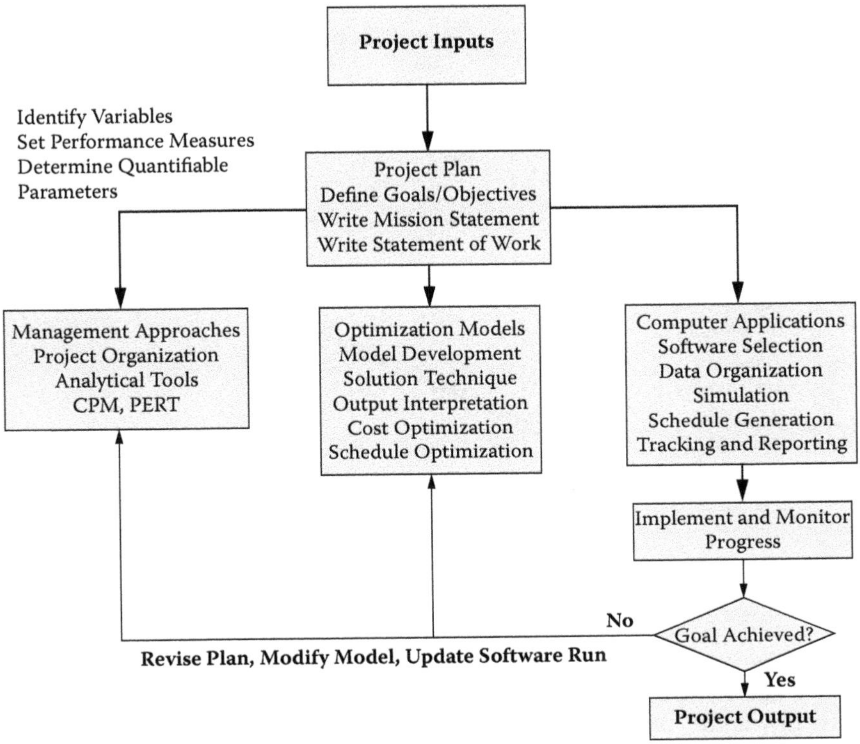

FIGURE 1.8
Flowchart of integrated project management.

The process starts with a managerial analysis of the project effort. Goals and objectives are defined, a mission statement is written, and a statement of work is developed; then the traditional project management approaches such as the selection of an organization structure are employed. Conventional analytical tools including the critical path method (CPM) and the precedence diagramming method (PERT) are then mobilized. Optimization models are used as appropriate. Some of the parameters to be optimized are cost, resource allocation, and schedule length. It should be understood that not all project parameters are amenable to optimization. The use of commercial project management software should start only after the managerial functions are completed. Some project management software includes built-in capabilities for planning and optimization.

A frequent mistake in project management is the rush to use project management software without first completing the planning and analytical studies required. Project management software should be used as a management tool, just as word processing is used as a writing tool. It is not effective to use word processing without first organizing the material to be written. Project

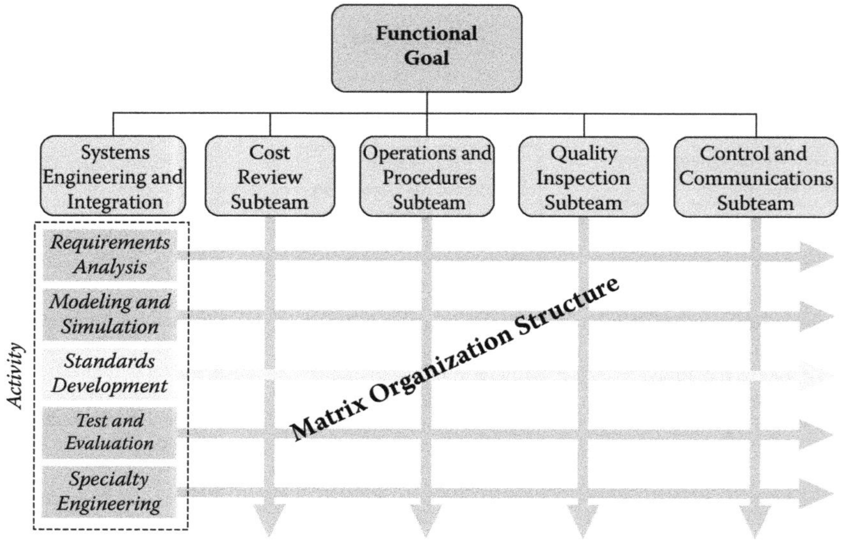

FIGURE 1.9
Example of matrix organization structure.

management is much more than the relevant software. If project management is carried out in accordance with the integration approach presented in the flowchart, the odds of success will increase. Of course, the flowchart should not be rigid. Flows and interfaces among the blocks in the flowchart may need to be altered or modified based on specific project needs.

2

Statistics for Project Control

> "Not everything that counts can be counted; and not everything that can be counted counts."
>
> **Albert Einstein**

The opening quote above typifies the difficulty and ambiguity associated with project control. Most attributes of a project that must be controlled are not within the realm of quantitative measurements because human elements affect many aspects of project execution. Achieving a workable level of project control requires a mix of both quantitative and qualitative approaches. If a project attribute is measurable, we will use quantitative metrics to assess performance. Often we must temper quantitative analysis with qualitative human judgment that makes project control nebulous but also offers opportunities to innovate and be creative in achieving control.

Modeling Project Environment

Conventional management takes a static view of a project environment, but rarely does any aspect of project management remain static. Dynamic changes arise during every project. The larger the project, the more dynamic variability can be expected. Statistical analysis permits adaptive decision making in a project environment. In practical real-world projects, decision factors interact dynamically and probabilistically. The occurrence level of each factor is uncertain and may depend on the occurrence levels of other factors. Statistical approaches can model and inform the best decision strategies for dynamic project scenarios.

Project modeling approaches take two forms. A *descriptive model* is used to describe a project environment characteristically so that a decision maker can better understand it and consequently make decisions that adequately align with the realities of the project. A *prescriptive model* for making decisions about a project environment is intended to serve as an adequate representation. Decisions made with such iconic models are believed to be applicable to actual project environments.

Quantitative modeling is used in both approaches and involves the application of mathematical and empirical knowledge to a problem environment.

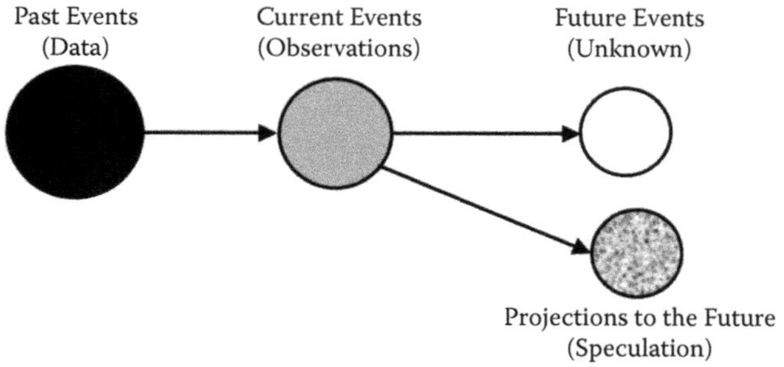

FIGURE 2.1
Conventional static modeling of project environment.

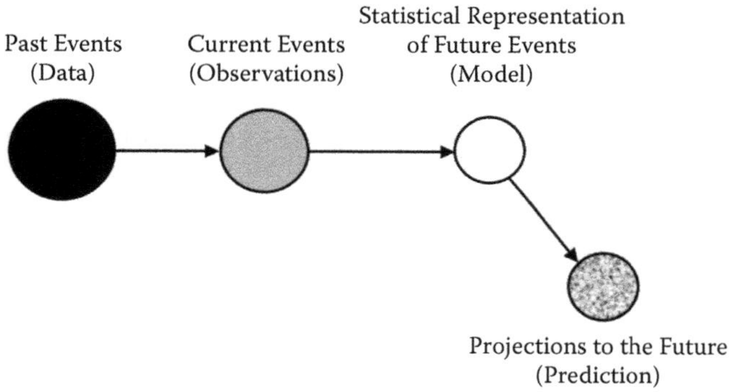

FIGURE 2.2
Dynamic statistical modeling of project environment.

A mathematical model consists of a set of mathematical formulae based on the validity of certain fundamental natural laws and various hypotheses relating to physical processes related to a problem. Projections are based on mathematical and statistical representations. Figure 2.1 illustrates the conventional static inference process. Past events (data) from a project are merged with current observations to project subjective future expectations for the project. Figure 2.2 shows a statistically based forecast of a project environment. Mathematical and statistical representations of a project environment are used to develop more robust forecasts.

Use of Statistics in Project Control

Clearly, it is impossible to guarantee that all aspects of a project are 100% correct. Control must be exercised to ensure that give-and-take processes keep a project on track. Control often requires the use of statistical inference, deduction, and conjecture. In an imperfect world, statistics guide decisions and performance. In the final analysis, statistics plays an important role in project analysis and control. Simple calculations such as average, variance, and sum can play big roles in setting control parameters for a project. The normal distribution is an important tool in statistical analysis of projects. The importance of applying statistical concepts to make good project decisions and improve project outcomes cannot be overemphasized. Statistical project control is a philosophy of understanding a project so that improvement actions can be pursued. Project improvement efforts are predicated on the following principles:

1. All project elements are handled via a system of interconnected processes.
2. Variation is present in all processes.
3. Understanding and reducing variations are key elements for project success.

Normal Distribution

Inferential statistics concerns measuring samples and then using the sample values to predict the values for a population. The measurements of samples are called *statistics*; measurements of population are called *parameters*. Some sample statistics are good predictors of their corresponding population parameters. Other sample statistics are not able to predict population parameters.

It is important to note that sample size will always be smaller than the population. The population size N cannot be predicted from the sample size n. The sample mode is not usually the same as the population mode. The sample median is also not necessarily a good predictor of the population median. The sample mean for a representative random sample from a population is a good *point estimate* of the population mean μ. The sample standard deviation s predicts the population standard deviation σ. The shape of the distribution of the sample is a good predictor of the shape of the distribution of a population. The ability to predict the shape of a population distribution by predicting the shape of the distribution of a good random sample is important. In many project situations, we will be predicting the population mean μ. Instead of predicting a single value of a project, an analyst will predict a range in which the population mean will likely be found.

Let us consider a project involving inter-city travel. Suppose for the purpose of estimating travel time (task time), we ask, "How long does it take to drive from Dayton, Ohio to Cleveland, Ohio?" A typical answer is "3 to 4 hours." A range estimate is needed because it is impossible to predict travel time with certainty because of unforeseen factors (e.g., traffic volume, average speed, weather and road conditions, traffic emergencies, accidents, road construction, etc.). Thus, a range estimate serves as an answer to the question. By so doing, we are implicitly using the concept of sampling from a population distribution.

Determining the appropriate range in which a population mean will be found depends on the shape of the distribution. A bimodal distribution is likely to need a larger range than a symmetrical bell-shaped distribution to ensure capture of the population mean. As a result of this realization, we must understand the shapes of distributions generated by different processes. The most important shape in statistics is the purely random distribution—the distribution noted when a coin is tossed numerous times and the number of times "heads" or "tails" appear is recorded. Fortunately and naturally, most random processes follow a bell-shaped curve. The normal distribution is the most ubiquitous bell-shaped distribution. In the experiment of tracking drive time, if we plot a histogram of the observed number of hours after many repeats of the driving experiment, we may generate a plot similar to Figure 2.3. Obviously, 0 hours of drive time means no travel at all. Thus, the observed number for 0 hours is 0. At the other end, driving 7 hours for a trip that normally takes 3 to 4 hours is not going to happen often. Thus, the 7-hour observations are few but not 0. Notice that the plot reveals s proportions instead of integers indicating the number of times a particular driving time (hours) was observed. The proportions eventually translate to probability values associated with each driving time.

If the number of driving experiments increases over a lot more observed drive times, the histogram plot and the fitted curve would become smoother and increasingly symmetrical. This is shown in Figure 2.4.

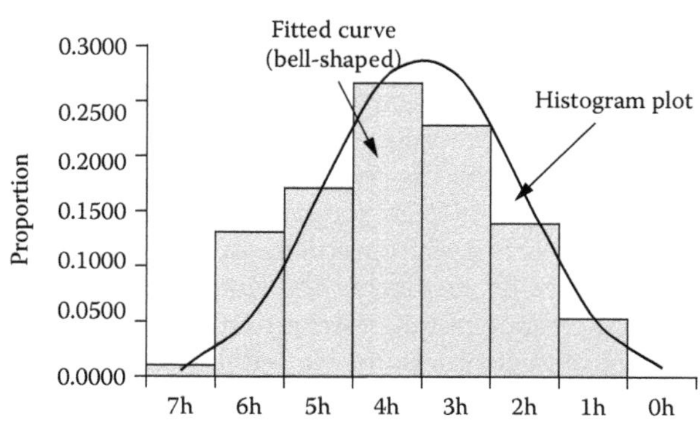

FIGURE 2.3
Histogram plot for drive time experiment.

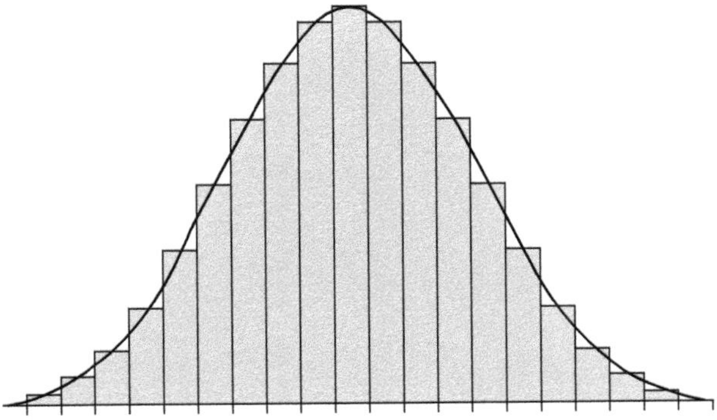

FIGURE 2.4
Bell-shaped curve fitted to larger histogram data.

If the number of trials is allowed to go to infinity, the resulting smooth curve looks like the curve in Figure 2.4 and can be described by a function. Statistical mathematicians would say that as the number of trials approaches infinity, the discrete distribution approaches a continuous distribution described by the function below.

$$f_x(y) = \frac{1}{\sqrt{2\pi\sigma}} e^{-(y-\mu)^2/2\sigma^2}$$

In the above function, σ is the population standard deviation, μ is the population mean, e is the base, and π is pi. The name of this function is the normal distribution or normal curve. If the function is graphed for a mean $\mu = 0$ and a population standard deviation $\sigma = 1$, then the graph in Figure 2.5 results. Figure 2.6 shows a plot of areas under the normal curve. The shape of the

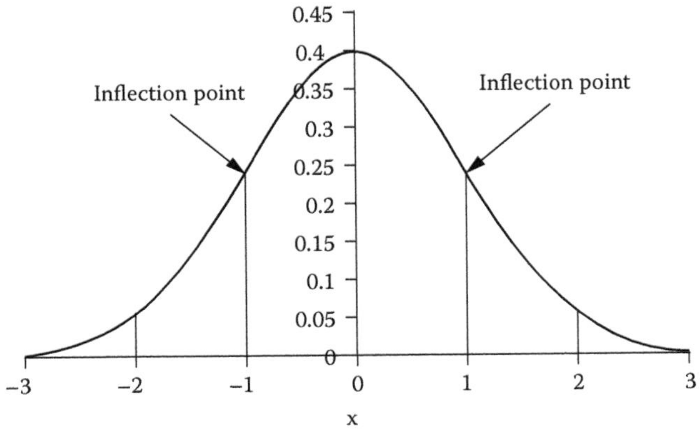

FIGURE 2.5
Plot of normal curve.

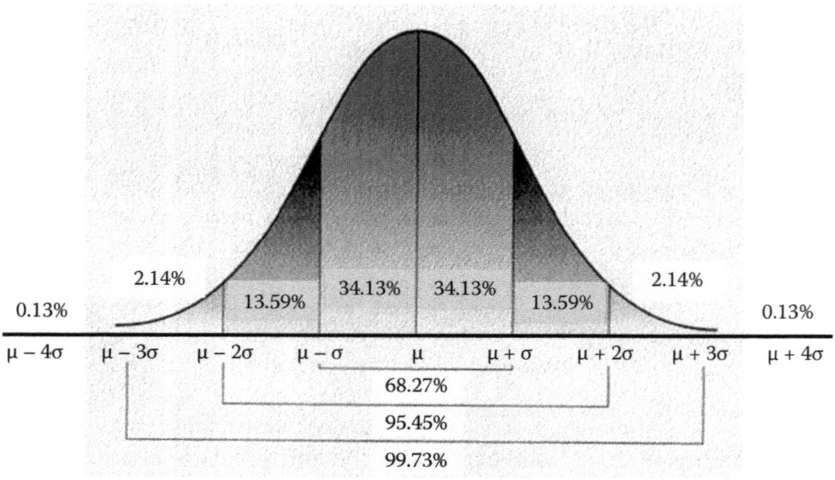

FIGURE 2.6
Areas under normal curve.

normal curve is affected by the standard deviation. Changes to the mean shift the normal curve horizontally. Because there is a mathematical equation for the normal distribution, the probabilities (i.e., areas under the curve) can be determined mathematically.

Relationship to Six Sigma

Although Six Sigma is covered in more detail in Chapter 8, it is timely to mention at this point, the relationship that the normal curve has to the popular Six Sigma technique. Six Sigma can be best described as a business process improvement approach that seeks to find and eliminate causes of defects and errors, reduce cycle times, reduce costs of operations, improve productivity, meet customer expectations, achieve higher asset utilization, and improve return on investment (ROI). Six Sigma is effective and appealing to management because of its focus on measurable bottom-line results. It uses a disciplined, fact-based approach to problem solving and can facilitate rapid project completion.

The Six Sigma name is based on a statistical measure that equates to 3.4 or fewer errors (e.g., in service functions) or defects (e.g., in production operations) per million opportunities. The area under the normal curve in Figure 2.6 bears out this claim as it shows the range within six sigma of the mean (i.e., $\pm 3\ \sigma$) corresponding to 99.73% coverage. The ultimate aspiration of all organizations that adopt Six Sigma philosophies is to have all critical processes, regardless of functional area, at a six sigma level of capability. The pursuit of Six Sigma has heightened the awareness of statistics among professionals in business and industry. The military and the government are also beginning to embrace Six Sigma programs in nonprofit service-oriented programs. With these developments, project management has emerged as a fertile platform for the application of statistical techniques.

A normal curve has the following properties:

Symmetry about $\mu = 0$.
Bell shape.
Highest probability at $\mu = 0$.
Approaches (asymptotic to) x-axis but never crosses it.
Numbers on x-axis represent standard deviations from mean.
Transition (inflection) points at $\mu \pm 1\sigma$.
Area under any portion of the curve is the probability that x is within that span.
Area under curve between $\mu - \sigma$ and $\mu + \sigma$ is 0.6826; thus the probability that an x value is between $\mu - \sigma$ and $\mu + \sigma = 68.26\%$.

Statistical Averages

In project control, an *average, central tendency* of a data set is a measure of the "middle" or "expected" value of the data set. Many different descriptive statistics can be chosen as measurements of the central tendency of the data items. These include the arithmetic mean, the median and the mode. Other statistical measures such as the standard deviation and the range are called measures of spread of data. An average is a single value meant to represent a list of values. The most common measure is the arithmetic mean, but there are many other measures of central tendency such as the median (used most often when the distribution of the values is skewed by small numbers with very high values). Typical examples measured are income distributions and house prices.

Arithmetic Mean

If n numbers are given and each number is denoted by a_i, where $i = 1, \ldots, n$, the arithmetic mean is the sum of the a_i's divided by n:

$$\bar{X} = \frac{1}{n}\sum_{i=1}^{n} a_i .$$

Although the computation to find an average is well known and commonly used, reviewing the logic behind the "average" calculation can provide a better understanding of why it is such a powerful tool in decision making. The arithmetic mean, often simply called the mean, of two numbers, such as 4 and 6, is obtained by finding a value A such that $4 + 6 = X + X$. We find that $X = (4 + 6)/2 = 5$. Switching the order of 4 and 6 does not change the resulting value obtained for X. The mean 5 is not less than the minimum 4 nor greater than the maximum 6. For example, if we increase the number of terms in the list for which we want an average, we find the arithmetic mean of 4, 6; 14 is found by solving for the value of X in the equation $4 + 6 + 14 = X + X + X$. We then solve for X to obtain $X = (4 + 6 + 14)/3 = 8$.

Changing the order of the three items in the list does not change the result: $X\,(6 + 14 + 4)/3 = 8$ and 8 is between 4 and 14. This summation method is easily generalized for lists with any number of elements. However, the mean of a list of integers is not necessarily an integer. For example, a statement such as "the average family has 2.35 children" represents a physical impossibility but conveys a practical expectation over a large number of trials. Thus, the statement can be restated as "the average number of children in the collection of families surveyed in the community is 2.35."

While an average value can appropriately represent a population, no specific family has 2.35 children. Likewise, if the average weight of a group of people is 175 pounds, there is no expectation that any individual in the group will actually weigh 175 pounds. In other words, the average value does not have to be an element of the data set it represents. Surprisingly, this subtle fact is not always understood by all decision makers and may lead to misinterpretation or misuse of what an average does or means. Despite its simplicity, the average is a powerful computational tool, but it is often misunderstood. Because many techniques of project planning and control depend on the calculations of averages, it is recommended that care be exercised in understanding the logic behind the concept.

An early sixteenth century meaning of the word *average* is "damage sustained at sea." The root is found in the Arabic word *awar,* in the Italian word *avaria,* in the French word *avarie,* and in the Dutch word *averij.* During that era, an *average adjuster* assessed insurable losses. Marine damage was classified as *particular average,* borne only by the owner of the damaged property, or *general average,* for which an owner could claim a proportional contribution from all the parties to the marine venture. The types of calculations used in adjusting general averages gave rise to the use of "average" to mean "arithmetic mean." We, thus, see that *average* has a rich history and practical utility. Other computational tools that are predicated on the logic of averages are discussed in the next sections.

Geometric Mean

The geometric mean of n numbers is obtained by multiplying all the numbers together and then calculating the nth root. In algebraic terms, the geometric mean of $a_1, a_2, \ldots, a_n$ is defined as shown below:

$$GM = \sqrt[n]{\prod_{i=1}^{n} a_i}$$

$$= \sqrt[n]{a_1 a_1 \ldots a_n}.$$

The geometric mean can be thought of as the antilog of the arithmetic mean of the logs of the numbers. As a computational example, the geometric mean of 4 and 11 is:

$$GM = \sqrt{4x11} = 3.873$$

For 4, 5, and 11 ($n = 3$), the geometric mean is:

$$\sqrt[3]{4x5x11} = \sqrt[3]{220} = 6.037.$$

Harmonic Mean

The harmonic mean for a set of numbers $a_1, a_2, \ldots, a_n$ is defined as the reciprocal of the arithmetic mean of the reciprocals of a_i's. That is,

$$HM = \frac{1}{\frac{1}{n}\sum_{i=1}^{n}\frac{1}{a_i}} = \frac{n}{\frac{1}{a_1} + \frac{1}{a_2} + \ldots + \frac{1}{a_n}}.$$

One possible application of harmonic mean calculation is for average speed. For example, if the out-bound speed from point A to point B is 45 miles per hour and the in-bound speed for returning from B to A is 55 miles per hour, the average speed is calculated as:

$$\frac{2}{\frac{1}{45} + \frac{1}{55}} = 49.50.$$

A well known inequality relationship of the arithmetic, geometric, and harmonic means for any set of positive numbers is

$$\text{HM} \le \text{GM} \le \text{AM}$$

$$\text{AM} \ge \text{GM} \ge \text{HM}.$$

Mode

The most frequently occurring number in a list is called the mode. The mode of a list of 1, 2, 2, 3, 3, 3, 4 is 3. A mode is not necessarily unique. For example, a list of 1, 2, 2, 3, 3, 5 has two modes: 2 and 3. The mode can be evaluated under the general method of defining averages by setting each member of a list equal to the most common value in the list if there is a most common value. This list is then equated to the resulting list with all values replaced by the same value. Since the components are already the same, no change is required. The mode is more meaningful and potentially useful if a list has many numbers and the frequency of the numbers progresses smoothly. For example, if in a group of 1,000 people, 30 people weigh 61 lbs., 32 people weigh 62 lbs., 29 people weigh 63 lbs., and all the other possible weights occur less frequently, 62 lbs. is the mode. The mode has the advantage that it can be used with non-numerical data (e.g., red cars are most common), while other averages cannot.

Median

The median is the middle number of a group in which the numbers are ranked in order. If a group contains an even number of numbers, the mean of the middle two numbers is taken as the mode. To find the median, order the list according to the magnitude of its elements and repeatedly remove the pair consisting of the highest and lowest values until one or two values are left. If one value is left, it is the median; if two values are left, the median is the arithmetic mean of the two values. This method reorders 1, 7, 3, 13 and orders it as 1, 3, 7, 13, then the 1 and 13 are removed and the 3 and 7 remain. Since 3 and 7 are two elements, the median is their arithmetic mean, $(3 + 7)/2 = 5$.

Average Percentage Return

The average percentage return is a special type of average commonly used in finance and economic analysis. It is an example of a geometric mean. For example, consider a duration of 2 years for a high-profile project. The estimated investment return in the first year is –10% and the estimated return in the second year is +60%. The average percentage return R can be obtained by solving $(1 - 10\%) \times (1 + 60\%) = (1 - 0.1) \times (1 + 0.6) = 1.44$. We now set up the equation relationship $(1 + R) \times (1 + R) = 1.44$, where R is the unknown annual percentage return. This equation simplifies to $R^2 + 2R - 0.44 = 0$ and yields $R = 0.2$ (20%).

This method can be generalized to examples utilizing durations that are not full years. The average percentage of a set of returns is a variation on the geometric average that provides the intensive property of a return per year corresponding to a list of percentage returns. For example, consider a period of a half of a year for which the return is –23% and a period of 2 and one half years for which the return is +13%. The average percentage return for the combined period is the single year return *R* that is, the solution of $(1 - 0.23)^{0.5} \times (1 + 0.13)^{2.5} = (1 + R)^{0.5} \times (1 + R)^{2.5}$ to yield R = 0.0600 (6.0%). The various average formulas are summarized below:

Arithmetic Mean: $$\bar{x} = \frac{1}{n}\sum_{i=1}^{n} x_i = \frac{1}{n}(x_1 + x_2 + \ldots + x_n)$$

Median: Middle value that separates the higher half from the lower half of a data set

Geometric median: Rotation invariant extension of the median for points in R^n

Mode: Most frequent value in a data set

Geometric mean: $$\left(\prod_{i=1}^{n} x_i\right)^{\frac{1}{n}} = \sqrt[n]{x_1.x_2 \ldots x_n}$$

Harmonic mean: $$\frac{n}{\frac{1}{a_1} + \frac{1}{a_2} + \ldots + \frac{1}{a_n}}$$

Quadratic mean: $$\sqrt{\frac{x_1^2 + x_2^2 + \ldots + x_n^2}{n}}$$

Generalized mean: $$\sqrt[p]{\frac{1}{n}\sum_{i=1}^{n} x_i^p}$$

Weighted mean: $$\frac{w_1x_1 + w_2x_2 + \ldots + w_nx_n}{w_1 + w_2 + \ldots + w_n}$$

Truncated mean: Arithmetic mean of data values after a certain number of the highest and lowest data values are discarded

Interquartile mean: Special case of truncated mean using the interquartile range

Midrange: $\frac{\max x + \min x}{2}$

Winsorized mean: Similar to truncated mean; extreme values are not deleted; they are set equal to the largest and smallest values retained

Annualization: $\left[\prod (1+R_i)^{t_i}\right]^{1/\sum t_i} - 1$

Other more sophisticated averages are the trimean, trimedian, and normalized mean. We can also create an average metric using the generalized *f-mean* of the form:

$$y = f^{-1}\left(\frac{f(x_1)+f(x_2)+\ldots+f(x_n)}{n}\right),$$

where f is any invertible function. The harmonic mean is an example of this approach using $f(x) = 1/x$, and the geometric mean is another, using $f(x) = \log x$. Another example, the exponential mean, uses the $f(x) = e^x$ function and is inherently biased toward higher values. However, this method for generating means is not general enough to capture all averages. A more general method for defining an average y takes any function of a list $g(x_1, x_2, \ldots, x_n)$, which is symmetric under permutation of the members of the list, and equates it to the same function with the value of the average replacing each member of the list: $g(x_1, x_2, \ldots, x_n) = g(y, y, \ldots, y)$.

This most general definition still captures the important property of all averages: the average of a list of identical elements is that element itself. The function $g(x_1, x_2, \ldots, x_n) = x_1 + x_2 + \ldots + x_n$ provides the arithmetic mean. The function $g(x_1, x_2, \ldots, x_n) = x_1 \cdot x_2 \cdot \ldots \cdot x_n$ provides the geometric mean. The function $g(x_1, x_2, \ldots, x_n) = x_1^{-1} + x_2^{-1} + \ldots + x_n^{-1}$ provides the harmonic mean.

Several other measures of central tendency can be developed by solving a variational problem in the sense of the calculus of variations, that is, minimizing variation from the center of the data set. Thus, given a measure of statistical dispersion, we seek a measure of central tendency that minimizes variation among all choices of center. Thus, standard deviation about the mean is lower than the standard deviation about any other point, and the maximum deviation about the midrange is lower than the maximum deviation about any other point. The uniqueness of this characterization of mean follows from convex optimization, which is beyond the scope of this book.

Data Patterns

The concept of an average can be applied to a stream of data or to a bounded set—the goal is to find a value about which recent data is in some way clustered. The stream may be distributed in time, as in samples taken by some data acquisition system from which we want to remove noise, or in space, as in pixels in an image from which we want to extract some property. An easy-to-understand and widely used application of average to a stream is the simple moving average in which we compute the arithmetic mean of the most recent N data items in the stream. To advance one position in the stream, we add $1/N$ times the new data item and subtract $1/N$ times the data item N places back in the stream.

Averages of Functions

The concept of calculating an average can be extended to functions. In calculus, the average value of a function f that can be integrated on an interval $[a,b]$, is defined as shown below:

$$\bar{f} = \frac{1}{b-a}\int_a^b f(x)dx.$$

The various computational approaches described above can be leveraged as metrics to assess project performance. With the appropriate measures, project parameters can be better controlled. Such parameters are drawn from time (schedule), cost (budget), and quality (performance) factors.

Statistical Thinking: Computational Examples

Using statistics means being able to reason statistically in the context of a decision problem relevant to project control. A few statistical reasoning and computations are provided here to illustrate how statistics play a role in a variety of problem scenarios.

Example 1

Suppose seven cards in a pile are numbered 1 through 7 to represent a project personnel assignment process. One card is drawn. The sum of the numbers on the remaining cards is 7. What is the number on the drawn card?

Solution: The sum of the numbers on all seven cards is 1 + 2 + 3 + 4 + 5 + 6 + 7, or 28. Removing a numbered card reduces the sum of the remaining cards by the number on that card. For example, if the card drawn were numbered 5, the sum of the remaining cards would be 28 – 5 = 23, a number whose units digit is 3. Similarly, if the card drawn is numbered 1, then the sum of the numbers on the remaining cards is 27, whose units digit is 7. Therefore, the drawn card was numbered 1.

Example 2

If a number is chosen at random from the set {–10, –5, 0, 5, 10}, what is the probability that it is a member of the solution set of both $3x - 2 < 10$ and $x + 2 > -8$?

Solution: If $3x - 2 < 10$ and $x + 2 > -8$, then $-10 < x < 4$. The numbers from the listed set that satisfy these inequalities are –5 and 0. The listed set contains five elements so the probability that a number chosen at random from the set will be a member of the solution set of both inequalities is 2/5.

Example 3

A six-sided number cube with faces numbered 1 through 6 is to be rolled twice. What is the probability that the number that comes up on the first roll will be less than the number that comes up on the second roll?

Solution: The outcome space for this experiment is the set of all ordered pairs (a, b), where a represents the first number that comes up, and b represents the second one. Since a and b can take any value from 1 through 6, the total possible number of outcomes is 36. The outcomes can be represented in a table as shown below:

a/*b*	1	2	3	4	5	6
1	(1,1)	(1,2)	(1,3)	(1,4)	(1,5)	(1,6)
2	(2,1)	(2,2)	(2,3)	(2,4)	(2,5)	(2,6)
3	(3,1)	(3,2)	(3,3)	(3,4)	(3,5)	(3,6)
4	(4,1)	(4,2)	(4,3)	(4,4)	(4,5)	(4,6)
5	(5,1)	(5,2)	(5,3)	(5,4)	(5,5)	(5,6)
6	(6,1)	(6,2)	(6,3)	(6,4)	(6,5)	(6,6)

Among all the pairs (a,b), 6 are pairs for which $a = b$ and for the 30 remaining pairs $a \neq b$. Notice that the number of pairs (a,b) for which $a < b$ is the same as the number of pairs (a,b) for which $a > b$. Therefore, the number of pairs (a,b) for which $a < b$ is 15 (i.e., half of 30). Then the probability that $a < b$ is equal to:

$$\frac{\text{number of pairs } (a,b) \text{ where } a < b}{\text{number of all possible outcomes}} = \frac{15}{36} = \frac{5}{12}$$

Example 4

Among 5 employees on a project, 3 are to be assigned offices and 2 are assigned cubicles. If 3 are men and 2 are women and those assigned to offices are to be chosen at random, what is the probability that the offices will be assigned to 2 men and 1 woman?

Solution: Of the 5 workers, 3 are to be assigned offices. This is an example of a combination problem. To find the number of ways of choosing 3 of the 5 workers, we can count the number of ways of selecting the workers one at a time and then divide by the number of times each group of 3 workers is repeated. Initially, there are 5 ways of choosing the first worker, then there will be 4 ways to choose the second worker, and 3 ways of choosing the third worker or a total of $5 \times 4 \times 3 = 60$ possibilities. Of these 60 possible selections, each distinct group of 3 workers will occur $3 \times 2 \times 1 = 6$ times. There are 3 possibilities for the first worker chosen from the group, 2 possibilities for the second worker chosen, and only 1 for the third worker. Therefore, there are $60/6 = 10$ different ways to choose 3 workers who get offices among the 5 workers. How many of these 10 possible groups of 3 workers consist of 2 men and 1 woman? From the 3 male workers, 2 can be chosen in 3 different ways. There are 2 possibilities for the female worker. Therefore, $3 \times 2 = 6$ of the groups of 3 workers consist of 2 men and 1 woman. Since there are 10 different ways the 3 workers who get offices can be chosen and 6 of these possible groups of 3 workers consist of 2 men and 1 woman, the probability that the offices will be assigned to 2 men and 1 woman is 6/10 or 3/5 (0.6).

Example 5

If a number is chosen at random from the set {–10, –5, 0, 5, 10}, what is the probability that it is a member of the solution set of both $3x - 2 < 10$ and $x + 2 > -8$?

Solution: If $3x - 2 < 10$ and $x + 2 > -8$, then $-10 < x < 4$. The numbers from the listed set that satisfy these inequalities are –5 and 0. The set contains five elements so the probability of a number chosen at random from the set being a member of the solution set of both inequalities is 2/5.

Example 6

Suppose for a manufacturing project we are interested in import and export analysis. If United States imports increased 20% and exports decreased 10 % during a certain year, the ratio of imports to exports at the end of the year was how many times the ratio at the beginning of the year?

Solution: Write the ratio of imports to exports as I/E. At the end of the year, imports were up 20%. The end-of-year imports can be expressed as 100% of the start-of-year imports *plus* 20% or 120% of I. At the end of the year, exports were down 10%. The end-of-year exports can be expressed as 100% of the start-of-year exports *minus* 10% or 90% of E. The ratio of imports to exports at the end of the year can be expressed as $1.20I/0.90E$, which is equivalent to 4/3 of I/E (i.e., 1.3333 I/E).

Example 7

This is a good Six Sigma problem. In a certain computer factory, 0.06% of all computers produced are defective. On the average, 3 computers will be defective among how many produced?

Solution: Recall that 0.06% means 0.06 of every 100, or 6 of every 10,000. The question asks for an equivalent ratio: 3 of how many? The ratio can be inferred from the 6 of 10,000 figure that can be reduced to 3 of 5,000.

Example 8

In a project to assess future STEM workforce recruitment potential, we are interested in the composition of senior and junior classes on a college campus. A class of 80 seniors consists of 3 males for every 5 females. The junior class contains 3 males for every 2 females. If the two classes are combined to include equal numbers of males and females, how many students are in the junior class?

Solution: Among the 80 seniors are 3 males for every 5 females. Thus 3/8 (30) of the seniors are males and 5/8 (50) are females. Among the juniors, 3/5 are males and 2/5 are females. If x denotes the total number of juniors, then $(3/5)x$ are males and $(2/5)x$ are females. The total number of senior and junior males is $30 + (3/5)x$. The total number of senior and junior females is $50 + (2/5)x$. The question states that these quantities are equal, so $30 + (3/5)x = 50 + (2/5)x$. Solving this gives $150 + 3x = 250 + 2x$, or $x = 100$. Thus, the junior class has 100 students.

Factors of Project Control

The three factors (time, budget, and performance) that form the basis of the operating characteristics of a project also help determine the basis for project control. Project control is the reduction of the deviation between actual performance and planned performance. This deviation reduction goal makes project control amenable to the application of statistical techniques. For example, we may be interested in the 95% confidence interval for a particular control action. To control a project, we must be able to measure performance. The ability to measure accurately is a critical aspect of control. Measurements are taken on each of the three components of project

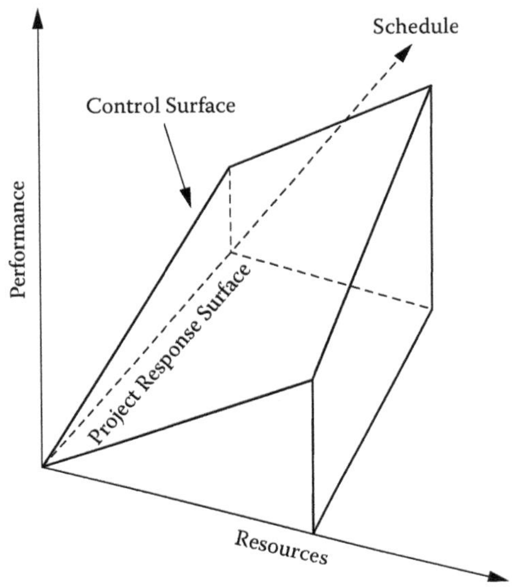

FIGURE 2.7
Project control boundary.

constraints: time, performance, and cost. These constraints are often in conflict and they cannot be fully satisfied simultaneously. The conflicting nature of the three constraints on a project is represented in Figure 2.7.

The higher the desired performance, the more time and resources (costs) will be required. It will be necessary to compromise one constraint in favor of another. Performance is the sole focus on certain projects. Time and costs may be of secondary importance in such projects. In other projects, cost reduction may be the main goal. Performance and time may be sacrificed to an extent in such projects. In some projects, schedule compression (time) is the ultimate goal. The specific nature of a project determines which constraints must be satisfied and which can be sacrificed. Some of the factors that can require control of a project are listed below.

1. Factors affecting time:
   Supply delays
   Missed milestones
   Delay of key tasks
   Changes in customer specifications
   Changes of due dates
   Unreliable time estimates
   Increased use of expediting
   Time-consuming technical problems

Impractical precedence relationships
New regulations that need time to implement

2. Factors affecting performance:
   Poor design
   Poor quality
   Low reliability
   Fragile components
   Poor functionality
   Maintenance problems
   Complicated technology
   Change in statement of work
   Conflicting objectives
   Restricted access to resources
   Employee morale
   Poor retention of experienced workforce
   Sicknesses and injuries
3. Factors affecting costs:
   Inflation
   New vendors
   Incorrect bids
   High labor cost
   Budget revisions
   High overhead costs
   Inadequate budget
   Increased scope of work
   Poor timing of cash flows

In project hierarchy, products are project dependent and processes are product dependent. The performance of a unit depends on the process from which the unit is made. Such a control hierarchy makes the control process more adaptive, dynamic, and effective. The basic elements of adaptive project control are:

- Continual tracking and reporting
- Modifying project implementation as objectives change
- Re-planning to deal with new developments
- Evaluating achievement versus objectives
- Documenting project success and failures as guides for the future

Examples of Statistics for Project Control

Team Formation and Combination

Consider the following project team formation and office allocation problem.

> Of 5 employees, 3 are to be assigned offices and 2 are to be assigned cubicles. If 3 of the employees are men and 2 are women and if those assigned offices are to be chosen at random, what is the probability that the offices will be assigned to 2 men and 1 woman?
>
> *Solution:* Of the 5 workers, 3 are to be assigned offices. This is an example of the use of combinations: to find the number of ways of choosing 3 of the 5 workers, we can count the number of ways of selecting the workers one at a time and then divide by the number of times each group of 3 workers will be repeated.
>
> There are 5 ways to choose the first worker to get an office; 4 ways of choosing the second worker, and 3 ways of choosing the third worker. This represents a total of $5 \times 4 \times 3 = 60$ possibilities. In these 60 possibilities, each distinct group of 3 workers will occur $3 \times 2 \times 1 = 6$ times. There are 3 possibilities for the first worker chosen from the group, 2 for the second worker chosen, and only 1 for the third. The answer is that $60/6 = 10$ different ways to choose 3 workers among the 5 to get offices.
>
> Now, consider this. How many of these 10 possible groups of 3 workers consist of 2 men and 1 woman? From the 3 male workers, 2 can be chosen in 3 different ways. There are two possibilities for the female worker. Therefore, $3 \times 2 = 6$ of the groups of 3 workers consist of 2 men and 1 woman. Since there are 10 different ways of choosing the 3 workers to get offices and 6 of these possible groups of 3 workers consist of 2 men and 1 woman, the probability that the offices will be assigned to 2 men and 1 woman is 6/10 or 3/5.

Statistics for Production Changeovers

Changeover times affect production times significantly. A reduction in changeovers is always desired, but the proper statistical approach must be followed. Factors in changeover times are set-ups, start-ups, cleaning for allergens, and product changeovers. The main ways to reduce changeover times are through training and adherence to standard operating procedures. However, statistical approaches also need to be understood. The formulas for set-up and run times are:

Set-up time = Total elapsed time until first good part is in production

Run time = Total amount of productive time for parts being produced

The goal is to reduce time needed for set-ups and start production more quickly. Set-up requirements vary from product to product, so a key should be made for each different set-up. Ensuring consistent set-up procedures for each product is based on specifications for the set-ups. Also, eliminating non-value added tasks should be considered. Some of these tasks can be accomplished outside production, for example, making spare parts accessible for quick changes, using no-fasten type clamps that eliminate fastening time and adjustments, and proper storage of materials. Most changeovers take 1 to 2 hours. On average, 25% of food facilities require more than 2 hours for a changeover.

Case Study

A food manufacturing company must perform an allergen clean-out every time it produces an item that contains nuts. In a given month, ten shifts are devoted to nut allergen products. The factory runs two 12-hour shifts daily; the production week is 5 days. The goal of the factory is to optimize the amount of run time while continuing to utilize at least two shifts per week for nut allergen production. An allergen clean-out takes 4 hours.

Solution: The factory could run all the allergens for an entire week (5 days, two shift operations) and then manufacture other products for the remainder of the month. If sales volumes will not allow all the allergens to be produced back to back and cannot support more than three allergen runs in a row, the solution would be to maximize the production of allergens in a row:

Week 1: Three shifts of allergens
Week 2: Three shifts of allergens
Week 3: Two shifts of allergens
Week 4: Two shifts of allergens

Total Amount of Downtime and Production Hours Are Summarized as Follows:

Allergen Shifts	Regular Shifts	Downtime Caused by Allergen Shifts	Total Hours of Regular Production	Total Hours of Allergen Production
3	7	4	84	36
3	7	4	84	36
2	8	4	96	24
2	8	4	96	24

The reduction of changeover time from 4 hours to 3 hours or less could be anticipated with training, standard operating procedures, brainstorming faster cleaning procedures, and quick changeover techniques utilizing SMED operations (single-minute exchange of dies). SMED allows production to be more efficient by improving flows. It also specifies that each item should be touched only once (one-touch exchange).

Specifications: Specifications are based on customer requirements. A customer will complain if a product does not have the right packaging, amount, consistency, flavor, or quality desired. Based on customer feedback, specifications are developed to ensure customer satisfaction. Most customers want the same types of products, enabling a manufacturer to determine a target goal and set specifications. The manufacturer should then ensure that its products fall within these specifications and have a tangible way to measure product against specifications before delivery to customers.

Customer dissatisfaction can arise from a number of situations, for example, a consumer is very loyal to a brand and purchases the same package of food regularly. One package contained eight pieces of chicken which seemed like a fair amount. The next package purchased contained five pieces of chicken (based on specification). Because the customer expected to receive eight pieces of chicken and received five, he was dissatisfied and his dissatisfaction could have led to a consumer complaint.

If consistency becomes an issue due to production machinery or methods, the root cause should be evaluated. Many production specifications are not applicable to the machinery or methods used. All commodities should be piloted before specifications are set to ensure consistency and reduce variation. Suppliers that provide raw commodities should be given guidelines for acceptable materials used in production and should be held accountable for meeting specifications within 99% tolerance. Suppliers should also pilot commodities and supplier colleges or training sessions should be held to ensure that they understand the requirements.

Another key issue is that the end production of a factory may not always look the same way after it is transported. Proper transportation methods and packaging materials are critical for maintaining commodities in appropriate conditions for end users. If product quality is affected by movement, the consumer will blame the producer for the quality issue.

If the product specifications indicate five pieces of the commodity per package, the specification for the production facility should be five to six pieces (four pieces will cause a warning; three pieces constitute an "out"). More than six pieces should set off a warning as well. If a weight requirement specification is 45 to 50 grams, a standard deviation of less than 2 should be met.

If the requirement for a cake is not to touch the packaging, the cakes should be stacked tightly within their cartons to minimize shifting during transportation.

General Control Steps

Parkinson's law states that a schedule will expand to fill available time and cost will increase to consume the available budget. Project control prevents a schedule from expanding out of control and assures that a project can be completed within budget. A recommended project control process is presented as follows:

Step 1: Determine the criteria for control; the specific aspects that will be measured should be identified.

Step 2: Set performance standards; they may be based on industry practice, prevailing project agreements, work analysis, forecasting, or other parameters.

Step 3: Measure actual performance based on a predetermined measurement scale. The measurement approach must be calibrated and verified for accuracy. Quantitative and non-quantitative factors may require different measurement approaches. This step also requires reliable project tracking and reporting tools. Project status, no matter how unfavorable, must be reported.

Step 4: Compare actual performance with the specified performance standard. The comparison should be objective and based consistently on the specified control criteria. Meetings should be held regularly to determine:

What has been achieved.

What remains to be done.

If the date of completion is not met, the date should not be changed without determining the root cause for failing to meet the date requirement.

Step 5: Identify unacceptable variance from expectation.

Step 6: Determine the expected impact of the variance on overall project performance.

Step 7: Investigate the source of the poor performance.

Step 8: Determine the appropriate control actions needed to overcome (nullify) the variance observed.

Step 9: Implement the control actions with total dedication.

Step 10: Ensure that the poor performance does not occur elsewhere on the project. The control steps can be carried out within the framework of the flowchart presented in Figure 2.8. The flow of capital, materials, and labor must be controlled throughout the project management process.

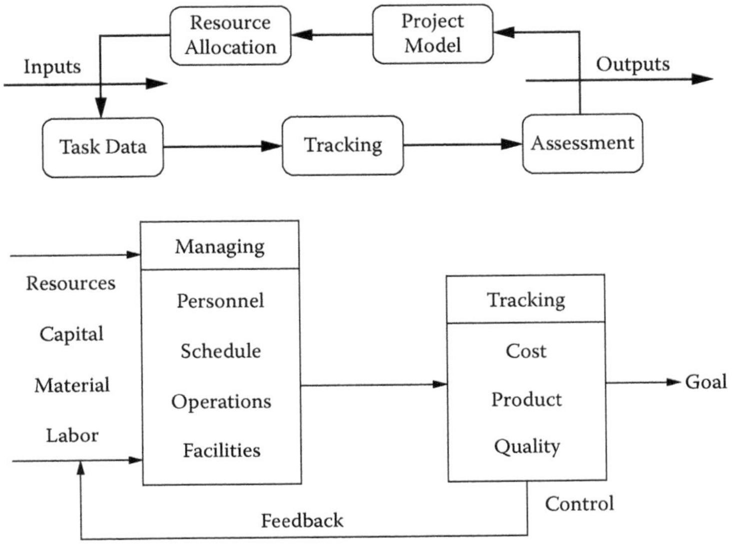

FIGURE 2.8
Flowchart for control process.

Formal and Informal Control

Informal control refers to the use of unscheduled or unplanned approaches to assess project performance along with informal control actions. Informal control requires unscheduled visits and impromptu queries to track progress. Informal control:

- Allows the project manager to learn more about project progress.
- Creates a surprise element that keeps workers on their toes.
- Precludes the temptation to generate "doctored" progress reports.
- Allows peers and subordinates to assume control roles.
- Facilitates prompt appraisal of latest results.
- Gives the project manager more visibility.

Formal control involves control of a project via formal and scheduled reports, consultations, and meetings. Formal control is typically more time consuming and it:

- Can be used by only a limited (designated) group of people.
- Reduces the direct visibility of the project manager.

May impede the implementations of control actions.
Encourages bureaucracy and "paper pushing."
Requires a rigid structure.

Despite its disadvantages, formal control can be effective for all types of projects and project responsibilities and accountability can be pursued in a structured manner. For example, standard audit questions may be posed to determine the current status of a project to establish strategies for future performance. Examples of suitable questions are

Where are we today?
Where were we expected to be today?
What are the prevailing problems?
What problems are expected in the future?
Where shall we be at the next reporting time?
What are the major results since the last project review?
What is the ratio of percent completion to budget depletion?
Is the project plan still the same?
What resources are needed for the next stage of the project?

A formal structured documentation of questions to ask can guide a project auditor to conduct project audits in a consistent manner. The standard question format makes it unnecessary for an auditor to guess or ignore certain factors that may be crucial for project control.

Schedule Control

The Gantt charts developed in the scheduling phase of a project can serve as benchmarks for measuring project progress. Project status should be monitored frequently. Symbols may be used to mark actual activity positions on a Gantt chart. Records showing differences between actual and expected status of an activity should be retained. The information should be conveyed to the appropriate personnel with clear indications of the required control actions. The more milestones or control points a project has, the easier the control function. The larger number allows for more frequent and distinct avenues for monitoring the schedule. Problems can be identified and controlled before they become more serious. However, more control points mean higher control costs.

Schedule variance magnitudes may be plotted on a time scale (e.g., daily). If the variance worsens, drastic actions may be needed. Temporary deviations that do not exert lasting effects on the project may not be causes for concern. Examples of control actions that may be needed to counter schedule delays are:

Job redesign
Productivity improvement
Revision of project scope
Revision of project master plan
Expediting or activity crashing
Elimination of unnecessary activities
Reevaluation of milestones or due dates
Revision of time estimates for pending activities

Project Tracking and Reporting

Tracking and reporting provide the avenues for monitoring and evaluating project progress. Computerized tools are useful for these purposes. Customized computer-based project tracking programs can be developed to monitor various aspects of project performance. An example is the tracking scheme shown in Table 2.1. The model tracks several parameters for each activity including:

Days of work planned for an activity
Planned costs
Number of days worked
Number of days credited
Percent completion expected
Actual percent completion
Actual cost
Amount of budget shortfall or surplus

The activity parameters used in the model and shown in Table 2.1 are explained next.

Workdays planned indicates the amount of work required for a task expressed in terms of the number of days required to perform the work. This is entered as an input by the user. In a more sophisticated model, the number of workdays required may be computed directly by the program based on a standard data base or internal estimation formula. Planned cost is the estimated cost of performing the full amount of work required by the activity. This can be entered as a user input or calculated by a computer program. Task weight represents the relative importance of an activity based on

TABLE 2.1

Data for Project Performance Monitoring

Project Start Date: 1-15-10
Current Date: 7-15-10
Project Name: Installation of New Product Line

Task Name	Planned Workdays	Planned Cost	Task Weight	Workdays Completed	Workdays Credited
Define problem	16	$100,000	0.1524	16	16
Interviews	38	100,000	0.3619	38	38
Network analysis	7	100,000	0.0667	7	7
Improve flow	7	150,000	0.0667	7	7
New process	7	200,000	0.0667	14	7
Expert system	30	42,000	0.2857	0	0
	105		1.000		75

Task Name	Expected % Completion	Expected Relative % Completion	Actual % Completion	Actual Relative % Completion	Actual Cost	Cost Deviation
Define problem	100	15.24	100	15.24	$59,000	($41,000)
Interviews	100	36.19	100	36.19	146,000	46,000
Network analysis	100	6.67	100	6.67	110,000	10,000
Improve flow	100	6.67	65	4.33	65,000	(85,000)
New process	100	6.67	10	0.67	216,000	16,000
Expert system	0	0	0	0		0
		71.43		0.6310		(54,000)

Planned project % completion = 71.43%
Project tracking index = –11.67%
Days ahead/behind schedule = –8.75 days' worth of work
Cost deviation = ($54,000) (Cost saving)

stated criteria. In Table 2.1, the task weight is computed on the basis of the work content of an activity relative to the overall project. The work content is expressed as days. Task weight is computed as:

$$\text{Task weight} = \frac{(\text{Workdays for activity})}{(\text{Total workdays for project})}$$

In the table, the total work content for the project is 105 days. The work content for task 2 is 38 days. Thus the weight for task 2 is 38/105 = 0.3619.

Workdays completed indicates the amount of work actually completed out of the total planned workdays as of the current date. This value is based on actual measurement of work accomplishment. This may be done by monitoring the total cumulative hours worked and dividing that total by the number of hours per day to obtain number of workdays. Workdays credited are the productive workdays among total workdays completed. For example, if no credit is given for set-up times, the workdays credited will be determined by subtracting all set-up times from workdays completed. For example, in Table 2.1, a total of 14 days were worked on task 5, but only 7 of the 14 were credited to the task.

The expected percent completion column in the table indicates the percentage of the required work expected to be completed by the current date computed as:

$$\text{Expected \% completion} = \frac{(\text{workdays completed})}{(\text{workdays planned})}$$

The expected relative percent completion column shows the percentage of work expected to be completed on a given activity relative to the total work of the project computed as:

$$\text{Expected relative \% completion} = (\text{expected \% completion}) \times (\text{task weight})$$

For example, task 1 is expected to have completed 16 of the 105 days for the project by the current date. That translates to 16/105 = 0.1524 or 15.24%.

Actual percent completion indicates the actual percentage of work completed by the current date. This value will normally be obtained by a direct observation of the goal of the task. It is possible that even though the number of planned workdays have been completed, the actual amount of work completed may not be worth the number of days worked, possibly due to low employee productivity. For example, if a certain amount of work is expected to take 10 days, an employee may work diligently on the task for 10 days but may accomplish only one-half the work required.

In Table 2.1, this is the case for tasks 4 and 5 as reflected in the actual percent completion column. Even though the required 7 days for task 4 have been completed, only 65% of the required work has been done. The amount of work yet to be done on task 4 is equivalent to 35% of 7 days (i.e., 2.45 days' worth of work behind schedule). Similarly, the amount of work remaining to be done on task 5 is 90% of 7 days (i.e., 6.3 days' worth of work behind schedule). Consequently, the total project is behind schedule by 2.45 days + 6.3 days = 8.75 days. Note that a project may be behind based on the schedule (number of days elapsed) or the amount of work accomplished. The 8.75 days represents work to be accomplished, that is, it will take 8.75 days to accomplish work required for tasks 4 and 5.

Actual relative percent completion indicates the relative percentage of work actually completed on a specific activity compared to other activities in the project. This is computed as

$$\text{Actual relative \% completion} = (\text{actual \% completion}) \times (\text{task weight})$$

Actual cost indicates the actual total amount spent on the activity as of the current date. Cost deviation denotes the difference between actual and planned costs. Cost overrun occurs when actual exceeds planned cost; cost saving is achieved when actual falls below planned cost. Brackets in Table 2.1 indicate cost savings. Tasks 2, 3, and 5 show cost overruns. In addition to the activity parameters explained above, other factors are associated with a project.

Planned project percent completion indicates the percent of a project expected to be completed by the current date computed as:

$$\text{Planned project \% completion} = \frac{(\text{workdays completed on project})}{(\text{total workdays planned})}$$

Table 2.1 shows an expected relative percent completion total of 71.43%. The project tracking index is a relative measure of the performance of a project based on the amount of work completed rather than the number of days elapsed. It is computed as:

$$\text{Index} = \frac{(\text{actual relative \% completion})}{(\text{expected relative \% completion})} - 1$$

If a project is performing better than expected (ahead of schedule), the index will be positive. The index will be negative if the project performance is below expectation. The index expressed as a percentage may be viewed as a measure of the criticality of the control action needed for the project. For the example in Table 2.1, the project tracking index is computed as (0.6310)/(0.7143) – 1 = –0.1167 or –11.67%, indicating that the project is behind schedule. This implies that 11.67% of the work expected by the current date has not been accomplished.

Days ahead or of behind schedule refers to the performance of the project schedule. It indicates the amount of work by which the project deviates from expectation. The deviation is expressed in terms of days of work. This measure is computed as:

$$\text{Schedule performance} = (\text{project tracking index}) \times (\text{planned days credited})$$

If this number is negative, the work accomplishment parameter of the project is behind schedule. If the number is positive, the project is ahead of schedule.

Keep in mind that a project may be on schedule but still be behind schedule in terms of actual work accomplishment. The measure used in the Baker model provides a better basis for evaluating project performance than the traditional approach of merely looking at the schedule. Entries at the end of Table 2.1 indicate that the project is behind schedule based on (–11.67%) × (75 days) = –8.75 days' worth of work. Recall that the same –8.75 figure was obtained earlier by adding the work deficiencies of individual activities. Even though this project has been credited with 75 workdays, the actual amount of work accomplished is 75 days – 8.75 days = 66.25 days.

Performance Control

Many project performance problems may not surface until a project has been completed. This makes performance control very difficult. Effort should be made to measure all the interim factors that may influence final performance. After-the-fact performance measurements are not effective for project control. Some performance problems may be revealed by time and cost deviations. When project time and costs are problems, an analysis of how the problems may affect performance should be made. Since project performance requirements usually relate to the performance of the end product, controlling performance problems may require changes of product specifications. Performance analysis involves checking key elements of a product such as those discussed next.

Scope

Is the scope reasonable based on the project environment? Can the required output be achieved with the available resources? The current wave of *downsizing* and *rightsizing* in industry may be an attempt to define proper scopes of operations.

Documentation

Is the requirement specification accurate?

Are statements clearly understood?

Requirements

Is the technical basis for the specification sound?

Are the requirements properly organized?

What are the interactions among specific requirements?
How does the prototype perform?
Is the raw material appropriate?
What is a reasonable level of reliability?
What is the durability of the product?
What are the maintainability characteristics?
Is the product compatible with other products?
Are the physical characteristics satisfactory?

Quality Assurance

Who is responsible for inspection?
What are the inspection policies and methods?
What actions are needed for nonconforming units?

Function

Is the product usable as designed?
Can the expected use be achieved by other means?
Is there any potential for misusing the product?

Careful evaluation of performance on the basis of these questions throughout the life cycle of a project should help identify problems early so that control actions may be initiated to forestall greater problems later.

Continuous Performance Improvement

Continuous performance improvement (CPI) is an approach to achieving a steady flow of improvement in a project. The approach is based on the concept of continuous process improvement used in quality management. The iterative decision processes in project management can benefit greatly from the use of CPI because it represents a practical method of improving performance in business, management, and technical operations. The approach is based on the following key points:

Early detection of problems
Incremental improvement steps
Project-wide adoption of CPI concept

Comprehensive evaluation of procedures
Prompt review of methods of improvement
Prioritization of improvement opportunities
Establishment of long-term improvement goals
Continuous implementation of improvement actions

A steering committee is typically set up to guide the improvement efforts. The typical functions of the steering committee with respect to performance improvement include:

Determination of organizational goals and objectives
Communications with personnel
Team organization
Administration of CPI procedures
Education and guidance for companywide involvement
Allocation of resources or recommendations for requirements

Figure 2.9 represents the conventional fluctuating approach to performance improvement. The process starts with a certain level of performance specified as the target to be achieved by time *T*. Without proper control, performance will gradually degrade until it falls below the lower control limit at

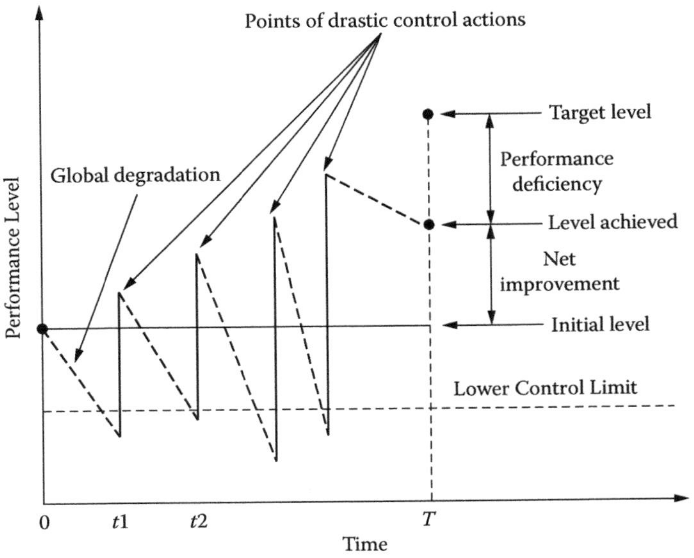

FIGURE 2.9
Conventional approach to control.

time t_1 when drastic efforts will be needed to raise the level. If control is neglected further, performance will undergo another gradual decline until it again falls below the lower control limit at time t_2. Again, costly measures will be needed to improve performance. This cycle of degradation and innovation may be repeated several times before time T is reached. At time T, a final attempt will be needed to suddenly raise performance to the target level but it may be too late at time T to achieve the target performance. The disadvantages of the conventional fluctuating approach to improvement include:

High cost of implementation
Need for drastic control actions
Potential loss of project support
Adverse effect on personnel morale
Frequent disruptions of project
Excessive focus on short-term benefits
Need for frequent and strict monitoring
Opportunity cost during degradation phase

Figure 2.10 represents the CPI approach. The process starts with the same initial quality level that is continuously improved in a gradual pursuit of the target performance level. Opportunities to improve are implemented immediately. The rate of improvement is not necessarily constant over the planning

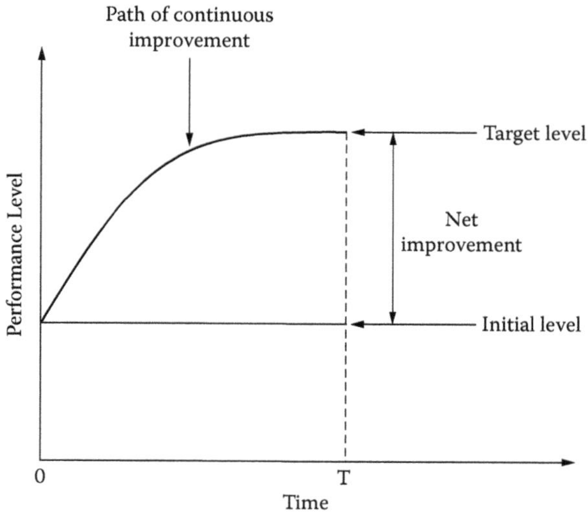

FIGURE 2.10
Continuous process of control.

horizon. Hence, the path of improvement is shown as a curve rather than a straight line. The important aspect of the CPI is that each subsequent performance level is at least as good as the one preceding it. The major advantages of CPI include

Better client satisfaction
Clear expression of project expectations
Consistent pace with available technology
Lower cost of achieving project objectives
Conducive environment for personnel involvement
Dedication to higher quality products and services

Cost Control

Several aspects of a project can contribute to its overall cost. These aspects must be tracked carefully to determine when control actions may be needed. Some of the important cost aspects of a project are:

Cost estimation approach
Cost accounting practices
Project cash flow management
Company cash flow status
Direct labor costing
Overhead rate costing
Incentives, penalties, and bonuses
Overtime payments

Controlling project cost requires management to address several key issues:

Proper planning to justify bases for cost elements
Reliable estimates of time, resources, and cost
Clear communication of project requirements, constraints, and available resources
Sustained cooperation of project personnel
Good coordination of project functions
Consistent policy for expenditures
Timely tracking and reporting of time, materials, and labor transactions

Periodic review of project progress

Revision of project schedule to adapt to prevailing project scenarios

Evaluation of budget depletion versus project progress

These items must be evaluated in an integrated control effort rather than as individual functions. The interactions among the various actions needed may be so unpredictable that success achieved at one level may be nullified by failure at another. Uncoordinated analysis makes cost control very difficult. Project managers must be alert and persistent when monitoring costs.

Information Required for Project Control

Complex projects require well-coordinated communication systems that can quickly reveal activity status. Reports on individual project elements must be tied together logically to facilitate managerial control. The project manager must have prompt access to individual activity data and overall status of the project. Three critical aspects of this function are the prevailing levels of communication, cooperation, and coordination. Project management information systems (PMISs) have evolved to solve the problems of monitoring, organizing, storing, and disseminating project information. Many PMIS commercial computer programs have been developed. The basic reporting elements in a PMIS are:

Financial reports

Project deliverables

Current project plan

Project progress reports

Material supply schedule

Client delivery schedule

Subcontract work and schedule

Project conference schedule and records

Graphical project schedule (Gantt chart)

Performance requirements evaluation plots

Time performance plots (plan versus actual)

Cost performance plots (expected versus actual)

Many standard forms have been developed to facilitate reporting. The availability of computerized systems led to a decline in use of manual project information systems.

Measurement Scales

Project control requires data collection, measurement, and analysis. A project manager will encounter different types of measurement scales, depending on specific items to be controlled. He or she may need data on project schedules, costs, performance levels, problems, and other factors. The various types of applicable data measurement scales are discussed next.

Nominal scale of measurement—A nominal scale is the lowest level of measurement scale. It classifies items into categories that are mutually exclusive and collectively exhaustive, that is, they do not overlap and cover all possible categories of the characteristics examined. For example, in the analysis of the critical path of a project network, each job is classified as critical or not critical. Gender, type of industry, job classification, and color are other examples of measurements on a nominal scale.

Ordinal scale of measurement—An ordinal scale is distinguished from a nominal scale by ordering of categories. An example is prioritizing project tasks for resource allocation. We know that first priority is above second, but we do not know how far above. Similarly, we know that better is preferred to good, but not by how much. In quality control, the ABC classification of items based on the Pareto distribution is an example of a measurement on an ordinal scale.

Interval scale of measurement—An interval scale is distinguished from an ordinal scale by utilizing equal intervals between the units of measure. The assignment of priority ratings to project objectives on a scale of 0 to 10 is an example of a measurement on an interval scale. A priority rating of 0 does not mean that an objective has absolutely no significance to a project team. Similarly, the scoring of 0 on an examination does not imply that a student knows absolutely nothing about the material covered by the examination. Temperature is a good example of an interval scale measurement. The zero point on a temperature scale is an arbitrary relative measure. Other examples of interval scales are IQ measurements and aptitude ratings.

Ratio scale of measurement—A ratio scale has the same properties as an interval scale but utilizes a true zero point. For example, an estimate of a zero time unit for the duration of a task is a ratio scale measurement. Other examples of items measured on a ratio scale are cost, time, volume, length, height, weight, and inventory level. Many items in a project management environment will be measured on a ratio scale.

Another important aspect of data analysis for project control involves the classification scheme used. Most projects will generate and use quantitative and qualitative data. Quantitative data describe the characteristics of items studied numerically. Qualitative data, on the other hand, are associated with object attributes that are not measured numerically. Most items measured on the nominal and ordinal scales will normally be classified as qualitative data; those measured on the interval and ratio scales will normally be classified into the quantitative data category.

The implication for project control is that qualitative data can lead to bias of control mechanisms because qualitative data are subject to the views and interpretations of the person using the data. Whenever possible, data for project control should be based on quantitative measurements.

Badiru and Whitehouse (1989) suggest a class referred to as *transient data*, defined as a volatile set of data used for one-time decision making and never needed again. An example may be the number of operators who show up at a job site on a given day. Unless there is a need to correlate day-to-day attendance records of operators, this information is relevant only for a given day. The project manager can make his decision for that day based on the number of available workers. Transient data need not be stored in a permanent database unless it may be needed for future analysis (forecasting, incentive programs, performance reviews).

Recurring data appear frequently enough to require permanent storage, for example, a file showing contract due dates that should be retained at least through the project life cycle. Recurring data may be further categorized into *static* and *dynamic data*. Recurring data that are static will retain their original parameters and values each time they are retrieved and used. Recurring data that are dynamic have the potential for taking on different parameters and values each time they are retrieved and used. Storage and retrieval considerations for project control should address certain questions:

What is the origin of the data?
How long will the data be maintained?
Who needs access to the data?
For what purpose will the data be used?
How often will the data be needed?
Is storage only for review purposes or must it be printed?
Will the data be used to generate reports?
In what formats will the data be needed?
How fast will the data need to be retrieved?
What security measures are needed for the data?

Data Determination and Collection

It is essential to determine what data to collect for project control purposes. Data collection and analysis are basic components of generating information for project control. The requirements for data collection are discussed next.

Choosing data—This involves selecting data on the basis of relevance, likelihood that they will be needed for future decisions, and whether they contribute to better decision making. The intended users should also be identified.

Collecting data—This identifies a suitable method of collecting the data and the sources from which the data will be collected. The collection method depends on the specific operation. The common methods include manual tabulation, direct keyboard entry, optical character reading, magnetic coding, electronic scanning, and now voice command. An input control may be used to confirm the accuracy of collected data. Examples of items to verify and control during data collection are:

- Relevance: whether the data are relevant to the prevailing problem. For example, data on personnel productivity may not be relevant for making a decision involving marketing strategies.
- Limitations: this check ensures that the data are within known or acceptable limits, for example, an overtime claim from an employee claiming to work more than 80 hours per week for several weeks in a row is well beyond ordinary limits.
- Critical value: This identifies a boundary point. Values below or above a critical value fall into different data categories. For example, the lower specification limit for a given characteristic of a product is a critical value that determines whether the product meets quality requirements.

Coding data—This technique represents data in a form useful for generating information. Coding should be compact and meaningful. The performance of information systems can be greatly improved if effective data formats and coding are designed into the system from the beginning.

Processing data—Data processing is the manipulation of data to generate useful information. Different types of information may be generated from a given data set, depending on processing method. The method should consider how the information will be used, who will use it, and what level of system response time is desired. If possible, processing controls should be used and may involve:

- Control total: check for the completeness of the processing by comparing accumulated results to a known total. An example is the comparison of machine throughput to a standard production level or the comparison of cumulative project budget depletion to a cost accounting standard.
- Consistency check: this determines whether processing yields the same results for similar data. For example, an electronic inspection device that suddenly shows a measurement ten times higher than the norm warrants an investigation of the input and the processing mechanisms.
- Scales of measurement: for numeric scales, specify units of measurement, increments, the zero point, and the range of values.

Using data—Use involves people. Computers can collect and manipulate data and generate information, but the ultimate decision rests with people. Decision making starts when information becomes available. Intuition, experience, training, interest, and ethics are only a few factors that determine how people use information. A piece of information used positively to further the progress of a project in one situation may also be used negatively in another case. To assure that data and information are used appropriately, computer-based security measures can be built into an information system. Project data may be obtained from several sources, for example:

Formal reports
Interviews and surveys
Regular project meetings
Personnel time cards or work schedules

The timing of data is also very important for project control purposes. The contents, level of detail, and frequency of data can affect the control process. An important aspect of project management is determining the data required to generate the information needed for project control. Tracking vast quantities of rapidly changing and interrelated data about project attributes can be very complicated. The major steps involved in data analysis for project control are:

Data collection
Data analysis and presentation
Decision making
Implementation of action

Data are processed to generate information that decision makers analyze when they make required decisions. Good decisions are based on timely and relevant information, which in turn is based on reliable data. The functions of data analysis for project control include:

Organizing and printing computer-generated information in a form usable by managers
Integrating different hardware and software systems to communicate in the same project environment
Incorporating new technologies such as expert systems into data analysis
Using graphics and other presentation techniques to convey project information

Proper data management will prevent misuse, misinterpretation, or mishandling. Data are needed at every stage in the life cycle of a project from problem identification through phase-out. Data may be needed to determine specifications, feasibility, resource availability, staff size, scheduling, project status, performance data, and phase-out plans. Data requirements include:

- Summary: a general summary of the information and decision for which the data is required along with the form in which the data should be prepared. The summary indicates the impact of the data requirements on organizational goals.
- Processing environment: the project for which the data is required, user personnel, and computer system to be used for processing the data. It should utilize a request system, authorization for use, relationships to other projects, and specify expected data communication needs and mode of transmission.
- Policies and procedures: specific methods that govern data handling, storage, and modification; procedures for implementing changes; and instructions for data collection and organization.
- Static data: segment of data used mainly for reference purposes and is rarely updated.
- Dynamic data: segment of data that is frequently updated based on changing circumstances.
- Frequency: expected frequency of data change for dynamic portion of the data, for example, quarterly. Data change frequency should be described in relation to the frequency of processing.
- Constraints: limitations on data requirements; may be procedural (based on corporate policy), technical (based on computer limitations), or imposed (based on project goals).
- Compatibility: ensuring that data collected for project control must be compatible with future needs.
- Contingency: security measures for use in cases of accidental or deliberate damage or sabotage affecting hardware, software, or personnel.

Data Analysis and Presentation

Data analysis encompasses various mathematical and graphical operations performed on data to elicit the inherent information the data contain. The manner in which project data is analyzed and presented can affect how the information is perceived by decision makers. The examples presented in this section illustrate how basic data analysis techniques can be used to convey important information for project control. In many cases, data provide answers to direct questions:

When is the project deadline?

Who are the people assigned to the first task?

How many resource units are available?

Are enough funds available for the project?

What were the quarterly expenditures on the project for the past 2 years?

Is personnel productivity low, average, or high?

Who is in charge of the project?

Answers to such questions constitute data of different forms or expressed on different scales. The resulting data may be qualitative or quantitative. Different techniques are available for analyzing different types of data. This section discusses some of the basic techniques for data analysis. The data presented in Table 2.2 illustrate data analysis techniques.

Raw data—Raw data consists of ordinary observations recorded for a decision variable or factor. Examples of factors for which data may be collected for decision making are revenues, costs, personnel productivity, task duration, project completion time, product quality, and resource availability. Raw data should be organized into a format suitable for visual review and computational analysis. Table 2.2 represents quarterly revenues from projects A, B, C, and D. For example, the data for quarter 1 indicate that project C yielded the highest revenue of $4,500,000 while project B yielded the lowest revenue of $1,200,000. Figure 2.11 presents the raw project revenue data as a line graph. Figure 2.12 presents the same information as a multiple bar chart.

Total revenue—This is a total or sum that indicates the overall effect of a specific variable. If $X_1, X_2, X_3, \ldots, X_n$ represent a set of n observations (e.g., revenues), the total is computed as:

$$T = \sum_{i=1}^{n} X_i .$$

In Table 2.2, the total revenue for each project is shown in the last column. The totals indicate that project C yielded the largest total revenue over the four quarters under consideration while project B produced the lowest. The

TABLE 2.2

Quarterly Revenues from Four Projects ($1,000)

Project	Quarter 1	Quarter 2	Quarter 3	Quarter 4	Row Total
A	3,000	3,200	3,400	2,800	12,400
B	1,200	1,900	2,500	2,400	8,000
C	4,500	3,400	4,600	4,200	16,700
D	2,000	2,500	3,200	2,600	10,300
Column total	10,700	11,000	13,700	12,000	47,400

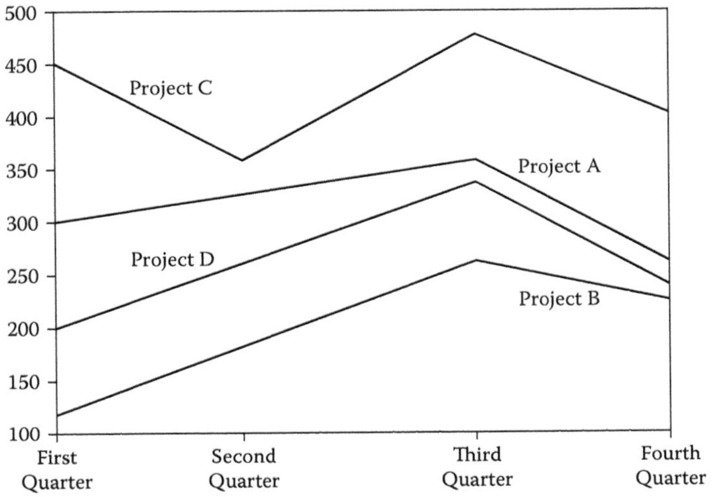

FIGURE 2.11
Line graph of quarterly project revenues.

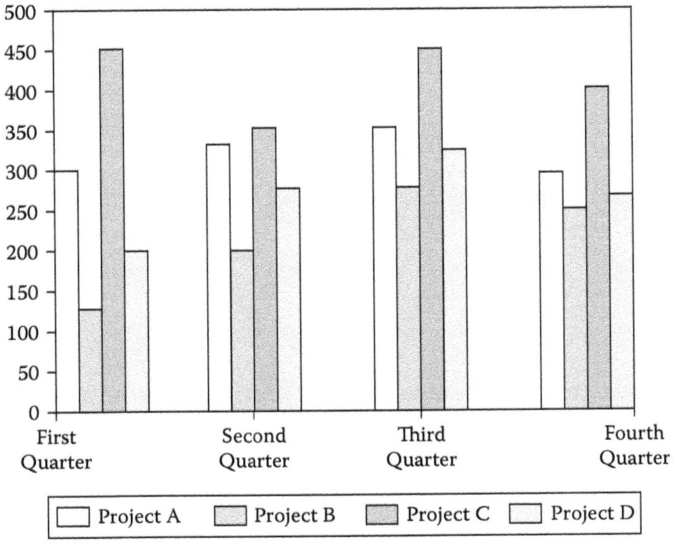

FIGURE 2.12
Multiple bar chart of quarterly project revenues.

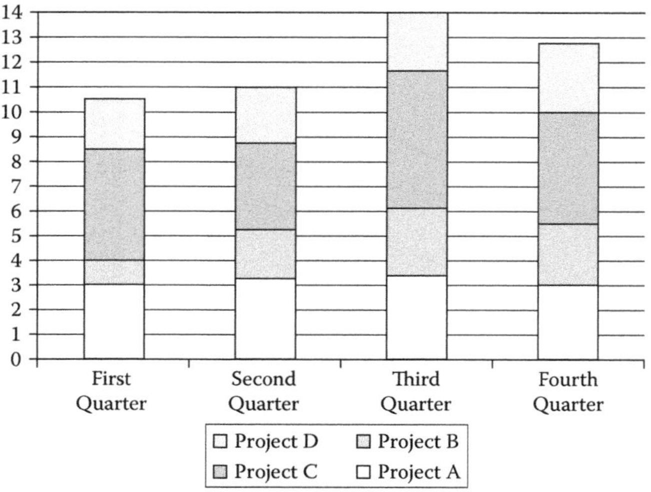

FIGURE 2.13
Stacked bar graph of quarterly total revenues.

last row shows the total revenue for each quarter. The totals show the largest revenue received in the third quarter; the first quarter produced the lowest total revenue. The total revenue for the four projects over the four quarters is shown as $47,400,000 in the last cell in the table. Figure 2.13 charts the quarterly total revenues as stacked bars. Each segment in a stack of bars represents the revenue contribution from a project. The total revenues for the four projects over the four quarters are also shown as a pie chart (Figure 2.14). The chart also shows percentages of the overall revenue contributed by each project.

Average revenue—This is one of the most common measures in data analysis. Given n observations (e.g., revenues), $X_1, X_2, X_3, \ldots, X_n$, the average of the observations is computed as:

$$\bar{X} = \frac{\sum_{i=1}^{n} X_i}{n}$$

$$= \frac{T_x}{n},$$

where T_x is the sum of n revenues. For our sample data, the average quarterly revenues for the four projects are:

$$\bar{X}_A = \frac{(3,000 + 3,200 + 3,400 + 2,800)(\$1,000)}{4}$$

$$= \$3,100,000$$

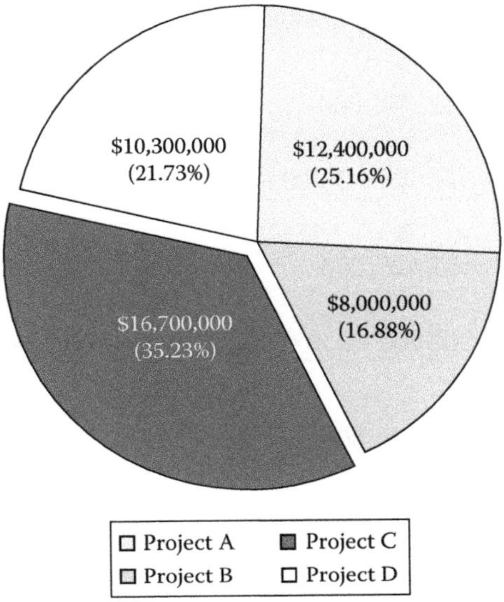

FIGURE 2.14
Pie chart of total revenue per project.

$$\bar{X}_B = \frac{(1,200+1,900+2,500+2,400)(\$1,000)}{4}$$

$$= \$2,000,000$$

$$\bar{X}_C = \frac{(4,500+3,400+4,600+4,200)(\$1,000)}{4}$$

$$= \$4,175,000$$

$$\bar{X}_D = \frac{(2,000+2,500+3,200+2,600)(\$1,000)}{4}$$

$$= \$2,575,000.$$

Similarly, the expected average revenues per project for the four quarters are:

$$\bar{X}_1 = \frac{(3,000+1,200+4,500+2,000)(\$1,000)}{4}$$

$$= \$2,675,000$$

$$\bar{X}_2 = \frac{(3,200 + 1,900 + 3,400 + 2,500)(\$1,000)}{4}$$

$$= \$2,750,000$$

$$\bar{X}_3 = \frac{(3,400 + 2,500 + 4,600 + 3,200)(\$1,000)}{4}$$

$$= \$3,425,000$$

$$\bar{X}_4 = \frac{(2,800 + 2,400 + 4,200 + 2,600)(\$1,000)}{4}$$

$$= \$3,000,000.$$

The above values are shown in a bar chart in Figure 2.15. The average revenue from any of the four projects in a given quarter is calculated as the sum of all the observations divided by the number of observations, that is:

$$\bar{\bar{X}} = \frac{\sum_{i=1}^{N} \sum_{j=1}^{M} X_{ij}}{K},$$

where N = number of projects; M = number of quarters; and K = total number of observations ($K = NM$). The overall average per project per quarter is

$$\bar{\bar{X}} = \frac{\$47,400,000}{16}$$

$$= \$2,962,500$$

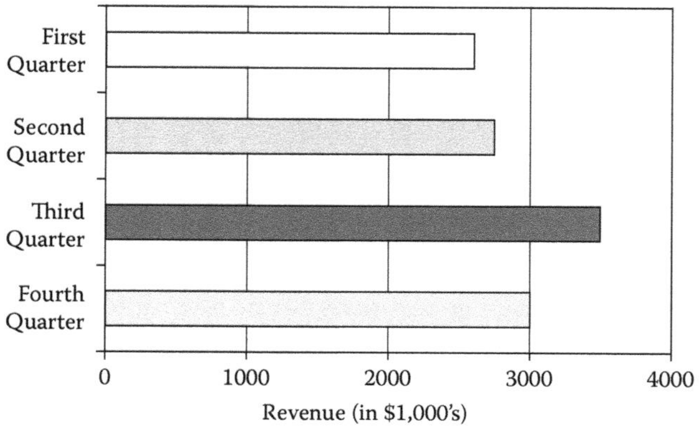

FIGURE 2.15
Average revenue per project for each quarter.

As a cross-check, the sum of the quarterly averages should equal the sum of the project revenue averages (equal to the grand total divided by four): (2,675 + 2,750 + 3,425 + 3,000)($1,000) = (3,100 + 2,000 + 4,175 + 2,575)($1,000) = $11,800,000 = $47,400,000/4. The cross-check procedure works because we have a balanced table of observations—four projects and four quarters. If we had only three projects, for example, the sum of the quarterly averages would not equal the sum of the project averages.

Median revenue—This is the value that falls in the middle of a group of observations arranged in order of magnitude. One-half of the observations are above the median and the other half are below it. The method of determining the median depends on whether the observations are organized into a frequency distribution. It is necessary to arrange unorganized data into increasing or decreasing order before finding the median. Given K observations (e.g., revenues), $X_1, X_2, X_3, \ldots, X_K$, arranged in increasing or decreasing order, the median is identified as the value in position $(K + 1)/2$ in the data arrangement if K is an odd number. If K is even, the average of the two middle values is considered the median. An example of sample data arranged in increasing order is 1,200, 1,900, 2,000, 2,400, 2,500, 2,500, 2,600, 2,800, 3,000, 3,200, 3,200, 3,400, 3,400, 4,200, 4,500, and 4,600. The median is calculated as (2,800 + 3,000)/2 = 2,900. Half the recorded revenues are expected to be above $2,900,000 and half are expected to be below that amount. Figure 2.16 is a bar chart of the revenue data arranged in increasing order. The median is anywhere between the eighth and ninth values in the ordered data.

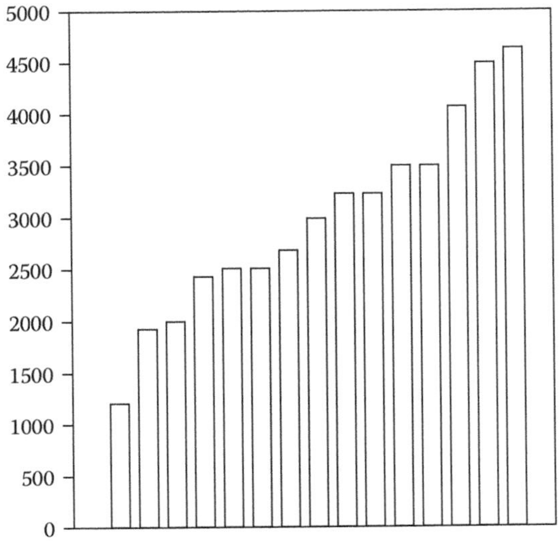

FIGURE 2.16
Bar chart of ordered data.

Quartiles and percentiles—The median is a position measure: its value is based on its position in a set of observations. Other measures of position are *quartiles* and *percentiles*. Three quartiles divide a set of data into four equal categories. The first quartile, denoted Q_1, is the value below which one-fourth of all the observations in the data set fall. The second quartile, denoted, Q_2, is the value below which two-fourths or one-half of all the observations in the data set fall. The third quartile, denoted Q_3, is the value below which three-fourths of the observations fall. The second quartile is identical to the median. It is technically incorrect to cite a fourth quartile because that implies a point within the data set below which all the data points fall: a contradiction! A data point cannot lie within the range of the observations and at the same time exceed all the observations including itself.

The concept of percentiles is similar to the concept of quartiles except that reference is made to percentage points. There are 99 percentiles that divide a set of observations into 100 equal parts. The X percentile is the value below which X percent of the data fall. The 99th percentile refers to the point below which 99% of the observations fall. The three quartiles discussed previously are regarded as the 25th, 50th, and 75th percentiles. A 100th percentile is technically incorrect. In performance ratings, for example, on an examination or product quality level, the higher the percentile the better. In many cases, recorded data are classified into categories that are not indexed to numerical measures. In such cases, other measures of central tendency or position are needed. One such measure is the mode.

Mode—This measure is defined as the value that has the highest frequency in a set of observations. When the recorded observations can be classified only into categories, the mode can be particularly helpful in describing the data. In a set of K observations (e.g., revenues), $X_1, X_2, X_3, \ldots, X_K$, the mode value occurs more than any other value in the set. Sometimes, the mode is not unique in a set of observations. For example, in Table 2.2, \$2,500, \$3,200, and \$3,400 all show the same number of occurrences. Each is a mode of the set of revenue observations. If a set of observations reveals a unique mode, the data are said to be unimodal. The mode is very useful in expressing central observations of qualitative characteristics such as color, marital status, or state of origin. The three modes in the raw data can be identified in Figure 2.16.

Range of revenue—The range is determined by the two extreme values in a set of observations. Given K observations (e.g., revenues), $X_1, X_2, X_3, \ldots, X_K$, the range is simply the difference between the lowest and the highest observations. This measure is useful when an analyst wants to know the extent of extreme variations in a parameter. The range of the revenues in our sample data is (\$4,600,000 to \$1,200,000) = \$3,400,000. Because of its dependence on only two values, a range tends to increase as sample size increases. Furthermore, it does not measure the variability of observations relative to the center of the distribution. This is why the standard deviation is normally used as a more reliable measure of dispersion than the range.

The variability of a distribution is generally expressed in terms of the deviation of each observed value from the sample average. If the deviations are small, the set of data is said to have low variability. The deviations provide information about the degree of dispersion in a set of observations. A general formula to evaluate the variability of data cannot be based on the deviations because some deviations are negative and some are positive and the sum of all the deviations is equal to zero. One possible solution is to compute the average deviation.

Average deviation—This measure is the average of the absolute values of the deviations from the sample average. Given K observations (e.g., revenues), $X_1, X_2, X_3, \ldots, X_K$, the average deviation of the data is computed as:

$$\bar{D} = \frac{\sum_{i=1}^{K} \left| X_i - \bar{X} \right|}{K}.$$

Table 2.3 shows how the average deviation is computed for our sample data. One aspect of the average deviation measure is that the procedure ignores the signs associated with deviations. Despite this disadvantage, its simplicity and ease of computation make it useful. In addition, a knowledge of the

TABLE 2.3

Computation of Average Deviation, Standard Deviation, and Variance

Observation Number (i)	Recorded Observation X_i	Deviation from Average	Absolute Value	Square of Deviation
1	3,000	37.5	37.5	1,406.25
2	1,200	−1762.5	1762.5	3,106,406.30
3	4,500	1537.5	1537.5	2,363,906.30
4	2,000	−962.5	962.5	926,406.25
5	3,200	237.5	237.5	56,406.25
6	1,900	−1062.5	1062.5	1,128,906.30
7	3,400	437.5	437.5	191,406.25
8	2,500	−462.5	462.5	213,906.25
9	3,400	437.5	437.5	191,406.25
10	2,500	−462.5	462.5	213,906.25
11	4,600	1637.5	1637.5	2,681,406.30
12	3,200	237.5	237.5	56,406.25
13	2,800	−162.5	162.5	26,406.25
14	2,400	−562.5	562.5	316,406.25
15	4,200	1237.5	1237.5	1,531,406.30
16	2,600	−362.5	362.5	131,406.25
Total	47,400.0	0.0	11,600.0	13,137,500.25
Average	2,962.5	0.0	725.0	821,093.77
Square root				906.14

average deviation helps explain the standard deviation, which is the most important measure of dispersion available.

Sample variance—This is the average of the squared deviations computed from a set of observations. If the variance of a set of observations is large, the data is said to have large variability. For example, large variability in the levels of productivity of a project team may indicate a lack of consistency or improper methods in the project functions. Given K observations (e.g., revenues), $X_1, X_2, X_3, \ldots, X_K$, the sample variance of the data is computed as:

$$s^2 = \frac{\sum_{i=1}^{K}(X_i - \bar{X})^2}{K-1}$$

The variance can also be computed by the following alternate formulas:

$$s^2 = \frac{\sum_{i=1}^{K}\left(X_i^2 - \left(\frac{1}{K}\right)\right)\left[\sum_{i=1}^{K} X_i\right]^2}{K-1}$$

$$s^2 = \frac{\sum_{i=1}^{K} X_i^2 - K\left(\bar{X}^2\right)}{K-1}$$

Using the first formula, the sample variance of the data in Table 2.3 is calculated as:

$$s^2 = \frac{13,137,500.25}{16-1}$$
$$= 875,833.33$$

The average calculated in the last column of Table 2.3 is obtained by dividing the total for that column by 16 instead of 16 – 1 = 15. That average is not the correct value of the sample variance. However, as the number of observations increases, the average as computed in the table will become a close estimate for the correct sample variance. Analysts distinguish the two values by referring to the average calculated in the table as the population variance when K is very large; they refer to the average calculated by the formulas above as the sample variance particularly when K is small. For our example, the population variance is given by:

$$\sigma^2 = \frac{\sum_{i=1}^{K}\left(X_i - \bar{X}\right)^2}{K}$$
$$= \frac{13,137,500.25}{16}$$
$$= 821,093.77$$

while the sample variance, as shown previously for the same data set, is given by:

$$\sigma^2 = \frac{\sum_{i=1}^{K}(X_i - \bar{X})^2}{K-1}$$
$$= \frac{13{,}137{,}500.25}{(16-1)}$$
$$= 875{,}833.33$$

Standard deviation—The sample standard deviation of a set of observations is the positive square root of the sample variance. The use of variance as a measure of variability has some drawbacks. For example, the knowledge of the variance is helpful only when two or more sets of observations are compared. Because of the squaring operation, the variance is expressed in square units rather than the original units of the raw data. To get a reliable feel for variability in the data, it is necessary to restore the original units by performing the square root operation on the variance. This is why standard deviation is a widely recognized measure of variability. Given K observations (e.g., revenues), $X_1, X_2, X_3, \ldots, X_K$, the sample standard deviation of the data is computed as:

$$s = \sqrt{\frac{\sum_{i=1}^{K}(X_i - \bar{X})^2}{K-1}}$$

As in the case of the sample variance, the sample standard deviation can also be computed by the following alternate formulas:

$$s = \sqrt{\frac{\sum_{i=1}^{K} X_i^2 - \left(\frac{1}{K}\right)\left[\sum_{i=1}^{K} X_i\right]^2}{K-1}}$$

$$s = \sqrt{\frac{\sum_{i=1}^{K} X_i^2 - K(\bar{X})^2}{K-1}}$$

Using the first formula, the sample standard deviation of the data in Table 2.3 is calculated as:

$$s = \sqrt{\frac{13{,}137{,}500.25}{(16-1)}}$$
$$= \sqrt{875{,}833.33}$$
$$= 935.8597$$

We can say that the variability in the expected revenue per project per quarter is $935,859.70. The population sample standard deviation is given by:

$$\sigma = \sqrt{\frac{\sum_{i=1}^{K}\left(X_i - \bar{X}\right)^2}{K}}$$

$$= \sqrt{\frac{13,137,500.25}{16}}$$

$$= \sqrt{821,093.77}$$

$$= 906.1423$$

while the sample standard deviation is:

$$s = \sqrt{\frac{\sum_{i=1}^{K}\left(X_i = \bar{X}\right)^2}{K-1}}$$

$$s = \sqrt{\frac{13,137,500.25}{(16-1)}}$$

$$= \sqrt{875,833.33}$$

$$= 935.8597$$

The results of data analysis can be reviewed directly to determine where and when project control actions may be needed. The results can also be used to generate control charts as discussed in the next section.

Control Charts

Control charts may be used to track project performance before deciding what control actions are needed. Control limits are incorporated into the charts to indicate when control actions should be taken. Multiple control limits may be used to determine various levels of control points. Control charts may be developed for costs, scheduling, resource utilization, performance, and other criteria.

Figure 2.17 illustrates periodic monitoring of project progress. Costs are monitored and recorded monthly. If costs were monitored more frequently (e.g., daily), the result would be a more rigid control structure. Of course, we

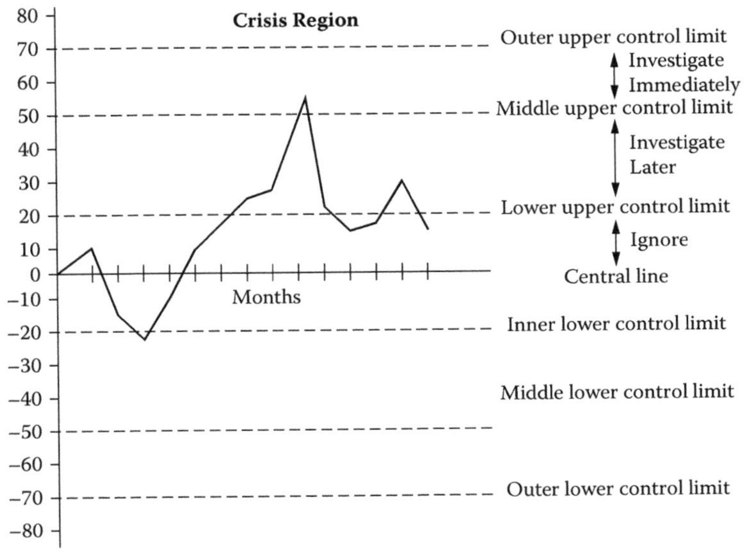

FIGURE 2.17
Control chart for project monitoring.

must decide whether the additional time needed for frequent monitoring is justified by the extra level of control provided. The control limits may be calculated with the same procedures used for X-bar and R charts in quality control or may be based on custom project requirements. In addition to drawing control charts to track costs, we can draw them for other measures of performance such as task duration, quality, or resource utilization.

Figure 2.18 shows a control chart for cumulative cost. The control limits on the chart are indexed to the project percent complete. Each percent complete point has a control limit that the cumulative project cost is not expected to exceed. A review of the control chart shows that the cumulative cost is out of control at the 10, 30, 40, 50, 60, and 80% completion points. Control actions should have been instituted right from the 10% completion point. If no control action is taken, the cumulative cost may continue to spiral out of control and eventually exceed the budget limit when the project is finished.

The information obtained from the project monitoring capabilities of project management software can be transformed into meaningful charts that can quickly identify when control actions are needed. A control chart can provide information about over-allocation of resources or unusually slow work progress. Figure 2.19 presents a chart for budget tracking and control. Starting at an initial level of $B, the budget is depleted in one of three possible modes. In case 1, the budget is depleted in such a way that a surplus is available at the end of the project cycle. This may be viewed as the ideal spending pattern. In case 2, the budget is depleted in an out-of-control pattern that

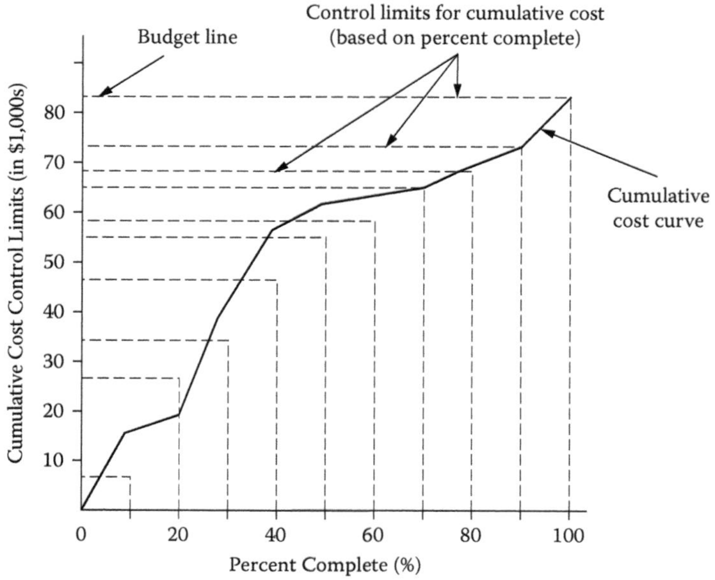

FIGURE 2.18
Control chart for cumulative cost.

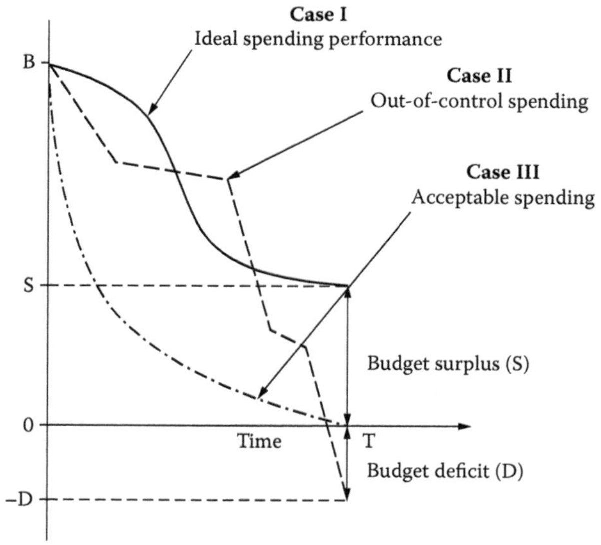

FIGURE 2.19
Chart for budget tracking and control.

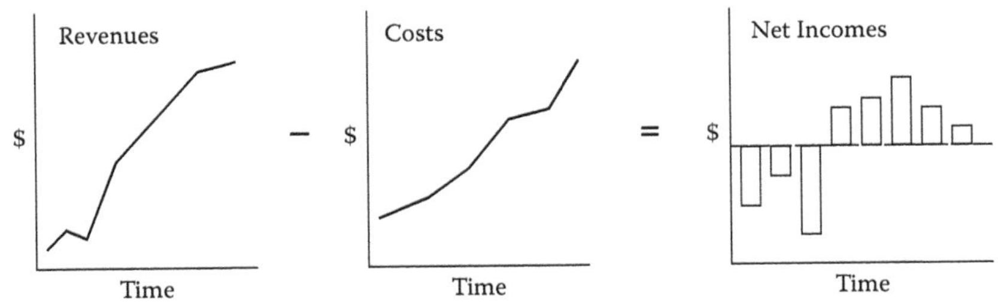

FIGURE 2.20
Review chart for revenue versus expenses.

leaves a deficit at the end of the project. In case 3, the budget is depleted at a rate that is proportional to the amount of work completed. In this case, the budget is zeroed out just as the project finishes. Intermediate control lines may be included in the chart so that needed control actions can be initiated at the appropriate times.

Figure 2.20 presents a chart for monitoring revenues versus expenses. As explained earlier, control limits may be added to a chart to determine when control actions should be taken.

Table 2.4 shows a report format for task progress analysis. The first three columns identify the task, its location, and its description. The activities column indicates the activities comprising the task. The next column denotes completed activities. Pending activities and past due activities are covered in the next two columns. The last column is intended to display comments about problems encountered. This control table helps focus control actions on specific activities in addition to overall project control actions. Table 2.5 presents a format for task-based time analysis. An additional feature is indicating percent completion for each task. The table also evaluates planned versus actual activities performed. Deviations from planned work should be explained.

Figure 2.21 is a graphical representation of task progress. This bar chart analysis would generate reports showing expected and actual completion times by task number, department, or project segment. Figure 2.22 is a bar chart showing project progress versus resource loading. This type of chart is useful for identifying the effects of resource allocation on percent completion.

Better performance can be achieved if more time and resources are available for a project. If lower costs and tighter schedules are desired, performance may have to be compromised and vice versa. From the view of the project manager, a project should be at the highest point along the performance axis. Of course, this represents an extreme case of getting something for nothing. From the view of project staff, the project should be at the point indicating highest performance, longest time, and most resources. This, of course, may be an unrealistic expectation since time and resources are typically in short supply. For project control, a feasible trade-off strategy must be developed.

TABLE 2.4
Task Analysis

Column 1	Column 2	Column 3	Column 4	Column 5	Column 6	Column 7	Column 8
Task #	Department	Description	Activities	Completed	Pending	Past Due	Comments

TABLE 2.5
Task-Based Time Analysis

Column 1	Column 2	Column 3	Column 4	Column 5	Column 6	Column 7	Column 8	Column 9
Task #	Department	Description	Expected Activity	Expected % Completed	Actual Activity	Actual % Completed	Deviation	Explain

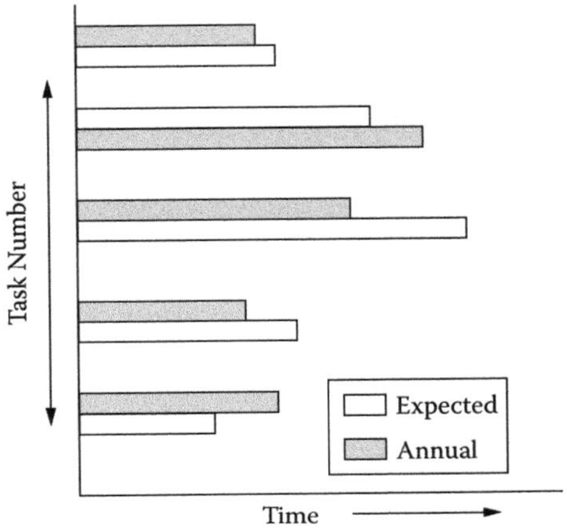

FIGURE 2.21
Graphical report on task progress.

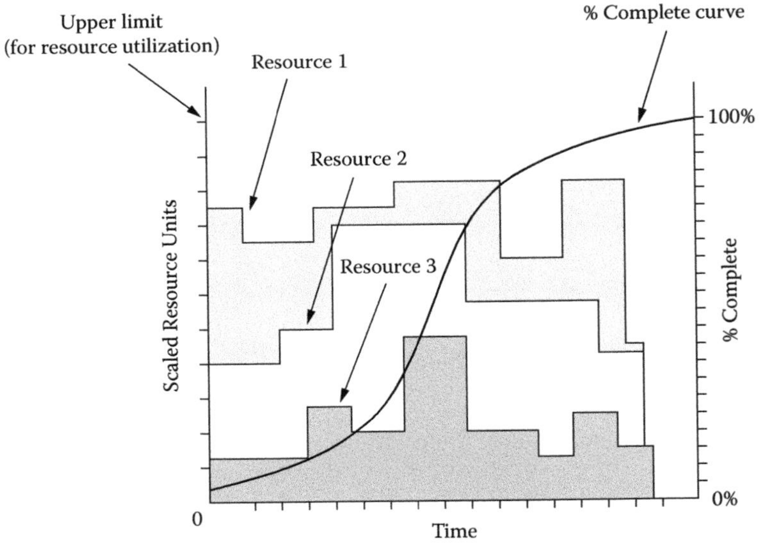

FIGURE 2.22
Resource loading versus project progress.

Although the control boundary is represented by a flat surface in Figure 2.7, it is obvious that the surface of the box will not be flat. If a multi-factor mathematical model is developed for the three factors, the nature of the response surface will vary depending on the specific interactions of the factors for a given project. Statistical control of a project requires that we recognize where risk and uncertainty will appear. Variability is a reality in project management. Uncertainty refers to the inability to predict the future accurately. Risk deals with the probability that something will happen and the loss that will result if it does happen. Risk and uncertainty may affect several project parameters including resource availability, activity durations, personnel productivity, budget, weather conditions, equipment failures, and cost. Statistical project control uses the techniques of probability and statistics to assess project performance and determine control actions.

3

Project Time Control

"On the flight of time,
What is the speed and direction of time?
Time flies; but it has no wings.
Time goes fast; but it has no speed.
Where has time gone? But it has no destination.
Time goes here and there; but it has no direction.
Time has no embodiment. It neither flies, walks, nor goes anywhere.
Yet, the passage of time is constant."

Adedeji Badiru

Project time, manifested in terms of project schedule, is a critical element of project management. To control time is to control the fate of a project. Project time management techniques constitute a major avenue to accomplishing goals and objectives in various government, business, military, industrial, and academic organizations. The techniques of project time management and control can be divided into three major categories:

- Qualitative managerial techniques
- Computational decision models
- Computer tools

This chapter focuses on computational network techniques for project schedule (time) control. Specifically, the network techniques of CPM (critical path method) and PERT (program evaluation and review technique) are covered. Network techniques emerged as a formal body of knowledge for project management during World War II and ushered in a golden era for operations research and quantitative modeling techniques. Badiru (1996) defines project management as the process of managing, allocating, and timing resources to achieve a given objective expeditiously. The phases of project management are:

- Planning
- Organizing
- Scheduling
- Controlling

The network techniques covered in this chapter apply primarily to the scheduling phase, although network planning is sometimes included in project planning. Every member of every organization needs project management to accomplish objectives. Consequently, the need for project management will continue to grow as organizations seek better ways to satisfy the constraints on:

- Schedules (time limitations)
- Costs (budget limitations)
- Performance (quality limitations)

The different tools of project management may change over time. Some will come and go, but the basic need to use network analysis to manage projects will remain constant. The network of activities in a project forms the basis for scheduling. CPM and PERT are the two most popular techniques for project network analysis. The precedence diagramming method (PDM) is also widely used, particularly for concurrent engineering applications. A project network is a graphical representation of the contents and objectives of the project. The basic project network analysis is typically implemented in three phases (network planning, network scheduling, and network control) that present the following advantages:

1. Advantages for communication:
   It clarifies project objectives.
   It establishes the specifications for project performance.
   It provides a starting point for more detailed task analysis.
   It presents a documentation of the project plan.
   It serves as a visual communication tool.
2. Advantages for control:
   It presents a measure for evaluating project performance.
   It helps determine what corrective actions are needed.
   It gives a clear message of what is expected.
   It encourages team interactions.
3. Advantages for team interaction:
   It offers a mechanism for a quick introduction to the project.
   It specifies functional interfaces on the project.
   It facilitates ease of application.
   It creates synergy between elements of the project.

Network planning is sometimes known as activity planning and involves the identification of relevant activities for a project. The required activities and their precedence relationships are determined in the planning phase. Precedence requirements may be determined on the basis of technological, procedural, or imposed constraints. The activities are then represented in the form of a network diagram. The two popular conventions for network diagrams are the activity-on-arrow (AOA) and the activity-on-node (AON). In the AOA approach, arrows represent activities and nodes represent starting and ending points of activities. In the AON approach, nodes represent activities and arrows represent precedence relationships. Time, cost, and resource requirement estimates are developed for each activity during network planning. Time estimates may be based on:

- Historical records
- Time standards
- Forecasting
- Regression functions
- Experiential estimates

Network scheduling is performed by using forward pass and backward pass computational procedures. These computations reveal the earliest and latest starting and finishing times for each activity. The slack time or float associated with each activity is determined by computations. The activity path with the minimum slack in the network is used to determine critical activities and also the duration of the project. Resource allocation, and time-cost trade-offs are other functions performed during network scheduling.

Network control involves tracking the progress of a project on the basis of the network schedule and taking corrective actions when needed—an evaluation of actual versus expected performance determines project deficiencies.

Critical Path Method

Precedence relationships in a CPM network fall into three major categories:

- Technical precedence
- Procedural precedence
- Imposed precedence

Technical precedence requirements arise from technical relationships among activities in a project. For example, in conventional construction, walls must be erected before the roof can be installed. Procedural precedence requirements are determined by policies and procedures that are often subjective and lack concrete justification. Imposed precedence requirements can be classified as resource-, project-, or environment-imposed. Resource shortages may require scheduling of one task before another. The current status of a project (e.g., percent completion) may determine that one activity should be performed before another. The environment of a project, for example, weather changes or the effects of concurrent projects, may determine precedence relationships of project activities.

The primary goal of a CPM analysis is the determination of the **critical path** that sets minimum completion time for a project. The computational analysis involves forward and backward pass procedures. The forward pass determines the earliest start and completion times for each activity in the network. The backward pass determines the latest start and completion times for each activity. Figure 3.1 depicts an activity network using the AON convention. The network is normally drawn from left to right. If this convention is followed, there is no need to use arrows to indicate the directional flow. The notations used for activity A in the network are explained below:

- *A* Activity Identification
- *ES* Earliest starting time
- *EC* Earliest completion time
- *LS* Latest starting time
- *LC* Latest completion time
- *t* Activity duration

During the forward pass analysis of a network, it is assumed that each activity will begin at its earliest starting time. An activity can begin as soon as its last predecessor activity is finished. The completion of the forward pass determines the earliest completion time of a project. The backward pass analysis

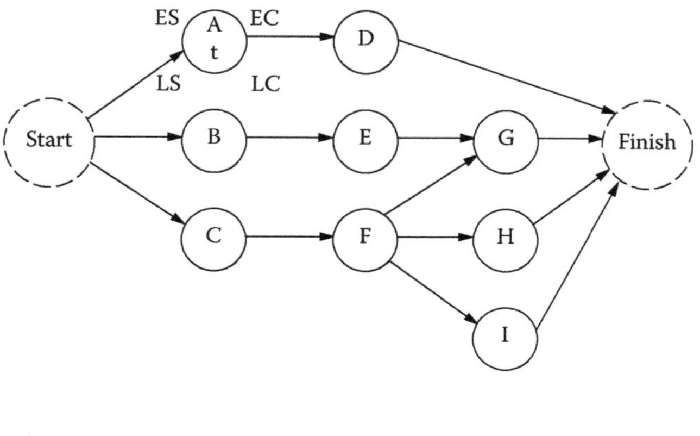

FIGURE 3.1
Activity network.

is a reverse procedure beginning at the latest project completion time and ending at the latest starting time of the first activity. The rules for implementing forward and backward pass analyses in CPM are presented below. The rules are implemented iteratively until the ES, EC, LS, and LC have been calculated for all nodes in the network.

Rule 1: Unless otherwise stated, the starting time of a project is set equal to time zero, that is, the first node in the network diagram has an earliest start time of zero.

Rule 2: The earliest start time (ES) for an activity is equal to the maximum of the earliest completion time (EC) of the immediate predecessor of the activity, that is, *ES = Maximum {Immediately Preceding EC}.*

Rule 3: The earliest completion time (EC) of an activity is its earliest start time plus its estimated duration, that is, *ES = ES + (Activity Time).*

Rule 4: The earliest completion time of a project is equal to the earliest completion time of the very last node in the network, that is, *EC of Project = EC of last activity.*

Rule 5: Unless the latest completion (LC) time of a project is explicitly specified, it is set equal to the earliest completion time of the project. This is called the *zero-project-slack* assumption, that is, *LC of Project = EC of Project.*

Rule 6: If a desired deadline is specified for the project, then *LC of Project = Specified Deadline.* Note that a latest completion time or deadline may be based on contract provisions.

Rule 7: The latest completion (LC) time for an activity is the smallest of the latest start times of the activity's immediate successors, that is, *LC = Minimum {Immediately Succeeding LSs}.*

Rule 8: The latest start time for an activity is the latest completion time minus the activity time, that is, *LS = LC-(Activity Time).*

CPM Example

Table 3.1 presents the data for an illustrative project. This network and extensions of it will be used for other computational examples in this chapter. The AON network for the example appears in Figure 3.2. Dummy activities are included in the network to designate single starting and ending points for the project.

Forward Pass

The forward pass calculations are shown in Figure 3.3. Zero is entered as the ES for the initial (dummy) node. Thus, EC for the starting node is equal to its ES. The ES values for the immediate successors of the starting node are set

TABLE 3.1
Data for Sample Project for CPM Analysis

Activity	Predecessor	Duration (Days)
A	—	2
B	—	6
C	—	4
D	A	3
E	C	5
F	A	4
G	B, D, E	2

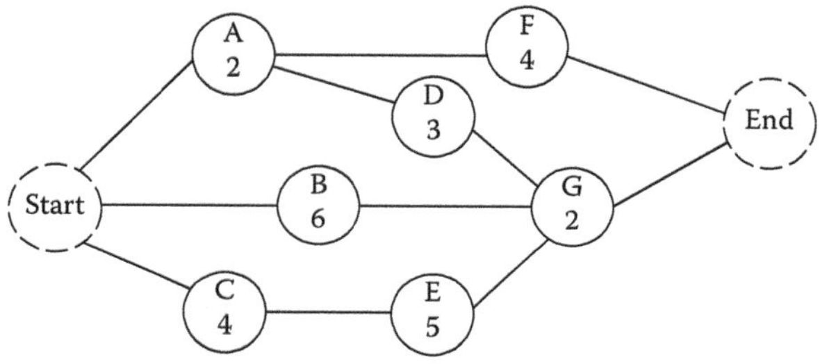

FIGURE 3.2
Project network for illustrative example.

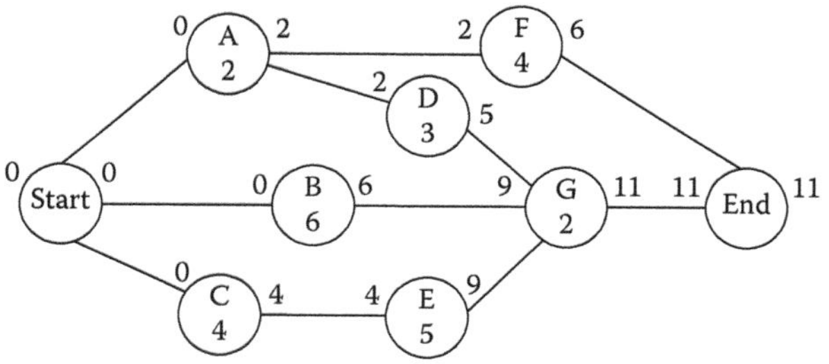

FIGURE 3.3
Forward pass analysis for CPM example.

equal to the EC of the start node and the resulting EC values are computed. Each node is treated as a start node for its successor or successors. However, if an activity has more than one predecessor, the maximum of the ECs of the preceding activities is used as the activity starting time. This happens in the case of activity G, whose ES is determined as Max {6,5,9} = 9. The earliest project completion time for the example is 11 days. Note that this is the maximum of the immediately preceding earliest completion times: Max {6,11} = 11. Since the dummy ending node has no duration, its earliest completion time is set equal to its earliest start time of 11 days.

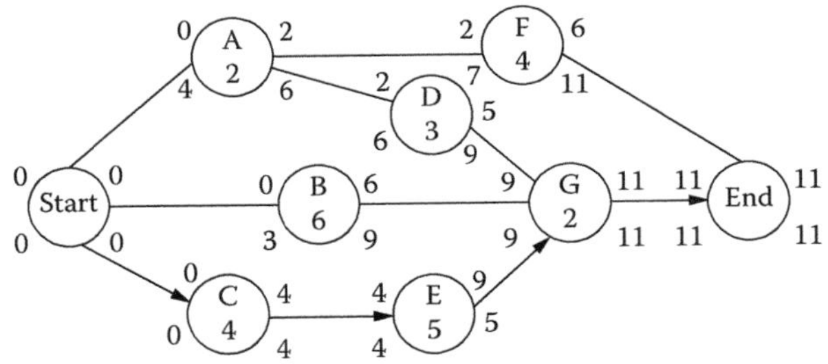

FIGURE 3.4
Backward pass analysis for CPM example.

Backward Pass

The backward pass computations establish the latest start (LS) and latest completion (LC) times for each node in the network. The results of the backward pass computations are shown in Figure 3.4. Since no deadline is specified, the latest completion time of the project is set equal to the earliest completion time. By back tracking and using the network analysis rules presented earlier, the latest completion and start times are determined for each node. Note that in the case of activity A with two successors, the latest completion time is determined as the minimum of the immediately succeeding latest start times, that is, Min {6,7} = 6. A similar situation occurs for the dummy starting node. In that case, the latest completion time of the dummy start node is Min {0,3,4} = 0. Since this dummy node has no duration, the latest starting time of the project is set equal to the node's latest completion time. Thus, the project starts at time 0 and is expected to be completed by time 11.

A project network usually involves several possible paths, a number of activities that must be performed sequentially, and some activities that may be performed concurrently. If an activity has ES and LS times that are not equal, its actual start and completion times may be flexible. The amount of flexibility of an activity is called slack. The slack time is used to determine the critical activities in a network as discussed below.

Determination of Critical Activities

The critical path is defined as one revealing the least slack in the network diagram. All the activities on the critical path are said to be critical. They can create bottlenecks in the network if they are delayed. The critical path is also the longest path in the network diagram. In some networks, particularly large ones, it is possible to have multiple critical paths. For a network with a large number of paths, it may be difficult to identify all the critical paths visually. The slack time of an activity is also known as its *float*. The four basic types of slack are described below:

Total slack (TS)—TS is the amount of time an activity may be delayed from its earliest starting time without delaying the latest completion time of the project. The total slack time of an activity is the difference between the latest completion time and the earliest completion time of the activity, or the difference between the latest starting time and the earliest starting time of the activity. TS = LC – EC or TS = LS – ES. TS is used to determine the critical activities in a project network. The critical activities are identified as those having the minimum total slack in the network diagram. If a network has only one critical path, all critical activities will be on that one path.

Free slack (FS)—FS is the amount of time an activity may be delayed from its earliest starting time without delaying the starting times of any of its immediate successors. Activity FS is calculated as the difference between the minimum earliest starting time of the activity's successors and the earliest completion time of the activity. FS = Min {Succeeding ESs} – EC.

Interfering slack (IS)—IS or interfering float is the amount of time by which an activity interferes with (or obstructs) its successors when its total slack is fully used. This measure is rarely used in practice. The interfering float is computed as the difference between the total slack and the free slack. IS = TS – FS.

Independent slack or independent float (IF)—Independent float or independent slack is the amount of float that an activity will always have regardless of the completion times of its predecessors or the starting times of its successors. IF = Max $\{0,(ES_j - LC_i - t)\}$, where ES_j is the earliest starting time of the preceding activity, LC_i is the latest completion time of the succeeding activity, and t is the duration of the activity whose independent float is calculated. Independent float takes a pessimistic view. It evaluates a situation whereby the activity is pressured from both sides—when its predecessors are delayed as late as possible and its successors are to be started as early as possible. Independent float is useful for conservative planning but is rarely used in practice. Despite its low level of use, independent float has practical implications for better project management. Activities can be buffered with independent floats as a way to handle contingencies.

In Figure 3.4, the total slack and the free slack for activity A are calculated, respectively, as

TS = 6 – 2 = 4 days and FS = Min {2,2} – 2 = 2 – 2 = 0. Similarly, the total slack and free slack for activity F are TS = 11 – 6 = 5 days and FS = Min {11} – 6 = 11 – 6 = 5 days.

Table 3.2 tabulates the results of the CPM example. It shows the earliest and latest times for each activity as well as the total and free slacks. The results indicate that the minimum total slack in the network is zero. Thus, activities

TABLE 3.2
CPM Analysis Results for Sample Project

Activity	Duration	ES	EC	LS	LC	TS	FS	Criticality
A	2	0	2	4	6	4	0	—
B	6	0	6	3	9	3	3	—
C	4	0	4	0	4	0	0	Critical
D	3	2	5	6	9	4	4	—
E	5	4	9	4	9	0	0	Critical
F	4	2	6	7	11	5	5	—
G	2	9	11	9	11	0	0	Critical

C, E, and G are identified as critical. Figure 3.4 highlights the critical path based on the following sequence of activities:

$$\text{Start} \rightarrow \text{C} \rightarrow \text{E} \rightarrow \text{G} \rightarrow \text{end}$$

The total slack for the overall project is equal to the total slack observed on the critical path. The minimum slack in most networks will be zero since the ending LC is set equal to the ending EC. If a deadline is specified for a project, then we would set the project's latest completion time to the specified deadline. In that case, the minimum total slack in the network would be TS_{Min} = (project deadline) – EC of last node.

This minimum total slack will then appear as the total slack for each activity on the critical path. If a specified deadline is lower than the EC at the finish node, the project will start with a negative slack—it will be behind schedule before it even starts. It may then become necessary to expedite some activities (i.e., crashing) to overcome the negative slack. Figure 3.5 shows an example with a specified deadline of 18 days after the earliest completion time of the last node in the network.

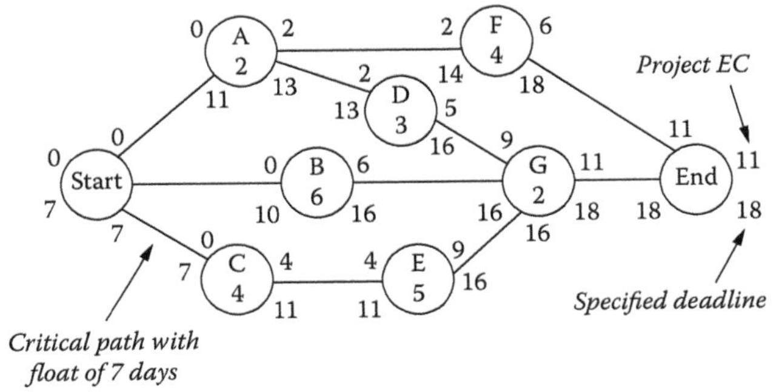

FIGURE 3.5
CPM network with deadline.

Using Forward Pass to Determine Critical Path

The critical path in CPM analysis can be determined only from the forward pass. This may be helpful if it is necessary to identify critical activities quickly without performing all the other calculations needed to determine the latest starting times, latest completion times, and total slacks. The steps for determining the critical path from the forward pass only are:

- Complete the forward pass in the usual manner.
- Identify the last node in the network as a critical activity.
- Work backward from the last node. Whenever a merge node occurs, the critical path will be along the path where the earliest completion time (EC) of the predecessor is equal to the earliest start time (ES) of the current node.
- Continue the backtracking from each critical activity until the project starting node is reached. Note that if the network has a single starting node or single ending node in the network, that node will always be on the critical path.

Subcritical Paths

A large network may contain paths that are near critical. Such paths require almost as much attention as the critical path since they exhibit the potential to become critical when changes occur in the network. Analysis of subcritical paths may help classify tasks into ABC categories on the basis of Pareto analysis. This type of analysis separates the few "vital" activities from the many "trivial" activities and permits more efficient allocations of resources. The principle of originated from the work of Italian economist Vilfredo Pareto (1848–1923) who discovered that most of the wealth in his country was held by a few individuals.

For project control purposes, the Pareto principle states that 80% of the bottlenecks are caused by only 20% of the tasks. This principle is applicable to many management processes. For example, in cost analysis, one can infer that 80% of the total cost is associated with only 20% of the cost items. Similarly, 20% of an automobile's parts cause 80% of the maintenance problems. In personnel management, about 20% of the employees account for about 80% of the absenteeism. For critical path analysis, 20% of the network activities will require 80% of the control efforts. The ABC classification based on the Pareto analysis divides items into three priority categories: A (most important), B (moderately important), and C (least important). Appropriate percentages (e.g., 20%, 25%, 55%) may be assigned to the categories. Pareto analysis allows a project manager to shift his or her focus only on the critical path to managing critical and near-critical tasks. The level of criticality of each path may be assessed by these steps:

Sort paths in increasing order of total slack.

Partition sorted paths into groups based on the magnitudes of their total slacks.

Sort the activities within each group in increasing order of their earliest starting times.

Assign the highest level of criticality to the first group of activities (e.g., 100%); this group represents the usual critical path.

Calculate the relative criticality indices for the other groups in decreasing order of criticality.

Define the following variables:

α_1 = minimum total slack in network

α_2 = maximum total slack in network

β = total slack for path whose criticality is to be calculated.

Compute the path's criticality as:

$$\lambda = \frac{\alpha_2 - \beta}{\alpha_2 - \alpha_1}(100\%).$$

This procedure yields relative criticality levels between 0 and 100%. Table 3.3 presents a hypothetical example of path criticality indices. The criticality

TABLE 3.3

Analysis of Subcritical Paths

Path No.	Activities on Path	Total Slack	Λ (%)	λ' (%)
1	A, C, G, H	0	100	10
2	B, D, E	1	97.56	9.78
3	F, I	5	87.81	8.90
4	J, K, L	9	78.05	8.03
5	O, P, Q, R	10	75.61	7.81
6	M, S, T	25	39.02	4.51
7	N, AA, BB, U	30	26.83	3.42
8	V, W, X	32	21.95	2.98
9	Y, CC, EE	35	17.14	2.54
10	DD, Z, FF	41	0	1.00

level may be converted to a scale between 1 (least critical) and 10 (most critical) by the expression below:

$$\lambda' = 1 + 0.09\lambda.$$

Gantt Chart

When the results of a CPM analysis are fitted to a calendar time, a project plan becomes a schedule. The Gantt chart is one of the most widely used tools for presenting a project schedule because it can show planned and actual progress. A time scale is indicated along the horizontal axis. Horizontal bars or lines representing activities are ordered along the vertical axis. As a project progresses, markers are made on the activity bars to indicate actual work accomplished. Gantt charts must be updated periodically to indicate project status. Figure 3.6 is a Gantt chart for our illustrative example using the earliest starting (ES) times from Table 3.2. Figure 3.7 is a Gantt chart for the example based on the latest starting (LS) times. Critical activities are indicated by shaded bars.

The Figure 3.6 chart is based on earliest starting time. Based on CPM network computations, activity F can be delayed from day 2 until day 7 (TS = 5) without delaying the overall project. Likewise, A, D, or both may be delayed by a combined total of 4 days (TS = 4) without delaying the project. If all the 4 days of slack are used up by A, D cannot be delayed. If A is delayed by 1 day, D can only be delayed up to 3 days without delaying G, which determines project completion. CPM computations also reveal that activity B may be delayed up to 3 days without affecting the project completion time.

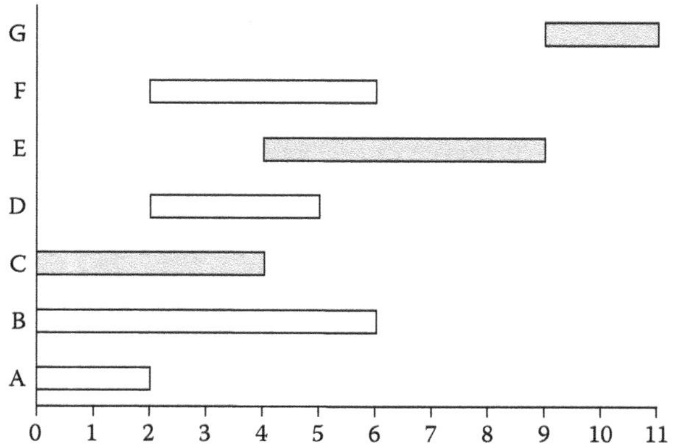

FIGURE 3.6
Gantt chart based on earliest starting times.

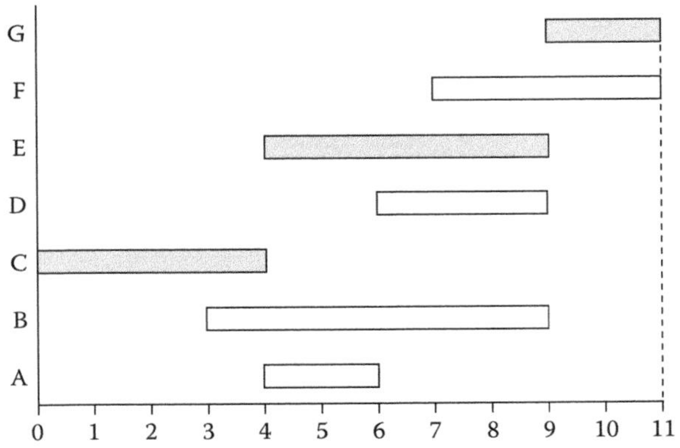

FIGURE 3.7
Gantt chart based on latest starting times.

Figure 3.7 shows activities are scheduled by their latest completion times. This represents a pessimistic case—activity slack times are fully used. No activity in this schedule can be delayed without delaying the project. Only one activity is scheduled over the first 3 days. This may be compared to the schedule in Figure 3.6, which lists three starting activities. The schedule in Figure 3.7 may be useful in situations that permit only a few activities to be scheduled in the early stages of the project, for example, shortages of personnel, lack of initial budget, time for project initiation, time for personnel training, allowance for learning period, or general resource constraints. Scheduling of activities based on ES times indicates an optimistic view. Scheduling on the basis of LS times represents a pessimistic approach.

Variations

A basic Gantt chart does not show precedence relationships among activities. It may be modified to show these relationships by coding appropriate bars, as shown by the cross-hatched bars in Figure 3.8. Other simple legends can be added to show which bars are related by precedence linking. The Figure 3.9 Gantt chart presents a comparison of planned and actual schedules. Note that two tasks are in progress at the time indicated in the figure. One of the ongoing tasks is unplanned. Figure 3.10 shows important milestones. The bars in Figure 3.11 represent a combination of related tasks.

Tasks may be combined for scheduling purposes or for conveying functional relationships required in a project. Figure 3.12 charts project phases. Each phase is further divided into parts. Figure 3.13 is a Gantt chart for multiple projects. Such charts are useful for evaluating resource allocation strategies. Resource loading over multiple projects may be needed for capital budgeting and cash flow analysis decisions. Figure 3.14 is a project slippage

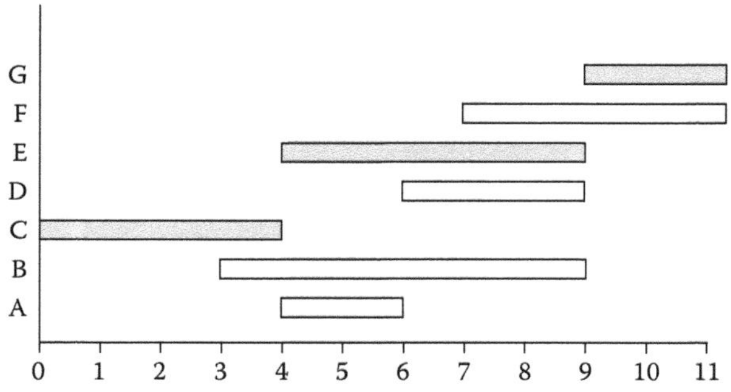

FIGURE 3.8
Coding of bars related by precedence linking.

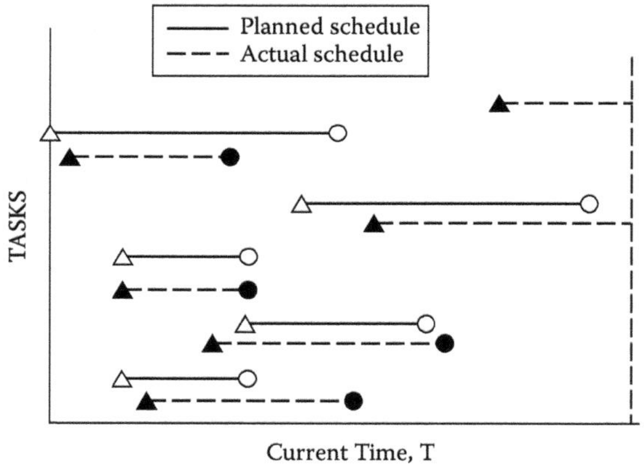

FIGURE 3.9
Progress monitoring Gantt chart.

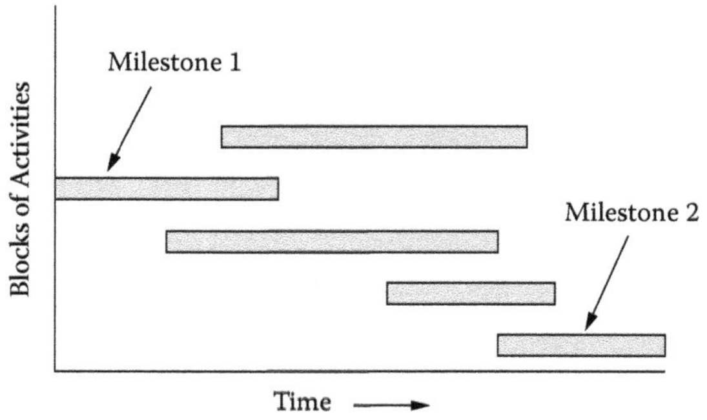

FIGURE 3.10
Milestone Gantt chart.

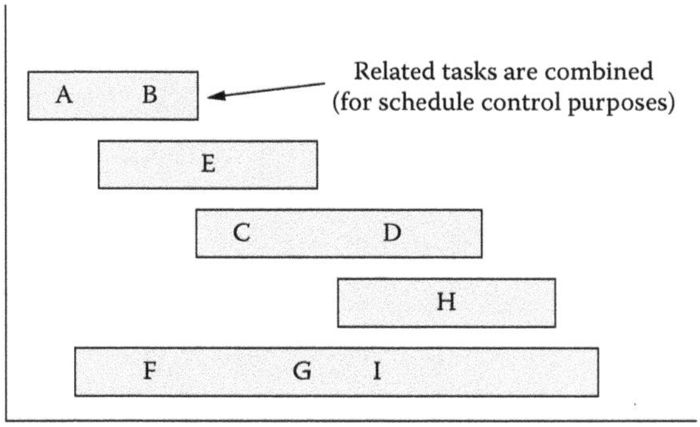

FIGURE 3.11
Task combination Gantt chart.

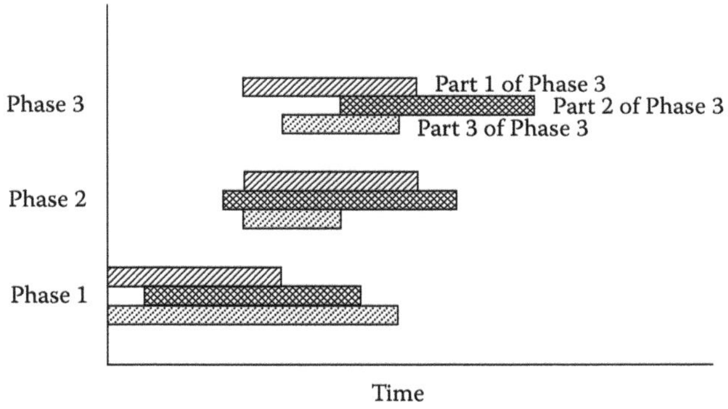

FIGURE 3.12
Phase-based Gantt chart.

chart useful for project tracking and control. Other variations of the basic Gantt chart may be developed for specific needs.

Activity Crashing and Schedule Compression

Schedule compression is a reduction of the length of a project network. Reduction is often accomplished by "crashing" activities. Crashing, also known as expediting, reduces activity durations, thereby reducing project duration. Crashing requires a trade-off between shorter task duration and higher task cost. Whether the total cost savings realized from reducing project duration justify the higher costs of reducing individual task durations must be determined. If a project involves delay penalties, it may be possible

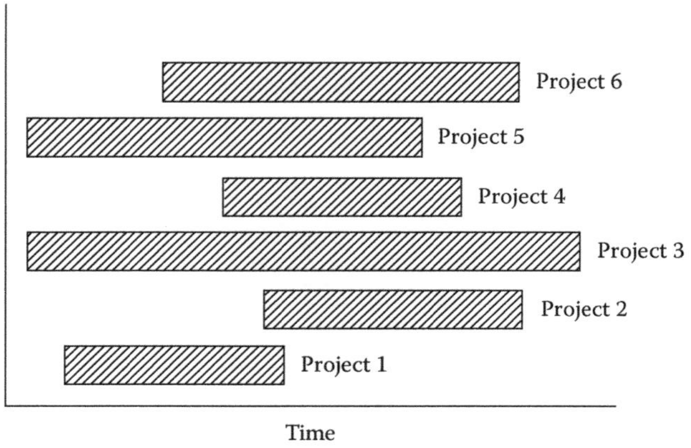

FIGURE 3.13
Multiple-project Gantt chart.

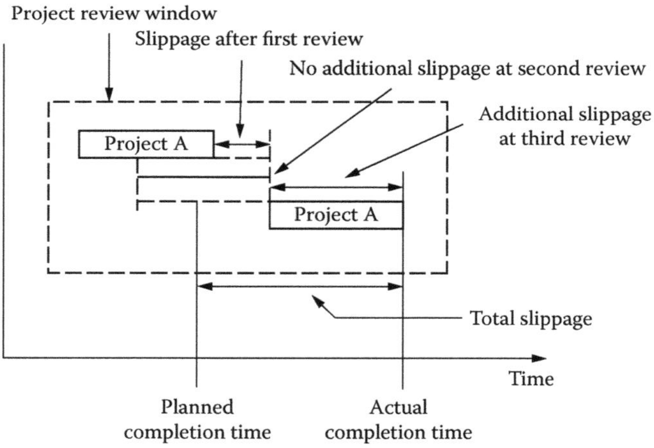

FIGURE 3.14
Project slippage tracking Gantt chart.

to reduce total project cost even though individual task costs are increased by crashing. If the potential cost saving by preventing a delay penalty is higher than the incremental cost of reducing the project duration, crashing is justified.

With conventional crashing, the more the duration of a project is compressed, the higher its total cost. The objective is to determine at what point to terminate further crashing in a network. **Normal task duration** is the time required to perform a task under normal circumstances. **Crash task duration**

TABLE 3.4

Normal and Crash Time and Cost Data

Activity	Normal Duration	Normal Cost	Crash Duration	Crash Cost	Crash Ratio
A	2	$210	2	$210	0
B	6	400	4	600	100
C	4	500	3	750	250
D	3	540	2	600	60
E	5	750	3	950	100
F	4	275	3	310	35
G	2	100	1	125	25
		$2,775		$3,545	

is the reduced time needed to perform a task when additional resources are allocated to it.

If each activity is assigned a range of time and cost estimates, several combinations of time and cost values will be associated with the overall project. Iterative procedures are used to determine the best time and cost combinations. Time and cost trade-off analyses may be conducted, for example, to determine the marginal cost of reducing the duration of a project by one time unit. Table 3.4 presents an extension of the data for the earlier example to include normal and crash times and costs for each activity. The normal duration of the project is 11 days, as seen earlier, and the normal cost is $2,775. If all the activities are reduced to their respective crash durations, the total crash cost of the project will be $3,545. In that case, the crash time based on CPM is 7 days. The CPM network for the fully crashed project is shown in Figure 3.15. Note that activities C, E, and G remain critical. The crashing of activities may result in a new critical path. The Gantt chart in Figure 3.16 shows a schedule of the crashed project using ES times. In practice, one would not crash all activities in a network. Rather, some heuristic would be

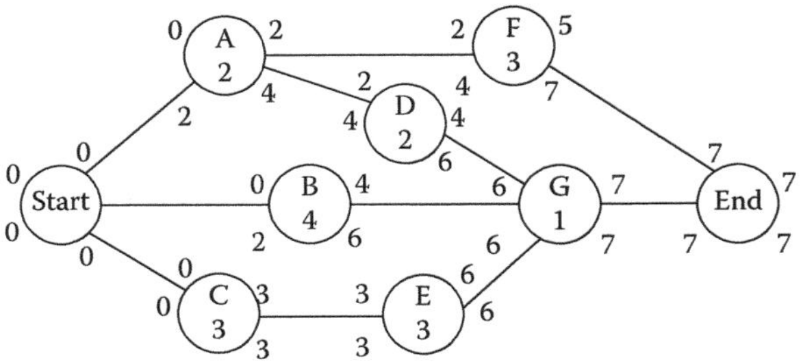

FIGURE 3.15
Fully crashed CPM network.

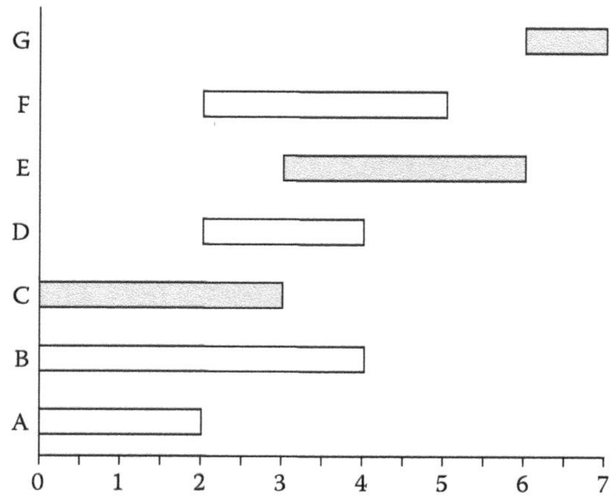

FIGURE 3.16
Gantt chart of fully crashed CPM network.

used to determine which activity should be crashed and by how much. One approach is to crash only the critical activities or those with the best ratios of incremental cost versus time reduction. The last column in Table 3.4 presents the respective ratios for the activities in our example. The crashing ratios are computed as:

$$r = \frac{\text{Crash cost} - \text{normal cost}}{\text{Normal duration} - \text{crash duration}}$$

This method of computing the crashing ratio gives crashing priority to the activity with the lowest cost slope. It is a common approach in CPM networks.

Activity G offers the lowest cost per unit time reduction of $25. If our approach is to crash only one activity at a time, we may decide to crash G first and evaluate the increase in project cost versus the reduction in project duration. The process can then be repeated for the next best candidate for crashing (activity F). After F has been crashed, activity D can then be crashed.

This approach is repeated iteratively in order of activity preference until no further reduction in project duration can be achieved or the total project cost exceeds a specified limit. A more comprehensive analysis is to evaluate all possible combinations of the activities that can be crashed. However, such a complete enumeration would be prohibitive, since a total of 2^c crashed networks would have to be evaluated. The c is the number of activities that can be crashed among n activities in the network ($c \leq n$). For our example, only six of the seven activities in the sample network can be crashed. Thus, a complete enumeration will involve $2^6 = 64$ alternate networks. Table 3.5 shows 7 of the 64 crashing options. Activity G, which offers the best crashing ratio, reduces the project duration by only 1 day. Even though activities F, D,

TABLE 3.5
Selected Crashing Options for CPM Example

Option No.	Activities Crashed	Network Duration	Time Reduction	Incremental Cost	Total Cost
1	None	11	—	—	2775
2	G	10	1	25	2800
3	G, F	10	0	35	2835
4	G, F, D	10	0	60	2895
5	G, F, D, B	10	0	200	3095
6	G, F, D, B, E	8	2	200	3295
7	G, F, D, B, E, C	7	1	250	3545

and B are crashed by a total of 4 days at an incremental cost of $295, they do not reduce project duration. Activity E is crashed by 2 days and generates a reduction of 2 days in project duration. Activity C, crashed by 1 day, further reduces project duration by 1 day. Activities that generate reductions in project duration were identified earlier as critical.

Figure 3.17 shows the crashed project duration versus the crashing options. Figure 3.18 plots total project cost after crashing versus the selected crashing options. As more activities are crashed, the project duration decreases and total project cost increases. If full enumeration were performed, Figure 3.17 would show additional points between the minimum possible project duration of 7 days (fully crashed) and the normal project duration of 11 days (no crashing). Similarly, the plot for total project cost (Figure 3.18) would contain additional points between the normal cost of $2,775 and the crash cost of $3,545.

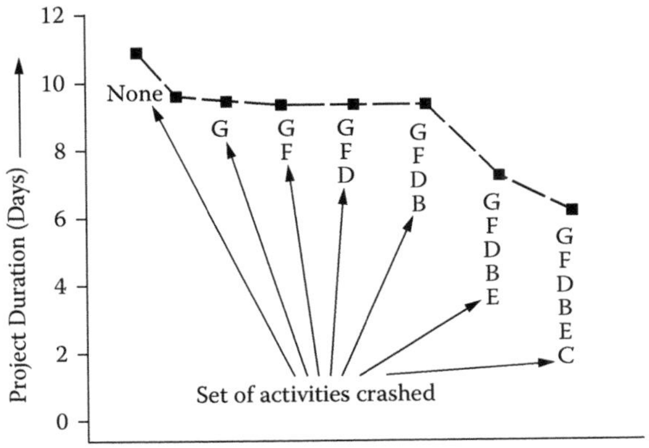

FIGURE 3.17
Duration as function of crashing options.

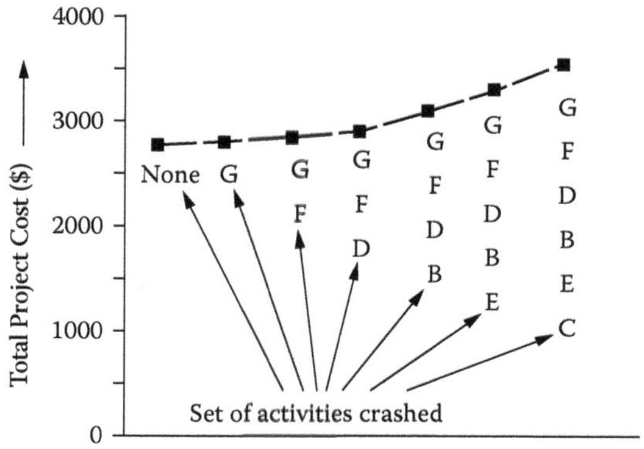

FIGURE 3.18
Project cost as function of crashing options.

Several other approaches exist for determining which activities to crash in a project network. Two alternate approaches are presented for computing the crashing ratio *r*. Let *r* = *criticality index* or let:

$$r = \frac{\text{Crash cost} - \text{normal cost}}{(\text{Normal duration} - \text{crash duration})\ (\text{Criticality index})}.$$

The first approach uses a *criticality index* criterion and gives crashing priority to the activity with the highest probability of being on the critical path. In deterministic networks, these are critical activities. In stochastic networks, an activity is expected to fall on the critical path only a percentage of the time. The second approach is a combination of the one used for the illustrative example and the criticality index. It reflects the process of selecting the least-cost expected value. The denominator of the expression represents the expected number of days by which the critical path can be shortened. For different project networks, different crashing approaches should be considered, and the one that best fits the nature of the network should be selected.

PERT Network Analysis

The program evaluation review technique (PERT) is an extension of CPM incorporating variability in activity durations into project network analysis. PERT has been used extensively and successfully. In real life, activities are often prone to uncertainties that determine the actual durations of activities. In CPM, activity durations are assumed to be deterministic. In PERT, the potential uncertainties in activity durations are accounted for by using three time estimates for each activity. The three time estimates represent the spread of the estimated activity duration. The greater the uncertainty of an activity, the wider the range of the estimates.

Estimates and Formulas

PERT uses three time estimates (optimistic, most likely, and pessimistic) to compute the expected duration and variance for each activity. The formulas are based on a simplification of the expressions for the mean and variance of a beta distribution. The approximation formula for the mean is a simple weighted average of the three time estimates, with the end points assumed to be equally likely and the mode four times as likely. The approximation formula for PERT is based on the recognition that most of the observations from a distribution will lie within plus or minus three standard deviations or a spread of six standard deviations. This leads to the simple method of setting the PERT formula for standard deviation equal to one-sixth of the estimated duration range. While there is no theoretical validation for these approximation approaches, the PERT formulas facilitate use and are presented below:

$$t_e = \frac{a + 4m + b}{6}$$

$$s = \frac{(b - a)}{6},$$

where a = optimistic time estimate; m = most likely time estimate; b = pessimistic time estimate $a < m < b$; t_e = expected time for activity; and s^2 = variance of duration of activity. After obtaining the estimate of the duration for each activity, the network analysis is carried out in the same manner illustrated for CPM. The major steps in PERT analysis are summarized below:

Obtain three time estimates a, m, and b for each activity.

Compute the expected duration for each activity using the formula for t_e.

Compute the variance of the duration of each activity from the formula for s^2.

Compute the expected project duration, T_e. As in the case of CPM, the duration of a project in PERT analysis is the sum of the durations of the activities on the critical path.

Compute the variance of the project duration as the sum of the variances of the activities on the critical path.

The variance of the project duration is denoted by s^2. Recall that CPM cannot compute variances of project duration because variances of activity durations are not computed. If a network has two or more critical paths, choose the one with the largest variance to determine the project duration and the variance of the project duration. Thus, PERT is pessimistic with respect to the variance of project duration when a network has multiple critical paths. For some networks, it may be necessary to perform a mean variance analysis to determine the relative importance of the multiple paths by plotting the expected project duration versus the path duration variance.

If desired, compute the probability of completing the project within a specified time period. This is not possible with CPM.

Modeling Activity Times

In practice, a question often arises as to how to obtain good estimates of *a*, *m*, and *b*. Several approaches can be used in obtaining the time estimates for PERT. Some of the approaches are:

- Estimate furnished by an experienced person
- Estimate extracted from standard time data
- Estimate obtained from historical data
- Estimate obtained from simple regression and/or forecasting
- Estimate generated by simulation
- Estimate derived from heuristic assumptions
- Estimate dictated by customer requirements

The pitfall of using estimates furnished by an individual is that they may be inconsistent because they are limited by the experience and personal bias of the person providing them. Individuals responsible for furnishing time estimates are usually not experts in estimation and generally have difficulty providing accurate PERT time estimates. They tend to select values of *a*, *m*, and *b* that are optimistically skewed and assign conservatively large values to *b*.

The use of time standards, on the other hand, may not reflect the changes in the current operating environment due to new technology, work simplification, new personnel, and other factors. Using historical data and forecasts is very popular because estimates can be verified and validated by records. Regression and forecasting present the danger of extrapolation beyond the data range used for fitting the models. If the sample size in a historical data set is sufficient and the data can be assumed to reasonably represent prevailing operating conditions, the three PERT estimates can be computed as follows:

$$\hat{a} = \bar{t} - kR$$

$$\hat{m} = \bar{t}$$

$$\hat{b} = \bar{t} + kR$$

where R = range of sample data; $\bar{t}$ = arithmetic average of sample data

$$k = 3/d_2$$

where d_2 = adjustment factor for estimating standard deviation of a population.

If $kR > \bar{t}$, set $a = 0$ and $b = 2\bar{t}$. The factor d_2 is widely tabulated in the quality control literature as a function of the number of sample points n. Selected values of d_2 are presented below.

n	5	10	15	20	25	30	40	50	75	100
d_2	2.326	3.078	3.472	3.735	3.931	4.086	4.322	4.498	4.806	5.015

As mentioned earlier, activity times can be determined from historical data. The procedure involves three steps:

Appropriate organization of historical data into histograms.

Determination of a distribution that reasonably fits the shape of the histogram.

Testing the goodness of fit of the hypothesized distribution via an appropriate statistical model.

The chi-square and Kolmogrov-Smirnov (K-S) tests are two popular methods for testing goodness of fit. Most statistical texts present the steps of goodness-of-fit tests.

Beta Distribution

PERT analysis assumes that the probabilistic properties of activity duration can be modeled by the beta probability density function. The beta distribution is defined by two end points and two shape parameters. It is used as a reasonable distribution to model activity times because it has finite end points and can assume a variety of shapes based on different parameters. While the true distribution of activity time will rarely be known, the beta distribution serves as an acceptable model. Figure 3.19 shows the beta distribution as a function of its shape parameters. The uniform distribution

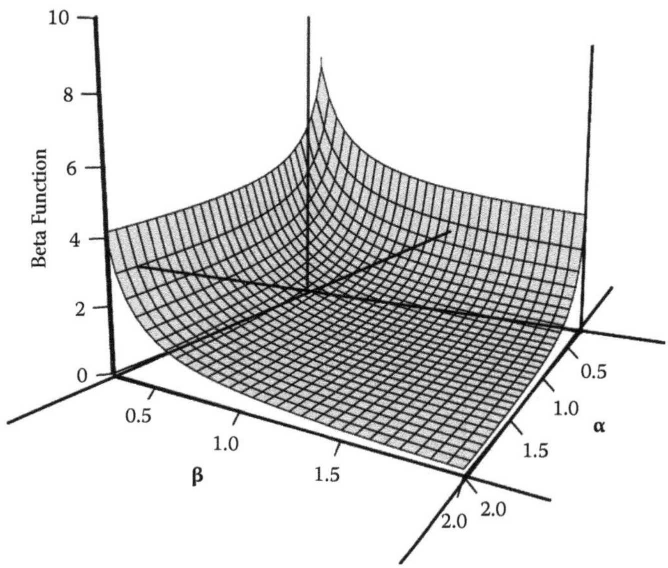

FIGURE 3.19
Beta distribution as function of shape parameters.

between 0 and 1 is a special case of the beta distribution with both shape parameters equal to one. The versatility of the beta distribution makes it effective for project time analysis and control.

The standard beta distribution is defined over the interval 0 to 1, while the general beta distribution is defined over any interval *a* to *b*. The general beta probability density function is given by:

$$f(t)=\frac{\Gamma(\alpha+\beta)}{\Gamma(\alpha)\Gamma(\beta)}\times\frac{1}{(b-a)^{\alpha+\beta-1}}\times(t-a)^{\alpha-1}(b-t)^{\beta-1}$$

for $a \le t \le b$ and $\alpha > 0$, $\beta > 0$. a = lower end point of distribution; b = upper end point of distribution; and α, β are shape parameters for the distribution. The mean, variance, and mode of the general beta distribution are defined as:

$$\mu = a+(b-a)\frac{\alpha}{\alpha+\beta}$$

$$\sigma^2 = (b-a)^2\frac{\alpha\beta}{(\alpha+\beta+1)(\alpha+\beta)^2}$$

$$m=\frac{a(\beta-1)+b(\alpha-1)}{\alpha+\beta-2}$$

The general beta distribution can be transformed into a standardized distribution by changing its domain from [*a*,*b*] to the unit interval, [0,1]. This is accomplished by using the relationship $t_s = a + (b-a)t_s$, where t_s is the standard beta random variable between 0 and 1. This yields the standardized beta distribution, given by:

$$f(t) = \frac{\Gamma(\alpha+\beta)}{\Gamma(\alpha)\Gamma(\beta)} t^{\alpha-1}(1-t)^{\beta-1}; \quad 0 < t < 1; \quad \alpha, \beta > 0.$$

The mean, variance, mode, and skewness of the standard beta distribution are defined as follow:

$$\mu = \frac{\alpha}{\alpha+\beta}$$

$$\sigma^2 = \frac{\alpha\beta}{(\alpha+\beta+1)(\alpha+\beta)^2}$$

$$m = \frac{\alpha-1}{\alpha+\beta-2}, \text{ if } \alpha, \beta > 1$$

$$e = \frac{2(\beta-\alpha)\sqrt{\alpha+\beta+1}}{\sqrt{\alpha\beta}(\alpha+\beta+2)}$$

A summary of the different shapes of the beta distribution, depending on the values of the two shape parameters, is presented below:

$\alpha = 1$, $\beta = 1$ is the uniform [0,1] distribution.
$\alpha < 1$, $\beta < 1$ is U-shaped.
$\alpha = 1/2$, $\beta = 1/2$ is the arcsine distribution.
$\alpha < 1$, $\beta \geq 1$ or $\alpha = 1$, $\beta > 1$ is strictly decreasing.
$\alpha = 1$, $\beta > 2$ is strictly convex.
$\alpha = 1$, $\beta = 2$ is a straight line.
$\alpha = 1$, $1 < \beta < 2$ is strictly concave.
$\alpha = 1$, $\beta < 1$ or $\alpha > 1$, $\beta \leq 1$ is strictly increasing.
$\alpha > 2$, $\beta = 1$ is strictly convex.
$\alpha = 2$, $\beta = 1$ is a straight line.
$1 < \alpha < 2$, $\beta = 1$ is strictly concave.
$\alpha > 1$, $\beta > 1$ is unimodal.
If $\alpha = \beta$, the density function is symmetric about 1/2.

Triangular Distribution

The triangular probability density function has been used as an alternative to the beta distribution for modeling activity times. Its three essential parameters are minimum value (a), mode (m) and maximum (b). It is defined mathematically as:

$$f(t) = \frac{2(t-a)}{(m-a)(b-a)}; \quad a \le t \le m$$

$$= \frac{2(b-t)}{(b-m)(b-a)}; \quad m \le t \le b$$

With mean and variance defined, respectively, as:

$$\mu = \frac{a+m+b}{3}$$

$$\sigma^2 = \frac{a(a-m)+b(b-a)+m(m-b)}{18}$$

Figure 3.20 represents the triangular density function. The three time estimates of PERT can be inserted into the expression for the mean of the triangular distribution to obtain an estimate of the expected activity duration. Note that in the conventional PERT formula, the mode (m) is assumed to carry four times as much weight as a or b when calculating the expected activity duration. By contrast, with triangular distribution, the three time estimates are assumed to carry equal weights.

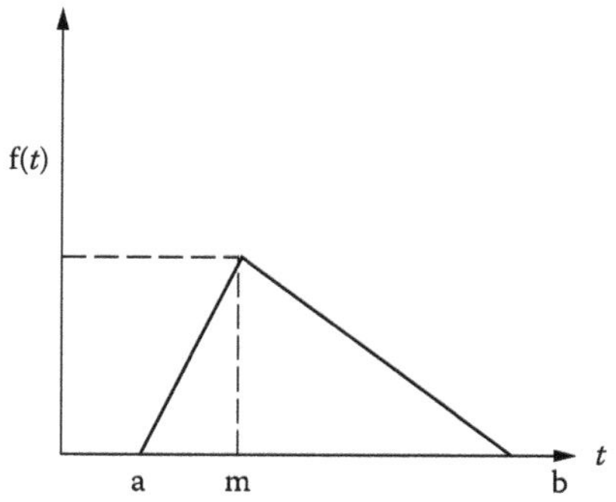

FIGURE 3.20
Triangular probability density function.

Uniform Distribution

For cases where only two time estimates instead of three are to be used for network analysis, the uniform density function may be assumed for activity times. This is acceptable for situations where extreme limits of an activity duration can be estimated and it can be assumed that the intermediate values are equally likely to occur. The uniform distribution is defined mathematically as:

$$f(t) = \frac{1}{b-a}; \quad a \leq t \leq b$$

$$= 0; \quad \text{otherwise},$$

with mean and variance defined, respectively, as:

$$\mu = \frac{a+b}{2}$$

$$\sigma^2 = \frac{(b-a)^2}{12}.$$

Figure 3.21 illustrates the uniform distribution for which the expected activity duration is computed as the average of the upper and lower limits of the distribution. The appeal of using only two time estimates a and b is that the estimation error due to subjectivity can be reduced and the estimation task simplified. Other distributions that have been explored for activity time modeling include the normal distribution, lognormal distribution, truncated exponential distribution, and Weibull distribution. After the expected activity durations have been computed, the analysis of the activity network is carried out as in the case of single-estimate CPM network analysis.

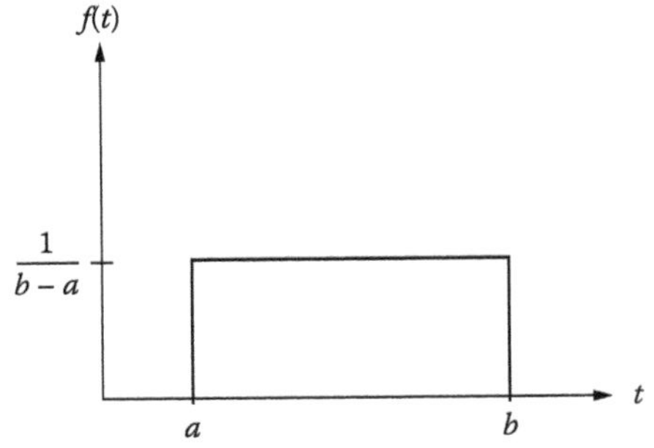

FIGURE 3.21
Uniform probability density function.

Statistical Analysis of Project Duration

Regardless of the distribution assumed for activity durations, the central limit theorem suggests that the distribution of the project duration will be approximately normal. The theorem states that the distribution of averages obtained from any probability density function will be approximately normally distributed if the sample size is large and the averages are independent.

Central Limit Theorem

Let $X_1, X_2, \ldots, X_N$ be independent and identically distributed random variables. The sum of the random variables is normally distributed for large values of N and is defined as:

$$T = X_1 + X_2 + \ldots + X_N$$

In activity network analysis, T represents the total project length as determined by the sum of the durations of the activities of the critical path. The mean and variance of T are expressed as:

$$\mu = \sum_{i=1}^{N} E[X_i]$$

$$\sigma^2 = \sum_{i=1}^{N} V[X_i]$$

where $E[X_i]$ = expected value of random variable X_i and $V[X_i]$ = variance of random variable X_i. When applying the central limit theorem to activity networks, note that the assumption of independent activity times may not always be satisfied. Because of precedence relationships and other interdependencies of activities, some activity durations may not be independent.

Probability Calculation

If a project duration T_e can be assumed to be approximately normally distributed based on the central limit theorem, the probability of meeting a specified deadline T_d can be computed by finding the area under the standard normal curve to the left of T_d. Figure 3.22 shows an example of a normal distribution describing project duration. Using the familiar transformation formula below, a relationship between the standard normal random variable z and the project duration variable can be obtained:

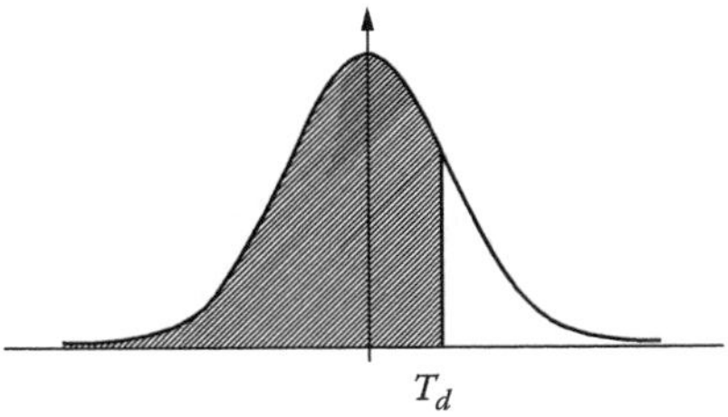

FIGURE 3.22
Area under normal curve.

$$z = \frac{T_d - T_e}{S},$$

where T_d = specified deadline; T_e = expected project duration based on network analysis; and S = standard deviation of project duration. The probability of completing a project by the deadline T_d is then computed as:

$$P(T \leq T_d) = P\left(z \leq \frac{T_d - T_e}{S}\right).$$

The probability is obtained from the standard normal table. Examples presented below illustrate the procedure for probability calculations in PERT.

PERT Network Example

Suppose we have the project data presented in Table 3.6. The expected activity durations and variances as calculated by the PERT formulas are shown in the last two columns of the table. Figure 3.23 shows the PERT network. Activities C, E, and G are shown as critical; the project completion time is

TABLE 3.6
Data for PERT Network Example

Activity	Predecessors	a	m	b	t_e	s^2
A	—	1	2	4	2.17	0.2500
B	—	5	6	7	6.00	0.1111
C	—	2	4	5	3.83	0.2500
D	A	1	3	4	2.83	0.2500
E	C	4	5	7	5.17	0.2500
F	A	3	4	5	4.00	0.1111
G	B, D, E	1	2	3	2.00	0.1111

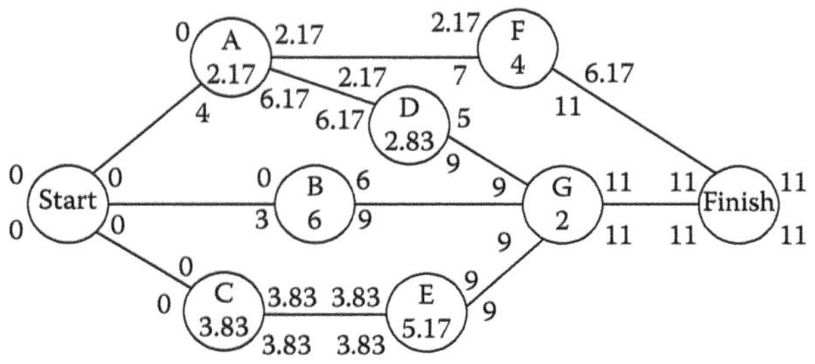

FIGURE 3.23
PERT network example.

11 time units. The probability of completing the project on or before a deadline of 10 time units (i.e., $T_d = 10$) is calculated as shown below:

$$T_e = 11$$

$$\begin{aligned} S^2 &= V[C] + V[E] + V[G] \\ &= 0.25 + 0.25 + 0.1111 \\ &= 0.6111 \end{aligned}$$

$$\begin{aligned} S &= \sqrt{0.6111} \\ &= .7817 \end{aligned}$$

$$\begin{aligned} P(T \le T_d) &= P(T \le 10) \\ &= P\left(z \le \frac{10 - T_e}{S}\right) \\ &= P\left(z \le \frac{10 - 11}{0.7817}\right) \\ &= P(z \le -1.2793) \\ &= 1 - P(z \le 1.2793) \\ &= 1 - 0.8997 \\ &= 0.1003. \end{aligned}$$

Thus, there is just over 10% probability of finishing the project within 10 days. By contrast, the probability of finishing the project in 13 days is calculated as:

$$P(T \leq 13) = P\left(z \leq \frac{13-11}{0.7817}\right)$$
$$= P(z \leq 2.5585)$$
$$= 0.9948.$$

This implies that there is over 99% probability of finishing the project within 13 days. Note that the probability of finishing the project in exactly 13 days will be zero. That is, $P(T = T_d) = 0$. If we desire the probability that the project can be completed within a certain lower limit (T_L) and a certain upper limit (T_U), let $T_L = 9$ and $T_U = 11.5$. Then,

$$P(T_L \leq T \leq T_u) = P(9 \leq T \leq 11.5)$$
$$= P(T \leq 11.5) - P(T \leq 9)$$
$$= P\left(z \leq \frac{11.5-11}{0.7817}\right) - P\left(z \leq \frac{9-11}{0.7817}\right)$$
$$= P(z \leq 0.6396) - P(z \leq -2.5585)$$
$$= P(z \leq 0.6396) - \left[1 - P(z \leq 2.5585)\right]$$
$$= 0.7389 - [1 - 0.9948]$$
$$= 0.7389 - 0.0052$$
$$= 0.7337.$$

That is, there is 73.4% chance of finishing the project within the specified range of duration.

Precedence Diagramming

The precedence diagramming method (PDM) was developed in the early 1960s as an extension of PERT/CPM network analysis. PDM permits mutually dependent activities to be performed partially in parallel instead of serially. The usual finish-to-start dependencies between activities are relaxed to allow activities to overlap to facilitate schedule compression. An example is the requirement that concrete should be allowed to dry for a number of days before holes are drilled for handrails, i.e., drilling cannot start until so

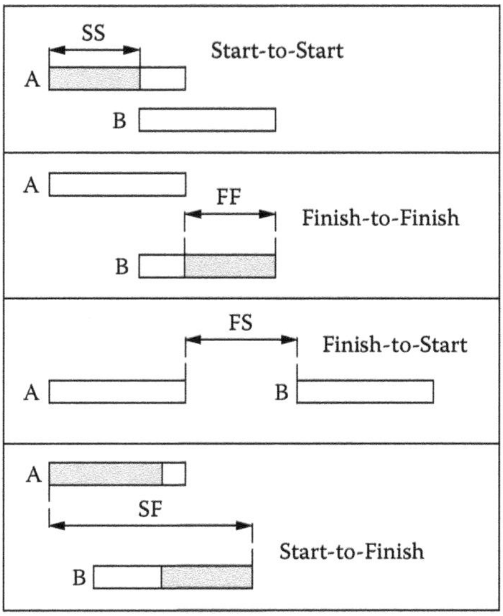

FIGURE 3.24
Lead-lag relationships in PDM.

many days elapse after the completion of concrete work. This is a finish-to-start constraint. The interval between the finishing time of the first activity and the starting time of the second activity is called the *lead-lag* requirement. Figure 3.24 represents the basic lead-lag relationships between activity A and activity B. The terminology in the figure is explained as follows.

- SS_{AB} (start-to-start) lead: Activity B cannot start until activity A has been in progress for at least SS time units.
- FF_{AB} (finish-to-finish) lead: Activity B cannot finish until at least FF time units after the completion of activity A.
- FS_{AB} (finish-to-start) lead: Activity B cannot start until at least FS time units after the completion of activity A. Note that PERT/CPM approaches use $FS_{AB} = 0$ for network analysis.
- SF_{AB} (start-to-finish) lead: There must be at least SF time units between the start of activity A and the completion of activity B.

The leads or lags may alternately be expressed as percentages rather than time units. For example, we may specify that 25% of the work content of activity A must be completed before activity B can start. If the percentage of work completed is used for determining lead-lag constraints, a reliable procedure must be used to estimate the percent completion. If the project work is analyzed properly using a work breakdown structure (WBS), it will be much easier to estimate percent completion by evaluating the work completed at the elementary task level.

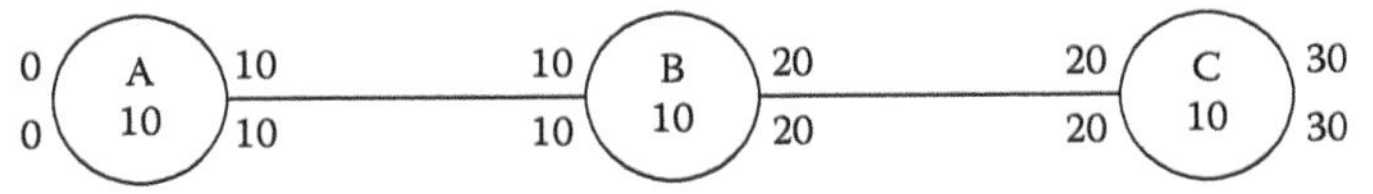

FIGURE 3.25
Serial activities in CPM network.

The lead-lag relationships may also be specified in terms of at-most relationships instead of at-least relationships. For example, we may have an at-most FF lag requirement between the finishing times of two activities. Splitting activities often simplifies the implementation of PDM, as will be shown later with examples. Some of the factors that determine whether an activity can be split are technical limitations, the morale of the person working on the split task, set-up times required to restart split tasks, difficulties of managing resources for split tasks, loss of consistency of work, and management policy about splitting jobs.

Figure 3.25 presents a simple CPM network consisting of three activities. The activities are to be performed serially and each has an expected duration of 10 days. The conventional CPM network analysis indicates that the duration of the network is 30 days. The earliest times and the latest times are as shown in the figure.

The Gantt chart for the example is shown in Figure 3.26. For comparison, Figure 3.27 shows the same network with lead-lag constraints, for example,

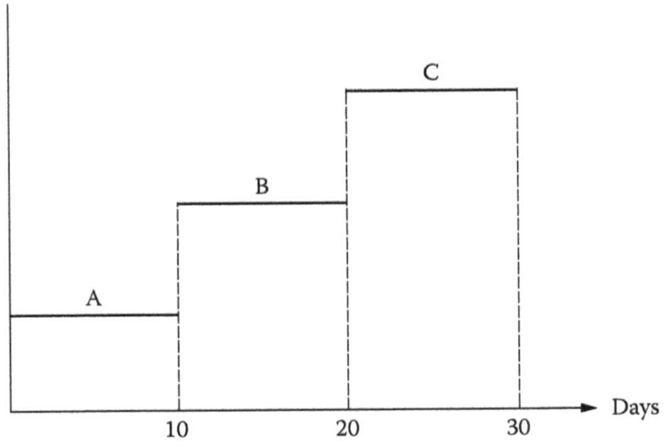

FIGURE 3.26
Gantt chart of serial activities in CPM example.

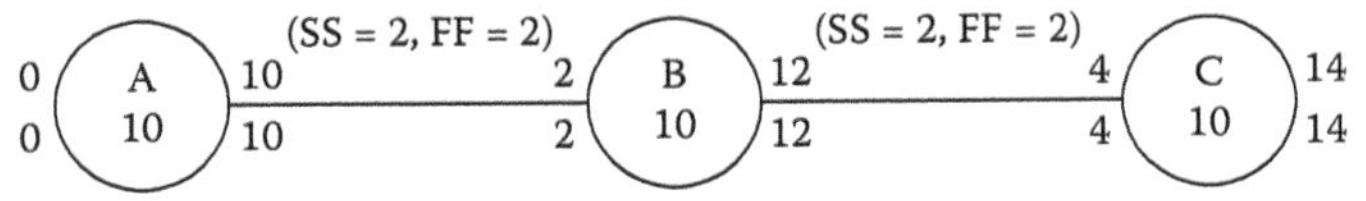

FIGURE 3.27
PDM network example.

an SS constraint of 2 days and an FF constraint of 2 days between activities A and B. Thus, activity B can start as early as 2 days after activity A starts, but it cannot finish until 2 days after the completion of A. In other words, *at least* 2 days must separate the finishing times of A and B. A similar precedence relationship exists between activities B and C. The earliest and latest times obtained by considering the lag constraints are indicated in Figure 3.27.

The calculations show that if B is started only 2 days after A is started, it can be completed as early as 12 days as opposed to the 20 days cited by conventional CPM. Similarly, activity C is completed at time 14—considerably less than the 30 days calculated by conventional CPM. The lead-lag constraints allow us to compress or overlap activities. Depending on the nature of the tasks, an activity does not have to wait until its predecessor finishes before it starts. Figure 3.28 is a Gantt chart incorporating lead-lag constraints. Note that a portion of a succeeding activity can be performed simultaneously with a portion of the preceding activity.

A portion of an activity that overlaps with a portion of another activity may be viewed as a distinct portion of the required work. Thus, partial completion of an activity may be evaluated. Figure 3.29 shows how each of the three activities is partitioned into contiguous parts. Although no break or termination of work in any activity is shown, the distinct parts (beginning and ending) can still be identified. This means that no splitting of the work content of any activity was done. The distinct parts are determined on the basis of the amount of work that must be completed before or after another activity, as dictated by the lead-lag relationships. In Figure 3.29, activity A is partitioned into A_1 and A_2. The duration of A_1 is 2 days because of the SS = 2 relationship between activities A and B. Since the original duration of A is 10 days, the duration of A_2 is then calculated to be 10 – 2 = 8 days.

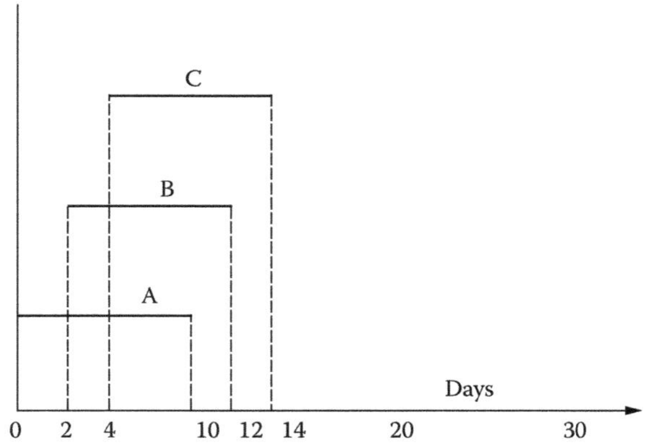

FIGURE 3.28
Gantt chart for PDM example.

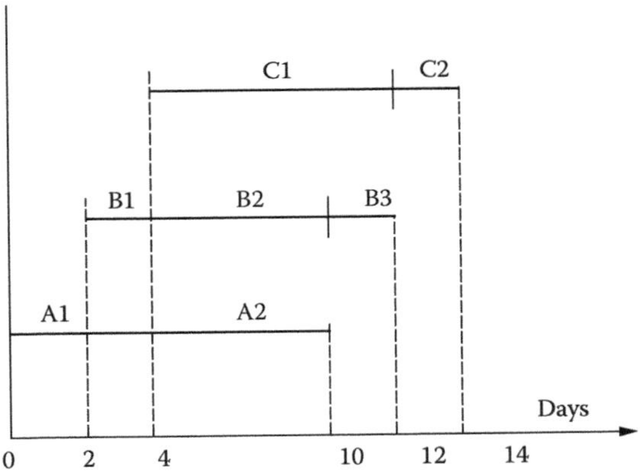

FIGURE 3.29
Partitioning of activities in PDM example.

Likewise, activity B is partitioned into B_1, B_2, and B_3. The duration of B_1 is 2 days because of an SS = 2 relationship between activities B and C. The duration of B_3 is also 2 days because of an FF = 2 relationship between activities A and B. Since the original duration of B is 10 days, the duration of B_2 is calculated as 10 – (2 + 2) = 6 days. In a similar fashion, activity C is partitioned into C_1 and C_2. The duration of C_2 is 2 days because of an FF = 2 relationship between activities B and C. Since the original duration of C is 10 days, the duration of C_1 is calculated as 10 – 2 = 8 days. Figure 3.30 shows a conventional CPM network for the three activities after they are partitioned into distinct parts. The conventional forward and backward passes reveal that all the activity parts are performed serially and no physical splitting of activities is involved. Note the three critical paths in Figure 3.30, each with

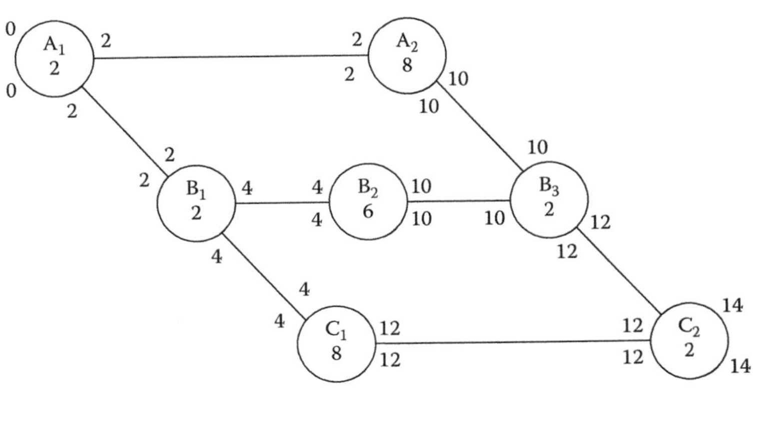

FIGURE 3.30
CPM network of partitioned activities.

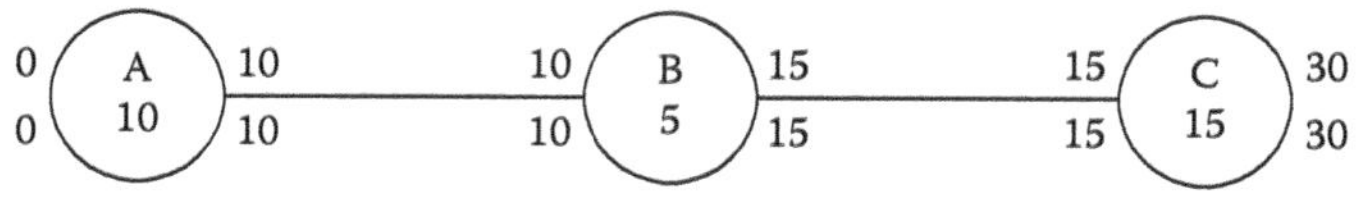

FIGURE 3.31
CPM example of serial activities.

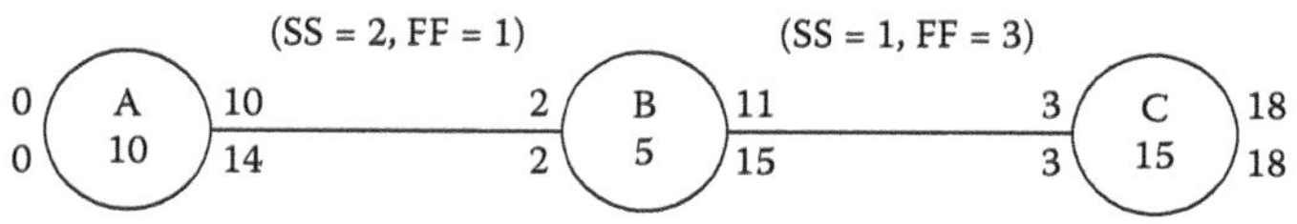

FIGURE 3.32
Compressed PDM network.

a length of 14 days. Note also that the distinct parts of each activity are performed contiguously.

Figure 3.31 shows an alternate example of three serial activities. The conventional CPM analysis shows that the duration of the network is 30 days. When lead-lag constraints are introduced into the network as shown in Figure 3.32, the network duration is compressed to 18 days.

In the forward pass computations in Figure 3.32, the earliest completion time of B is time 11 because of an FF = 1 restriction between activities A and B. Since A finishes at time 10, B cannot finish until at least time 11. Even though the earliest starting time of B is time 2 and its duration is 5 days, its earliest completion time cannot be earlier than time 11. Also note that C can start as early as time 3 because of the SS = 1 relationship between B and C. Thus, given a duration of 15 days for C, the earliest completion time of the network is 3 + 15 = 18 days. The difference between the earliest completion times of C and B is 18 – 11 = 7 days, which satisfies the FF = 3 relationship between B and C.

In the backward pass, the latest completion time of B is 15 (18 – 3= 15) due to the FF = 3 relationship between B and C. The latest start for B is time 2 (3 – 1 = 2) since an SS = 1 relationship exists between B and C. If we are not careful, we may erroneously set the latest start time of B to 10 (15 – 5 = 10), but that would violate the SS = 1 restriction between B and C. The latest completion time of A is 14 (15 – 1 = 14) because of an FF = 1 relationship between A and B.

All the earliest times and latest times at each node must be evaluated to ensure that they conform to all the lead-lag constraints. When computing earliest start or earliest completion times, the smallest possible value that satisfies the lead-lag constraints should be used. By the same reasoning, when computing the latest start or completion times, the largest possible value that satisfies the lead-lag constraints should be used.

Manual evaluations of the lead-lag precedence network analysis can become very tedious for large networks. A computer tool may be needed to

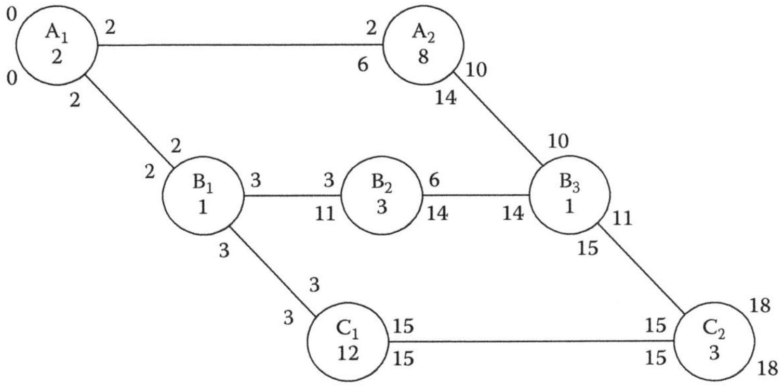

FIGURE 3.33
CPM expansion of second PDM example.

implement PDM. If manual analysis must be done for PDM computations, it is suggested that the network be partitioned into more manageable segments that may be relinked after the computations are completed. The expanded CPM network in Figure 3.33 was based on the precedence network in Figure 3.32. Activity A is partitioned into two parts, activity B is partitioned into three parts, and activity C is partitioned into two parts. The forward and backward passes show that only the first parts of activities A and B are on the critical path. Both parts of activity C are on the critical path.

Figure 3.34 shows the corresponding earliest-start Gantt chart for the expanded network. Looking at the earliest start times, activity B is physically split at the boundary of B_2 and B_3 in such a way that B_3 is separated from B_2

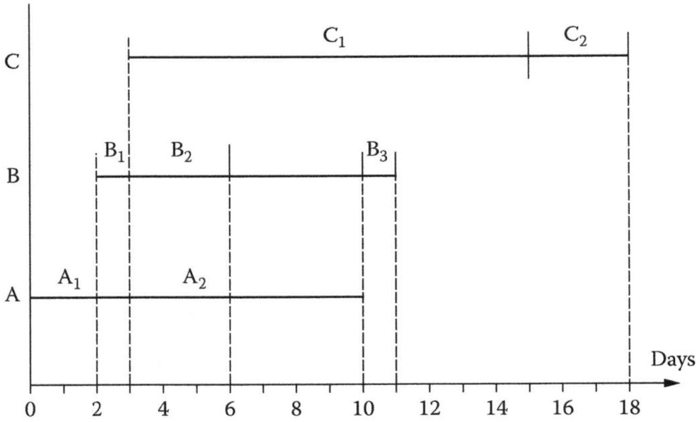

FIGURE 3.34
Compressed PDM schedule based on ES times.

by 4 days. This implies that work on activity B is temporarily stopped at time 6 after B_2 is finished and is not started again until time 10. Note that despite the 4-day delay in starting B_3, the entire project is not delayed because B_3 (the last segment of activity B) is not on the critical path. In fact, B_3 has a total slack of 4 days. In a situation like this, the duration of B can actually be increased from 5 days to 9 days without adverse effects on project duration. However, increasing the duration of an activity may have negative implications for cost, personnel productivity, and/or morale.

If physical splitting of activities is not possible, the best option available in Figure 3.34 is to stretch the duration of B_3 to fill the gap from time 6 to time 10. An alternative is to delay the starting time of B_1 until time 4 to consume the 4-day slack at the beginning of activity B. Unfortunately, delaying the starting time of B_1 by 4 days will delay the overall project by 4 days because B_1 is on the critical path as shown in Figure 3.33. The project analyst will need to evaluate the appropriate trade-offs between splitting, delaying activities, increasing activity durations, and incurring higher project costs. The prevailing project scenario should be considered when making such trade-off decisions.

Figure 3.35 shows the Gantt chart for the compressed PDM schedule based on latest start times. In this case, it will be necessary to split activities A and B even though the total project duration remains the same at 18 days. If activity splitting is not an option, we can increase the duration of activity A from 10 to 14 days and the duration of B from 5 to 13 days without adversely affecting the entire project duration. The benefit of precedence diagramming is that the ability to overlap activities facilitates flexibility in manipulating individual activity times and reducing project duration.

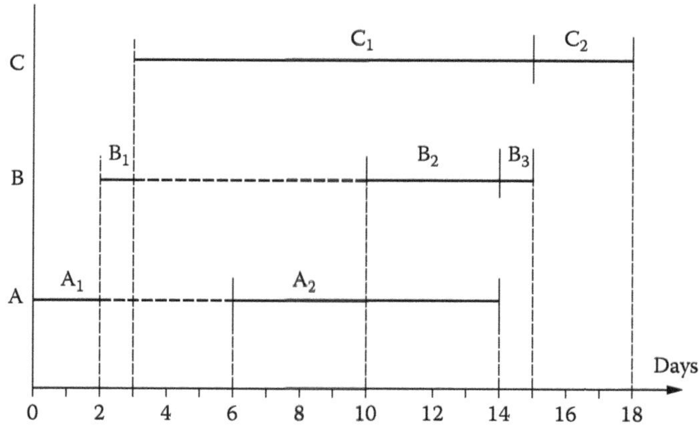

FIGURE 3.35
Compressed PDM schedule based on LS times.

Reverse Criticality in PDM Networks

Care must be exercised when working with PDM networks because of the potential for misuse or misinterpretation. Because of the lead and lag requirements, activities that have no slacks may appear to have generous slacks. Also, reverse critical activities may decrease project duration when their durations are increased. This may happen when the critical path enters the completion of an activity through a finish lead-lag constraint. Also, if a finish-to-finish dependency and a start-to-start dependency are connected to a reverse critical task, a reduction in the duration of the task may actually lead to an increase in the project duration.

Figure 3.36 illustrates this anomalous situation. The finish-to-finish constraint between A and B requires that B should finish no earlier than 20 days. If the duration of task B is reduced from 10 days to 5 days, the start-to-start constraint between B and C forces the starting time of C to shift forward by 5 days, thereby resulting in a 5-day increase in project duration.

Such anomalies can occur without being noticed in large PDM networks. One safeguard against their adverse effects is to make only one activity change at a time and document the resulting effect on the network structure and duration. The following categorizations are used to describe unusual characteristics of activities in PDM networks.

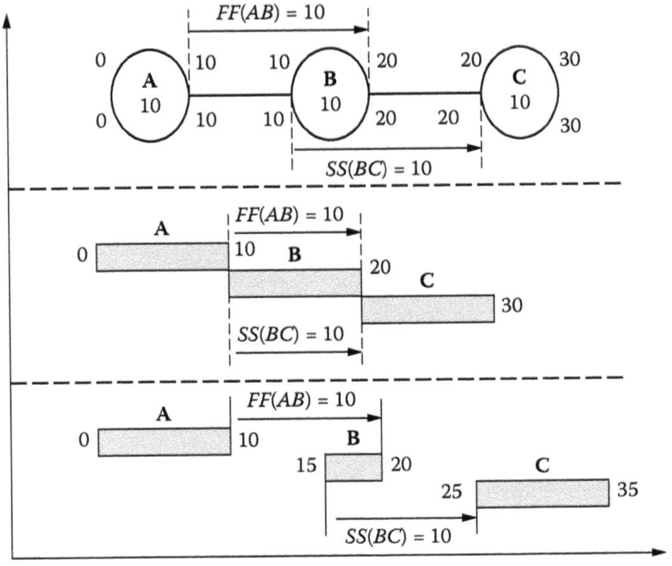

FIGURE 3.36
Reverse critical activity in PDM network.

Normal Critical (NC): An activity for which the project duration shifts in the same direction as the shift in the duration of the activity.

Reverse Critical (RC): An activity for which the project duration shifts in the reverse direction to the shift in the duration of the activity.

Bi-Critical (BC): An activity for which the project duration increases as a result of any shift in the duration of the activity.

Start Critical (SC): An activity for which the project duration shifts in the direction of the shift in the start time but is unaffected (neutral) by a shift in the overall duration of the activity.

Finish Critical (FC): An activity for which the project duration shifts in the direction of the shift in the finish time but is unaffected (neutral) by a shift in the overall duration of the activity.

Mid-Normal Critical (MNC): An activity whose mid portion is normal critical.

Mid-Reverse Critical (MRC): An activity whose mid portion is reverse critical.

Mid-Bi-Critical (MBC): An activity whose mid portion is bi-critical.

A computer-based decision support system can facilitate the integration and consistent usage of all the relevant information in a complex scheduling environment. Task precedence relaxation assessment and resource-constrained heuristic scheduling constitute examples of problems suitable for computer implementation. Examples of pseudo-coded heuristic rules for computer implementation are shown below:

```
IF:    Logistical conditions are satisfied
THEN:  Perform the selected scheduling action

IF:    condition A is satisfied and
       condition B is false and
       evidence C is present or
       observation D is available
THEN:  precedence belongs in class X

IF:    precedence belongs to class X
THEN:  activate heuristic scheduling procedure Y
```

A computer model helps a decision maker develop a task sequence that fits the needs of concurrent scheduling. Based on user input, the model will determine task precedence, establish precedence relaxation strategy, implement task scheduling heuristics, match the schedule to resource availability, and present a recommended task sequence to the user. The user can perform what-if analysis by making changes in the input data and conducting sensitivity analysis. The computer implementation can be achieved in an

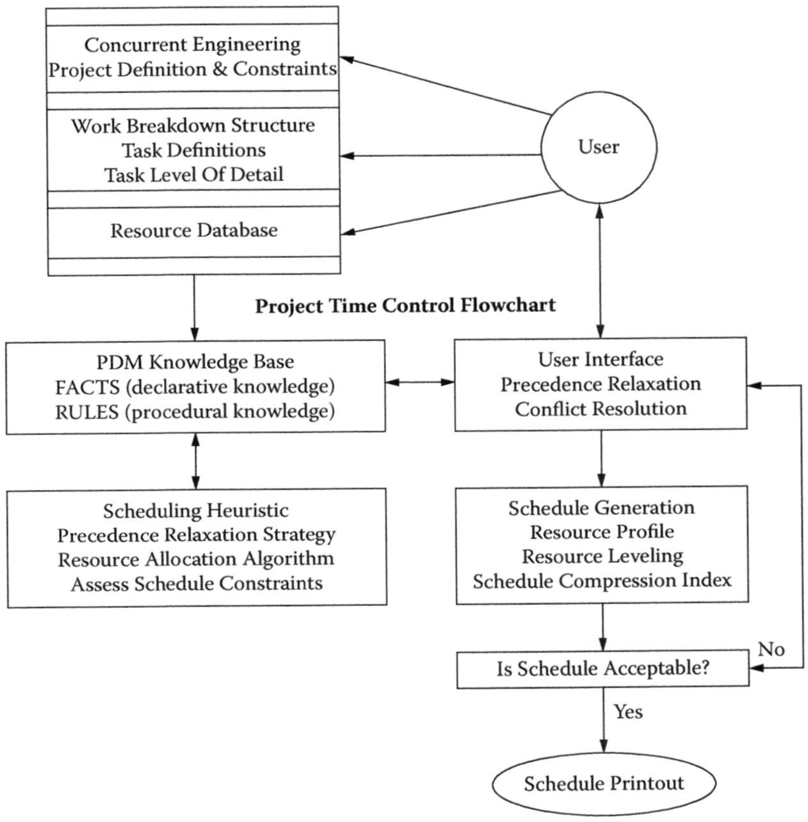

FIGURE 3.37
Decision support model for PDM.

interactive environment as shown in Figure 3.37. The user will provide task definitions and resource availabilities with appropriate precedence requirements. At each stage, the user is prompted to consider potential points for precedence relaxation. Wherever schedule options exist, they will be presented for the user's consideration and approval. The user will have the opportunity to make final decisions about task sequence.

This chapter presented the basic techniques of activity network analysis for project time control. In business and industry, project management is rapidly becoming one of the major tools for accomplishing goals. Engineers, managers, and operations research professionals are increasingly required to participate on teams assigned to complex projects. Network analysis is frequently utilized in quantitative assessments of projects. The computational approaches explained in this chapter can aid project analysts in developing effective project schedules and determining the best ways to exercise project control when needed.

References

Badiru, Adedeji B. (1996). *Project Management in Manufacturing and High Technology Operations*, 2nd ed. John Wiley & Sons, New York.

Badiru, Adedeji B. and P. S. Pulat (1995). *Comprehensive Project Management: Integrating Optimization Models, Management Principles, and Computers.* Prentice-Hall, Englewood Cliffs, NJ.

4

Project Performance Control

"If you can't measure it, you can't control it."

Management Anecdote

Inasmuch as measurement of performance is not always possible, we must still attempt to measure project characteristics so that we can apply control actions where needed. The quote above is a testament to the need to measure whatever we can, wherever we can.

The conventional functions of management include planning, organizing, leading, directing, and controlling. Somewhere along the lines running through this spectrum of management lies the function of measuring. Although the functions are not executed in a strictly ordered sequence, measuring must come before control action can be taken.

In complex project environments, the human cause and effect relationships make it imperative to be cautious about the accuracy of measured factors. We should realize that, although measurement is necessary, it is not sufficient for properly chaperoning a project toward success. Human sentiments, emotions, and interpersonal reactions play roles in project outcomes. We must thus make allowances when interpreting project control data. Project types vary greatly and can range from manufacturing, construction, education delivery, and customer service to medical surgery. Techniques that have been developed to achieve performance management and control in one application domain are equally applicable, albeit in modified versions, to other project management domains.

Importance of Work Performance Measurement

Innovation and advancement owe their sustainable foundation to some measurement scale (Badiru 1988, 2008). We must measure work before we can improve it. This means that we *still* need work measurement. One common thread in any project is the goal of achieving improvement to a product, process, service, or result. How can we improve a task if we don't know how long it takes us to accomplish it in the first place? Therein lies the efficacy of work measurement. The Six Sigma, Lean, TQM, 5s, theory of constraints, supply networking, and other techniques all utilize the same underlying principles of improving work elements. Even in personal pursuits or projects, time and motion analyses subconsciously play roles in charting the path to effective actions. The most successful leaders are those whose visions contain elements of time, motion toward action, and deliberate work standards.

At an industrial conference, a speaker mentioned the pursuit of improvement in the deposition characteristics of nanoparticles in microelectronics manufacturing. That is an example of work and process measurement with the goal of achieving improvement. Industrial engineers are often called upon to facilitate the integration and coordination of cutting-edge technologies. Responses to such calls cannot succeed without a basic understanding and measurement of time, motion, and performance of the technologies. Interoperability metrology may sound more erudite, but it still measures how work occurs across technologies.

The fast changing operating environments that we now face necessitate interdisciplinary improvement projects that may involve physicists, material scientists, chemists, engineers, bioscientists, and a host of other disciplines. They may all use different measurement scales and communicate in different jargons, but a common thread of work measurement permeates their collective improvement efforts. If we review the widely accepted definitions of industrial engineering, we will see references to energy measurement, cost measures, productivity measurement, scientific measurements, and systems capability assessment. These present convincing confirmation that work measurement has re-emerged in more palatable concatenations of words.

Work measurement must remain a core competency of project managers. Measurement of productivity, human performance, and resource consumption are essential components of achieving organizational goals and increasing profitability.

For example, trade-offs must be exercised among the triple constraints of production represented in terms of time, cost, and performance. Proper work measurement is required to ensure that the trade-offs occur within feasible regions of the operating environment. Work rate analysis, centered on work measurement, helps identify areas where operational efficiency and improvement can be pursued. In many production settings, workers encounter production breaks that require an analysis of the impact on production output. Some breaks are standard and scheduled; others are nonstandard and unscheduled. In both cases, the workstation is subject to work rate slowdown (ramp-down) and work rate pickup (ramp-up), respectively, before and after breaks. These impacts are subtle and barely noticed unless a formal engineered work measurement is put in place.

Although the conventional techniques of project management do not involve work sampling, it is the authors' opinion that some elements and principles of work sampling are useful for clarifying the processes and requirements of project performance control. Traditional work sampling is applied to production management and control. The basic requirement of work sampling is task observation. Without resorting to the rigid process of work sampling, any project control endeavor requires task observation. As a result, project management is akin to production management and the techniques used in both functions are complementary.

Basic Techniques of Performance Work Sampling

Ray (2006) presents work sampling as a very useful technique for measuring work that can be applied in wide varieties of situations. In its simplest form, it is used by shop supervisors and foremen to estimate idle times of machines. For example if a supervisor notices an idle machine during two of ten observations, he would estimate that the machine was idle 20% of the time. Other names, such as activity sampling, ratio delay, snap reading, and occurrence sampling are occasionally used to indicate work sampling. The technique is particularly useful for measuring indirect work and service activities and is suitable where the stop watch method is not acceptable or possible. Work sampling has become a standard tool for measuring indirect and service jobs.

It is based on the laws of probability and determines the proportions of total time devoted to the various components of a task. The probability that an event will or will not occur is based on statistical binomial distribution. When the number of observations is large, the binomial distribution can be approximated to the normal distribution. The binomial probability of x occurrences is calculated as follows:

$$b(x) = {}_nC_x p^x q^{(n-x)},$$

where p = probability of x occurrences; $q = 1 - p$ = probability of no x occurrence; n = number of observations; and ${}_nC_x$ = combination of x items from n items. The normal distribution is used instead of the binomial distribution in work sampling for computational convenience. The normal distribution of a proportion p has an average value of ratio equal to p and standard deviation defined as:

$$\sigma_p = \sqrt{\frac{p(1-p)}{n}},$$

where p = proportion of occurrence of an event and n = number of observations. Suppose the desired absolute accuracy desired is designated as A, and the relative accuracy for the proportion p is designated s. We will then have the relationship:

$$\begin{aligned} A &= sp \\ &= z\sigma_p, \end{aligned}$$

where z is the standard normal variate. Absolute accuracy indicates the range within which the value of p is expected to lie; it represents the closeness of the ratio p to the true ratio p. For example, if $p = 30\%$ and relative accuracy = 10%, then $A = (0.3) \times (0.1) = 0.03$ or 3%. For a study where the true p value = 30%, a ±3% absolute accuracy level indicates that the calculated value of p will be between $[(30 \pm (30 \times 0.03)] = 30 \pm 0.75$ or between 29.25 and 30.75%.

Performance Confidence Interval

Confidence interval denotes the long-term probability that the ratio p will be within the accuracy limits. The concept relies on relative frequency interpretation of probability. Thus, 95% confidence means that among a large number of samples, 95 out of 100 will contain the true value of p. The probability value is given by the proportion of the area under the normal curve included by a number of standard deviation (z). The usual confidence levels and corresponding z values are summarized in Table 4.1.

Sample Size Calculations

The number of observations or sample size can be determined as follows:

$$A = z\sigma_p = z\sqrt{\frac{p(1-p)}{n}}$$

or

$$n = \frac{z^2}{A^2}[p(1-p)],$$

TABLE 4.1

Confidence Levels for Standard Normal Curve

Confidence Level (%)	z Standard Deviation
68.27	$\pm 1.00\ \sigma$
90.00	$\pm 1.64\ \sigma$
95.00	$\pm 1.96\ \sigma$
95.45	$\pm 2.00\ \sigma$
99.73	$\pm 3.00\ \sigma$

where A is usually 0.05 or 5% for industrial work; p = percentage of total work time during which a component occurs; and z = number of standard deviations depending on confidence level desired. The following sample calculation was used to determine the sample size required to achieve a desired accuracy on a milling machine:

Relative accuracy desired = 0.05%

Confidence level desired = 95%

Preliminary study indicated p = 30%

For this accuracy $A = (0.05) \times (0.3) = 0.015$; $z = 1.96$; $n = (z^2/A^2) \times [(p(1-p)]$
$= (1.96^2/0.015^2) \times (0.30 \times 0.70) = 3{,}585.49$ or about 3,586 observations.

To be statistically acceptable, it is essential that a work sampling procedure provides each individual moment during observation an equal chance of being observed. The observations must be random, unbiased, and independent so that the assumption of the binomial theory of constant probability of event occurrence is attained. Hence, it is essential that observations are taken at random times during a work sampling study.

Control Charts for Performance Control

Control charts are extensively used in quality control work to identify when a system has gone out of control. The same principle is used to control the quality of work sampling studies. The 3σ limit is normally used in work sampling to set the upper and lower limits of control. First the value of p is plotted as the center line of a p chart. The variability of p is then found for the control limits.

Example

For p value = 0.3 and sample n = 500, $\sigma = \sqrt{p(1-p)/n} = 0.0205$ and $3\sigma = 0.0615$. The limits are then 0.3 ± 0.0615 or 0.2385 and 0.3615 as shown in Figure 4.1. On the second Friday, the calculated value of p based on observations fell beyond limits, indicating the need for investigation and initiation of a corrective action.

Plan for Typical Work Sampling Study

Determine objective

Identify elements of study

Conduct a preliminary study to estimate ratio percentages

Determine desired accuracy and confidence levels

Determine required number of observations

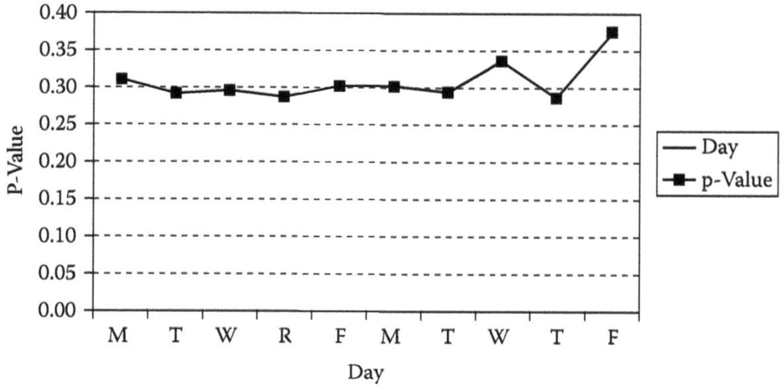

FIGURE 4.1
Control chart on daily basis.

Schedule random trip times and routes
Design observation form
Collect sampling data
Summarize results

Applications of Work Sampling

Work sampling is most suitable for determining (a) machine utilization, (b) allowances for unavoidable delays, and (c) work standards for direct and indirect work. The technique is particularly well suited for determining standards for indirect and service work, for example, in project management. Each type of application will be illustrated by an example (Ray 2006).

Machine Utilization

ABC Company was concerned about the utilization of forklift trucks and wanted to determine the average idle times of forklift trucks in its plant. A 5-week work sampling study was conducted. The data collected are presented in Table 4.2. The desired confidence level and relative accuracy were 95% and 5.0% respectively. Forklift truck utilization = (1500/2000) × 100% = 75%. Idle percentage = (500/2000) × 100% = 25%. Assuming a desired confidence level of 95%, the accuracy of idle percentage is determined as follows:

$$A = s \times 0.25 = 1.96 \times \sqrt{p*(1-p)/n} = 1.96 \times \sqrt{0.25*0.75/2000} = 0.0190.$$

TABLE 4.2
Work Sampling Summary Sheet

Element	Observations	Total
Working	11111 11111 11111 11111 … … … … … … … … …	1500
Idle	11111 11111 11111 … … … … … … … … … … …	500
		2000

s = relative accuracy = (0.0190/0.25) × 100% = 7.60%. The desired accuracy required was 5%. The number of observations required to achieve the accuracy was:

$$0.05 \times 0.25 = 1.96 \times \sqrt{0.25 * 0.75 / n}$$

or

$$n = (1.96/(0.05 \times 0.25)2 \times (0.05 \times 0.25) = 3{,}074.$$

The number of observations per day during the preliminary study was 2000/25 = 80. The same group was assigned to collect additional sampling data at 80 observations per day. Hence, the additional days required for the study to achieve the desired accuracy was (3,074 – 2,000)/80 = 13.4 days or about 14 days.

Allowances for Personal and Unavoidable Delays

Suppose the industrial engineering department of the ABC Company wanted to determine the allowances for personal reasons and unavoidable delays for the machine shop. The steps are explained below.

Identify elements of study—The three elements were: (a) working, (b) personal delay, and (c) unavoidable delay.

Design observation form—Observation forms are task specific to each study. For the ABC study, a sampling form was based on elements required to be observed: (a) working, (b) personal activity such as drinking water, and (c) unavoidable delay such as a summons to a supervisor's office. Each tally mark indicated one observation of the corresponding element. Additional items may be included on a work sampling form to reflect future requirements.

Preliminary study to estimate ratio percentages—The study was conducted to estimate the approximate percentage values for the elements cited in Table 4.3.

Confidence level and accuracy—The desired levels of confidence and accuracy were 95% and 0.05 %, respectively.

Sample size—Preliminary estimates were used to determine the sample size required. The smallest value of the percentage occurrence was used for computation to ensure the desired level or higher accuracy for all elements.

TABLE 4.3

Pilot Work Sampling Summary (ABC Company)

Element	Observations	Total Observations	Occurrence (%)
Working	1111 1111	102	85
Personal activity	1111	6	5
Unavoidable delay	1111 1111	12	10
Total		120	100

$A = 0.05 \times 0.05 = 0.0025$; $z = 1.96$; $n = (z/A)^2 \times (p) \times (1 - p) = (1.96/0.0025)^2 \times (0.05) \times (0.95) = 29{,}197$. Ten workers operated similar machines. Hence, the number of observations per trip was 10. The number of trips required was 29,197/10 = 2,920. One work sampler could make 5 trips per hour or 40 trips per day on an 8-hour shift. If three persons were available for sampling, they together could make 120 trips per day. Hence the duration required was 2,920/120 = 24.33 days or about 5 weeks.

Trip scheduling—A random number table was used to schedule 40 random trips per day for each observer. In addition, the routes of the observers were changed randomly each day.

Collecting Sampling Data

The observers were trained to be objective and not anticipate observations. Each trip followed a randomly selected route at a random time. Video cameras may be used to minimize bias when collecting data as the camera records all ongoing activities accurately. Table 4.4 summarizes the results of the ABC study. Certain unavoidable delays should be based exclusively on observations of work element data alone (24,730) if an unavoidable delay under study was highly dependent on work time, as in the case of fatigue.

Determining Work Standards

For the ABC study, work standards for the machine shop were developed. During the study period, the machine shop produced 100,000 pieces of fan motor shaft. Company policy set a fatigue allowance of 8% in the machine shop. Five workers in the section operated similar machines. The performance

TABLE 4.4

Summary of Sampled Data for ABC Study

Work Element	Total Observations	Occurrence (%)
Working	24,730	24,730/29,200 = 0.8469 = 84.69 % ≌ 84.7%
Personal delay	1,470	1,470/29,200 = 0.503 = 5.03 % ≌ 5.0%
Unavoidable delay	3,000	3,000/29,200 = 0.1027 = 10.27 % ≌ 10.3%
Total	29,200	

rating was 110%, determined by estimating the pace of work periodically during the sampling study and considered reasonable.

Total work time = (5 weeks) × (5 days/week) × (8 hours/day) × (5 operators) = 1,000 man hours

Observed time/piece = 1,000 × 60/100,000 = 0.6 minutes/piece

Normal time = 1.10 × 0.6 = 0.66 minutes

Total allowance = 5 + 10.3 + 8.0 = 23.30%

Standard time/piece = 0.66 × (1.233) = 0.814 minutes/piece of motor shaft

Learning Curve Analysis for Performance Control

Manufacturing progress function, also known as learning curve, is a major topic of interest in operations analysis and improvement in industry, where work measurement serves as a basis for productivity improvement. Industrial innovation and advancement owe their sustainable foundations to work measurement processes. We must measure work accurately if we want to improve it. Work measurement accuracy requires an assessment of the impact of learning during a production cycle.

This chapter presents the analytic framework for extending univariate learning curves to multivariate analysis. Multivariate functions are needed to account for multiple factors that can influence learning in industrial operations. In many operations, subtle, observable, quantitative, and qualitative factors intermingle to compound the productivity analysis problem. Such factors are best modeled within a multivariate learning curve framework. Industrial innovation and advancement are the results of measurement scales. We must measure work before we can improve it as a factor in evaluating performance. Standards obtained via work measurement provide information critical to success. The information covers:

Scheduling
Staffing
Line balancing
Materials requirement planning
Compensation
Costing
Employee evaluation

For all these data types, the effects of learning in the presence of multiple factors should be considered before operational decisions are made.

Effects of Learning

> "The illiterate of the 21st century will not be those who cannot read and write, but those who cannot learn, unlearn, and relearn."
>
> **Alvin Toffler**

Manufacturing progress function, also known as learning curve or experience curve, is a major topic of interest in operations analysis and cost estimation. Learning in the context of manufacturing operations management is the improved productivity obtained from repetition. Several studies have confirmed that human performance improves with reinforcement or frequent repetition. Reducing operation processing times achieved through learning curve effects can directly translate to cost savings for manufacturers and improved morale for employees. Learning curves are essential for setting production goals, monitoring progress, reducing waste, and improving efficiency. The applications of learning curves extend well beyond conventional productivity analysis.

Typical learning curves present the relationship between cumulative average production cost per unit and cumulative production volume based on the effect of learning. For example, an early study by Wright (1936) disclosed an 80% learning effect, which indicates that a given operation is subject to a 20% productivity improvement each time the production quantity doubles. This productivity improvement phenomenon is illustrated later in this chapter. With information about expected future productivity levels, a learning curve can serve as a predictive tool to develop time estimates for tasks that are repeated within a production cycle.

Manufacturing progress functions are applicable to all aspects of manufacturing planning and control and directly useful for project time control. Multivariate models are significant for cost analysis because they can help account for the multivariate influences on learning and thus facilitate a wider scope of learning curve analysis for manufacturing operations. A specific example of a bivariate manufacturing progress function is presented. Since production time and cost are inherently related, both terms are often used interchangeably as the basis for manufacturing progress function analysis. For consistency, cost is used as the basis for discussion. Manufacturing progress function and learning curve are used interchangeably.

In many manufacturing operations, tangible, intangible, quantitative, and qualitative factors intermingle to compound the productivity analysis problem. Consequently, a more comprehensive evaluation methodology such as a multivariate learning curve can be used for productivity analysis. Some of the specific analyses that can benefit from the results of multivariate learning curve analysis include cost estimation, work design and simplification, breakeven analysis, manpower scheduling, make or buy decisions, production planning, budgeting and resource allocation, and management of productivity improvement programs.

Univariate Learning Curve Models

The conventional univariate learning curve model presents several limitations in practice. Since the first formal publication of learning curve theory by Wright (1936), several alternative propositions relate to the geometry and functional forms of learning curves. Some of the most common models are:

Log-linear model
S-curve
Stanford-B model
DeJong's learning formula
Levy's adaptation function
Glover's learning formula
Pegel's exponential function
Knecht's upturn model
Yelle's product model
Multiplicative power model

Log-Linear Model

The log-linear model (Wright 1936) is often called the conventional learning curve model based on the principle that improvement in productivity is constant (constant slope) as output increases. The two basic forms of the log-linear model are the average cost function and the unit cost function. The average cost model is more popular. It specifies the relationship between the cumulative average cost per unit and cumulative production. The relationship indicates that cumulative cost per unit will decrease by a constant percentage as the cumulative production volume doubles. The model is expressed as

$$C(x) = C_1 x^b,$$

where $C(x)$ = cumulative average cost of producing x units; C_1 = cost of first unit; x = cumulative production count; and b = learning curve exponent (i.e., constant slope of learning curve on log-log paper). When linear graph paper is used, the log-linear learning curve is a hyperbola of the form shown in Figure 4.2. On log-log paper, the model is represented by the following straight line equation:

$$\log C_x = \log C_1 + b \log x,$$

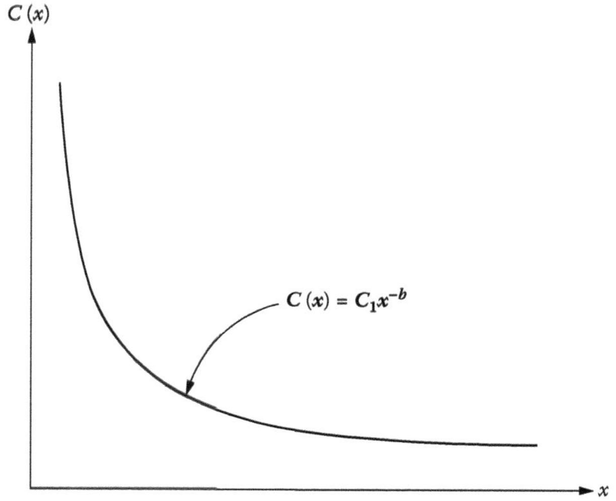

FIGURE 4.2
Log-linear learning curve model.

where b is the constant slope of the line. The expression for the learning rate percent p is derived by considering two production levels in which one level is double the other. For example, given the levels x_1 and x_2 (where $x_2 = 2x_1$), we have the following expressions:

$$C_{x_1} = C_1(x_1)^b$$

$$C_{x2} = C_1(2x_1)^b.$$

The percent productivity gain is then computed as:

$$p = \frac{C_1(2x_1)^b}{C_1(x_1)^b} = 2^b.$$

In general, the learning curve exponent can be computed as:

$$b = \frac{\log C_{x_1} - \log C_{x_2}}{\log x_1 - \log x_2},$$

where x_1 = first production level; x_2 = second production level; C_{x1} = cumulative average cost per unit at first production level; and C_{x2} = cumulative average cost per unit at second production level.

Expression for Total Cost: Using the basic cumulative average cost function, the total cost of producing x units is computed as:

$$TC_x = (x)(C_x) = C_1 x^{(b+1)}.$$

Expression for Unit Cost: The unit cost of producing the xth unit is given by:

$$UC_x = C_1 x^{(b+1)} - C_1 (x-1)^{(b+1)}.$$

Expression for Marginal Cost: The marginal cost of producing the xth unit is given by

$$MC_x = \frac{d[TC_x]}{dx}$$
$$= (b+1) C_1 x^b .$$

Multivariate Learning Curve Model

The learning rate of employees is often influenced by other factors that may be within the control of the organization. The conventional learning curve model is developed as a function of production level only. Nonetheless, other factors apart from cumulative production volume can influence how fast, how far, and how well a production operator learns within a given time horizon. The overall effect of learning may be influenced by several factors including skill, experience, prior training, and concurrent training levels, design changes, methods improvement, material substitutions, changes in tolerance levels, task complexity, degree of external interference affecting the task and/or operator, level of competence with available tools, and prior experience with related job functions (skill transfer).

Model Formulation

To account for the multivariate influence on learning, a model was developed to facilitate a wider scope of learning curve analysis for manufacturing operations. The general form of the hypothesized model is:

$$C_x = K \prod_{i=1}^{n} c_i x_i^{b_i} ,$$

where C_x = cumulative average cost per unit for a given set of x factor values (x = vector of specific values of independent variables); K = model parameter;

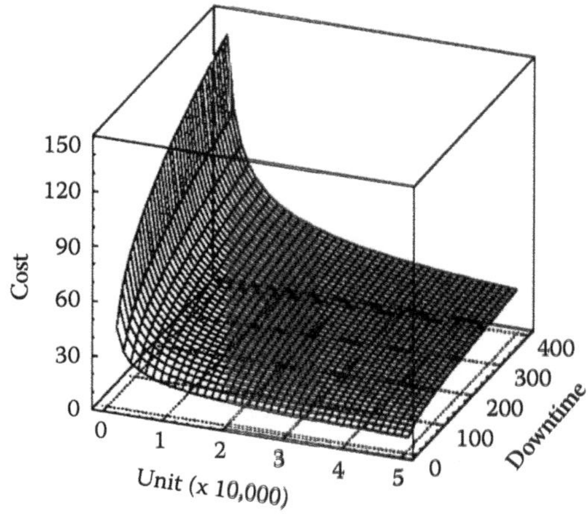

FIGURE 4.3
Response surface for bivariate learning curve model.

C_1 = cost of first unit of product; x_i = specific value of *i*th factor; n = number of factors in model; c_i = coefficient for *i*th factor; and b_i = learning exponent for *i*th factor. Figure 4.3 shows a generic graphical representation of a two-factor learning curve model based on the aforementioned formulation.

Bivariate Model

This section presents a computational experience with a learning curve model containing two independent variables: cumulative production (x_1) and cumulative training time (x_2). The inclusion of training time as a second variable in the model is reasonable because many factors that influence learning can be expressed as training-dependent variables. With the two-factor model, the expected cost of production can be estimated on the basis of cumulative production and cumulative training time.

Training time in this illustration refers to the amount of time explicitly dedicated to training an operator for his or her job functions. Previous learning curve models considered increases in productivity due to increases in production volumes. Little consideration has been given to the fact that production levels are time-dependent and influenced by other factors. For a two-factor model, the mathematical expression for the learning curve is hypothesized to be:

$$C_{x_1x_2} = Kc_1x_1^{b_1}c_2x_2^{b_2},$$

where C_x = cumulative average cost per unit for a given set of x factor values; K = intrinsic constant; c_1 = coefficient of factor 1; c_2 = coefficient of factor 2; x_1 =

TABLE 4.5

Multivariate Learning Curve Data

Treatment	Observation	Cumulative Average Cost ($)	Cumulative Production (Units)	Cumulative Training Time (Hours)
1	1	120	10	11
	2	140	10	8
2	3	95	20	54
	4	125	20	25
3	5	80	40	100
	6	75	40	80
4	7	65	80	220
	8	50	80	150
5	9	55	160	410
	10	40	160	500
6	11	40	320	660
	12	38	320	600
7	13	32	640	810
	14	36	640	750
8	15	25	1280	890
	16	25	1280	800
9	17	20	2560	990
	18	24	2560	900
10	19	19	5120	1155
	20	25	5120	1000

specific value of factor 1; x_2 = specific value of factor 2; b_1 = learning exponent for factor 1; and b_2 = learning exponent for factor 2.

A set of real data compiled from a manufacturing company is shown in Table 4.5 and used to illustrate the procedure for fitting and diagnostically evaluating a multivariate learning curve function. Two data replicates are used for each of the ten combinations of cost and time values. Observations are recorded for the number of units representing double production volumes. Data replicates were obtained by recording observations from identical set-ups of the process studied. The process involved the assembly of electronic components in communication equipment. The cost and time values were rounded off to the nearest whole numbers in accordance with strategic planning reports of the company. The two-factor model is represented in logarithmic scale to facilitate the curve fitting procedure:

$$\begin{aligned}\log C_x &= \left[\log K + \log(c_1 c_2)\right] + b_1 \log x_1 + b_2 \log x_2 \\ &= \log a + b_1 \log x_1 + b_2 \log x_2,\end{aligned}$$

where a represents the combined constant in the model such that:

$$a = (K)(c_1)(c_2).$$

Using statistical analysis software, the following model was fitted for the data:

$$\log C_x = 5.70 - 0.21(\log x_t) - 0.13(\log x_2),$$

which transforms into the multiplicative model:

$$C_x = 298.88 x_1^{-0.21} x_2^{-0.13},$$

where $a = 298.88$ [i.e., $\log(a) = 5.70$; C_x = cumulative average cost per unit; x_1 = cumulative production units; and x_2 = cumulative training time in hours.

Since $a = 298.88$, we have $(Kc_1c_2) = 298.88$. If two of the constants K, c_1, or c_2 are known, the third can be computed. The constants may be determined empirically from the analysis of historical data. Figure 4.4 shows the response surface for the multiplicative form of the fitted bivariate learning curve example. A visual inspection of the response surface plot indicates that the cumulative average cost per unit is sensitive to changes in both independent variables—cumulative units and training time. It does appear, however, that the sensitivity to cumulative units is greater than the sensitivity to training time. Diagnostic statistical analyses indicate that the model is a good fit for the data. The 95% confidence intervals for the parameters in the model are shown in Table 4.6.

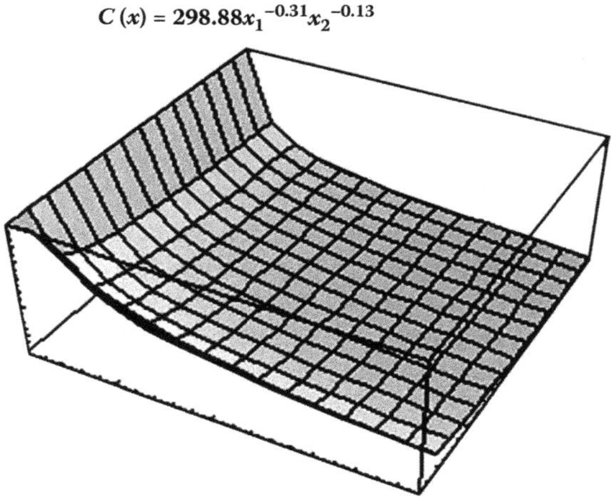

FIGURE 4.4
Bivariate learning curve example.

TABLE 4.6

95% Confidence Intervals for Model Parameters

Parameter	Estimate	Lower Limit	Upper Limit
log(a)	5,7024	5,4717	5,9331
b_1	–0.2093	–0.2826	–0.1359
b_2	–0.1321	–0.2269	–0.0373

TABLE 4.7

ANOVA for Full Regression of Learning Curve Model

Source	Sum of Squares	df	Mean Square	F-Ratio	P-Value
Model	7.41394	2	3.70697	248.778	0.0000
Error	0.253312	17	0.00149007		
Total	7.66725	19			

Note: R-squared = 0.966962; R-adjusted (for degree of freedom) = 0.963075; standard error of estimate = 0.122069.

TABLE 4.8

Further ANOVA for Variables in Model Fitted

Source	Sum of Squares	df	Mean Square	F-Ratio	P-Value
Log (units)	7.28516187	1	7.2851619	488.91	0.000
Log (time)	0.12877864	1	0.1287786	8.64	0.0092
Model	7.41394052	2			

The results of analysis of variance for the full regression are presented in Table 4.7. The P-value of 0.0000 in the table indicates a highly significant regression fit. The R-squared value of 0.966962 indicates that almost 97% of the variabilities in cumulative average cost are explained by the terms in the model. Table 4.8 shows the breakdown of the model component of the sum of squares. Based on the low P-values shown in the table, units and training time contributed significantly to the multiple regression model. Also, based on the sum of squares, production units account for most of the fit in this particular bivariate learning curve model.

The correlation matrix for the estimates of the coefficients in the model is shown in Table 4.9. Note that units and training time are negatively correlated and the constant is positively correlated with production units and negatively correlated with training time. The strong negative correlation (–0.9189) between units and training time confirms that training time decreases as production volume increases (i.e., learning curve effect). Variables that are statistically independent will have expected correlations of zero. As expected, Table 4.9 does not indicate zero correlations.

TABLE 4.9

Correlation Matrix for Coefficient Estimates

Source	Constant	Log (Units)	Log (Time)
Constant	1.0000	0.3654	−0.6895
Log (units)	0.3654	1.0000	−0.9189
Log (time)	−0.6895	−0.9189	1.0000

A residual plot of the modeling indicates a fairly random distribution of the residuals, suggesting a good fit. The following additional results are obtained for the residual analysis: residual average = 4.66294, coefficient of skewness = 0.140909, coefficient of kurtosis = −0.557022, and Durbin-Watson statistic = 1.21472. A normal probability plot of the residuals was also analyzed. The fairly straight line fit in the plot indicates that the residuals are approximately normally distributed—an essential condition for a good regression model. A perfectly normal distribution would plot as a straight line in a normal probability plot.

Given a production level of 1,750 units and a cumulative training time of 600 hours, the multivariate model indicates an estimated cumulative average cost per unit:

$$C_{(1750.600)} = (298.88)\left(1750^{-0.21}\right)\left(600^{-0.13}\right)$$

$$= 27.12.$$

Similarly, a production level of 3,500 units and training time of 950 hours yields the following cumulative average cost per unit:

$$C_{(3500.950)} = (298.88)\left(3500^{-0.21}\right)\left(950^{-0.13}\right)$$

$$= 22.08.$$

R-squared = 0.966962; R-adjusted (for degree of freedom) = 0.963075; standard error of estimate = 0.122069.

An important application of the bivariate model involves addressing problems such as the following:

Situation: The standards department of a manufacturing plant set a target average cost per unit of $12.75 to be achieved after 1,000 hours of training. Find the cumulative units that must be produced to achieve the target cost.

Analysis: Using the bivariate model previously fitted, the following relationship is obtained:

$$\$12.75 = (298.88)\left(X^{-0.21}\right)\left(1000^{-0.31}\right),$$

which yields the required cumulative production of $X = 46.444$ units.

Based on the large number of cumulative production units required to achieve the expected standard cost, the standards department may want to review the cost standard. The $12.75 may not be achievable if the product under consideration has a limited market demand. The flat surface of the learning curve model as units and training time increase implies that more units will be needed to achieve additional cost improvements. Thus, even though an average cost of $22.08 can be obtained at a cumulative production level of 3,500 units, it takes several thousand additional units to bring the average cost down to $12.75 per unit.

Comparison to Univariate Model

The bivariate model permits us to simultaneously analyze the effects of production level and training time on average cost per unit. If a univariate function is used with the conventional log-linear model, we would be limited to a relationship between average cost and cumulative production. This would deprive us of the full range of information we could have obtained about process productivity. This fact is evidenced by using only the second and third columns of the data in Table 4.9 as shown in the following. The univariate log-linear model is represented as

$$C_x = C_1 x^b,$$

where the variables are as defined previously, The second and third columns of Table 4.9 are fitted to the aforementioned model to yield:

$$C_x = 240.03X^{(-0.3032)},$$

with $R^2 = 95.02\%$. Thus, the learning exponent, b is equal to –03032 and the corresponding percent learning p is 81.05%. This indicates that a productivity gain of 18.95% is expected whenever the cumulative production level doubles. Although we obtained a good fit for the univariate model, the information about training time is lost. For example, the univariate model indicates that the cumulative average cost would be $33.84 per unit when cumulative production reaches 640 units. By contrast, the bivariate model shows that the cumulative average cost is $32.22 per unit at that cumulative production level with a cumulative training time of 810 hours. This lower cost is due to the additional effect of training time. Thus, the bivariate model provides a more accurate picture of the interactions of factors associated with the process.

Potential Applications

A multivariate learning curve model extends the conventional analysis of the effect of learning to a more robust decision-making tool for operations management. Learning curves can be used to plan manpower needs; set labor standards; establish prices; analyze make or buy decisions; review wage incentive payments; evaluate organizational efficiency; develop quantity sales discounts; evaluate employee training needs; evaluate capital equipment alternatives; predict future production unit costs; create production delivery schedules; and develop work design strategies. The parameters that affect learning rate are broad in scope. Unfortunately, practitioners do not fully comprehend the causes of and influences on learning rates. Learning is typically influenced by many factors in a manufacturing operation. Such factors may include:

Governmental factors:
- Industry standards
- Educational facilities
- Regulations
- Economic viability
- Social programs
- Employment support services
- Financial and trade supports

Organizational factors:
- Management awareness
- Training programs
- Incentive programs
- Employee turnover rate
- Cost accounting system
- Labor standards
- Quality objectives
- Pool of employees
- Departmental interfaces
- Market and competition pressures
- Obsolescence of machinery and equipment
- Long range versus short range goals

Product-based factors:
- Product specifications
- Raw material characteristics
- Environmental impacts of production
- Delivery schedules
- Numbers and types of constituent operations
- Product diversity

Assembly requirements
Design maturity
Work design and simplification
Level of automation

Multivariate learning curve models offer robust tools to simultaneously evaluate several factors quantitatively. Decisions based on multivariate analyses are generally more reliable than decisions based on single-factor analyses. The next sections present brief outlines of potential applications of multivariate learning curve models for performance analysis, management, and control.

Design of Training Programs

As shown in the bivariate model, the need for employee training can be assessed with the aid of a multivariate learning curve model. Projected productivity gains are more likely to be met if more of the factors influencing productivity can be included in the conventional analysis.

Manufacturing Economic Analysis

Results of multivariate learning curve analysis are important for various types of economic assessments in a manufacturing environment. The declining state of manufacturing in many countries has been a subject of much discussion in recent years. A reliable methodology for analyzing costs for specific operations is essential to the full exploitation of the recent advances in technology. Manufacturing economic analysis is the process of evaluating manufacturing operations on a cost basis. In manufacturing systems, many tangible, intangible, quantitative, and qualitative factors intermingle to compound the cost analysis problem. Consequently, a more comprehensive evaluation methodology such as a multivariate learning curve can be very useful.

Breakeven Analysis

The conventional breakeven analysis assumes that variable cost per unit is constant. Conversely, learning curve analysis recognizes the potential reduction in variable cost per unit due to the effects of learning. Based on the multiple factors involved in manufacturing, multivariate learning curve models should be investigated and adopted for breakeven cost analysis.

Make or Buy Decisions

Make or buy decisions can be enhanced by considering the effects of learning on items that are manufactured in-house. Make or buy analysis involves a choice between the cost of producing an item and the cost of purchasing it. Multivariate learning curves can provide the data for determining the accurate cost of producing an item. A make or buy analysis can be coupled with breakeven analysis to determine when to make or buy a product.

Manpower Scheduling

Considering the effects of learning in a manufacturing environment can lead to more accurate analyses of manpower requirements and accompanying schedules. In integrated production, where parts move sequentially from one station to another, the effect of multivariate learning curves can become even more applicable. The allocation of resources during production scheduling should not be made without considering the effect of learning (Liao 1978).

Production Planning

The overall production planning process can benefit from multivariate learning curve models. Preproduction planning analysis of the effects of multivariate learning curves can identify areas where better and more detailed planning may be needed. The more preproduction planning that is done, the better the potential for achieving the productivity gains produced by learning.

Labor Estimating

For manufacturing activities involving operations at different stations, several factors interact to determine the learning rates of workers. Multivariate curves can be useful for developing accurate labor standards and complement conventional work measurement studies.

Budgeting and Resource Allocation

Budgeting or capital rationing is a significant effort in any manufacturing operation. Multivariate learning curve analysis can provide a management guide for allocating resources to production operations on a more equitable basis.

Impacts of Multivariate Learning Curves

The learning curve phenomenon has been of interest to researchers and practitioners for many years. The variety of situations to which learning curves are applicable has necessitated the development of various functional forms for the curves. The conventional view of learning curves considers only one factor at a time as the major influence on productivity improvement. However, in today's integrated manufacturing environment, it is obvious that several factors interact to activate and perpetrate productivity improvement.

This chapter presented a general approach for developing multivariate learning curve models. Multivariate models are useful for various types of analyses in a manufacturing environment, for example economic analysis, breakeven analysis, make or buy decisions, manpower scheduling, production planning, budgeting, resource allocation, labor estimating, and cost optimization.

Multivariate learning curves can generate estimates of expected costs, productivity, process capability, work load composition, system response time, and other parameters. Such estimates can be valuable to decision makers for manufacturing process improvement and work simplification functions. The availability of reliable learning curve estimates can enhance the communication interfaces among different groups in an organization. Multivariate learning curves can provide mechanisms for the effective cost-based implementation of design systems, facilitate systems integration of production plans, improve supervisory interfaces, enhance design processes, and provide cost information to strengthen the engineering and finance interface. The productivity of human resources is becoming a major concern in many industries. Thus, a multivariate learning curve model can contribute significantly to better utilization of manpower resources.

The multiplicative model used in the illustrative example presented in this chapter is only one of several possible expressions that can be investigated for multivariate learning curve models. Further research and detailed experiments with alternate functional forms are necessary to formalize the proposed methodology. It is anticipated that such further studies will be of interest to practitioners and researchers.

Fast-changing operating environments that we now face require interdisciplinary improvement projects that may involve physicists, material and industrial engineers, chemists, engineers, bioscientists, and other disciplines. They may all use different measurement scales and jargons, but the common thread of work measurement permeates collective improvement efforts. We should not shy away from work measurement because of its inglorious (and false) past. It should be embraced as the forerunner of new approaches to pursuing improvement in industrial operations.

Work measurement, under whatever trendy name we choose, must remain a core competency of industrial engineering. Measurements of productivity, human performance, and resource consumption are essential components of achieving organizational goals and increasing profitability. Work rate analysis centered on work measurement helps identify areas where operational efficiency and improvement can be pursued. In many production settings, production breaks require an analysis of their impacts on production output. Whether production breaks are standard and scheduled or nonstandard and unscheduled, work stations are subject to work rate slowdown (ramp-down) and pickup (ramp-up) before and after breaks. These impacts are subtle and hard to see without a formal engineered work measurement technique. Practitioners, researchers, students, and policy makers in industrial engineering must embrace and promote industrial work measurement as a tool of productivity improvement.

Summary of Cost, Time, and Performance Formulas

This section summarizes common formulas useful for cost, time, and performance assessment.

Average time to perform task—Based on learning curve analysis, the average time required to perform a repetitive task is calculated as

$$t_n = an^{-b},$$

where t_n = cumulative average time resulting from performing a task n times; t_1 = time required to perform the task the first time; and k = learning factor for the task (usually known or assumed). The parameter k is a positive real constant whose magnitude is a function of the type of task performed. A large value of k would cause the overall average time to drop quickly after only a few repetitions of the task. Thus, simple tasks tend to have large learning factors. Complex tasks tend to have smaller learning factors, thereby requiring several repetitions before significant time reduction can be achieved.

Calculating learning factor—If a learning factor is not known, it may be estimated from time observations by

$$k = \frac{\log t_1 - \log t_n}{\log n}.$$

Calculating total time—Total time, T_n, to complete a task n times, if the learning factor and initial time are known, is obtained by multiplying the average time by the number of times:

$$T_n = t_1 n^{(1-k)}.$$

Determining time for nth performance of task—The formula is:

$$x_n = t_1(1-k)n^{-k}.$$

Determining limit of learning effect—The limit of learning effect indicates the number of times of performance of a task at which no further improvement is achieved. This is often called the improvement ratio and is represented as

$$n \geq \frac{1}{1 - r^{1/k}}.$$

Determining improvement target—It may be desirable to achieve a certain level of improvement after so many performances of a task based on a certain learning factor k. Suppose that it takes n_1 trials to achieve a certain average time performance y_1 and we want to calculate how many n_2 trials would be needed to achieve a given average time performance y_2. The formula is:

$$n_2 = n_1 y_1^{1/k} y_2^{-1/k} = n_1 (y_2 / y_1)^{-1/k} = n_1 r^{-1/k},$$

where r is the time improvement factor.

Calculation of number of machines required to meet output—The number of machines to achieve a specified total output is calculated as:

$$N = \frac{1.67t(O_T)}{uH},$$

where N = number of machines; t = processing time per unit (minutes); O_T = total output per shift; u = machine utilization ratio (decimal); and H = hours worked per day (8 times number of shifts).

Calculation of machine utilization—Machine idle times adversely affect utilization. The fraction of the time a machine is productively engaged is the utilization ratio calculated:

$$u = \frac{h_a}{h_m},$$

where h_a = actual hours worked and h_m = maximum hours the machine could work. The *percent utilization* = 100% times u.

Calculation of output to allow for defects—To allow a certain fraction of defects in total output, calculate starting output:

$$Y = \frac{X}{1 - f},$$

where Y = starting output; X = target output; and f = fraction defective.

Calculation of machine availability—The percent of time that a machine is available for productive work is calculated as:

$$A = \frac{o - u}{o}(100\%),$$

where o = operator time and u = unplanned downtime.

References

Badiru, Adedeji B. (2008). Long live work measurement. *Industrial Engineer*, Vol. 40, No. 3, p. 24.

Badiru, Adedeji B. (1988). *Project Management in Manufacturing and High Technology Operations*. John Wiley & Sons, New York.

Ray, Paul (2006). Work sampling. In *Handbook of Industrial and Systems Engineering*, Badiru, A. B., Ed. Taylor & Francis, Boca Raton, FL.

Wilson, J. R. and N. Corlett, Eds. (2005). *Evaluation of Human Work*, 3rd ed. Taylor & Francis, New York.

Wright, T. P. (1936). Factors affecting the cost of airplanes. *Journal of Aeronautical Science*, 3, 122–128.

5

Project Cost Control

"Complex acquisition processes do not promote program success—they increase costs, add to schedule and obfuscate accountability."

2006 Report of Defense Advanced Research Projects Agency (DARPA)

As cautioned by the opening quote above, complexity impedes project control, particularly along the metrics of cost, schedule, and quality. Every project needs procurement and acquisition processes. If those processes are too complex, they negate project control endeavors. With the ever changing economy, cost control is one of the most important steps of controlling any type of project. Low investment projects with high returns of process control and efficiency improvements are desired. High investment opportunities are still desired if the returns on investment opportunities are below the 2-year mark.

Determining return on investment (ROI) is important for investing in capital projects, but they normally have higher push-backs than non-capital projects. To begin cost control on a project, it is important to define the scope of the project, the investment, the return on investment, and the failure modes. Utilizing failure modes and effects analysis (FMEA) is one major way to start a cost control project. The details for constructing an FMEA are explained in Chapter 8. The basic steps are:

1. Define process steps.
2. Define functions.
3. Define potential failure modes.
4. Define potential effects of failure.
5. Define the severity of a failure.
6. Define the potential mechanisms of failure.
7. Define current process controls.
8. Define the occurrence of failure.
9. Define current process control detection mechanisms.
10. Define the ease of detecting a failure.
11. Multiply severity, occurrence, and detection to calculate a risk priority number (RPN).
12. Define recommended actions.

13. Assign actions with key target dates to responsible personnel.
14. Revisit the process after actions have been taken to improve it.
15. Recalculate RPNs with the improvements.

The biggest difficulty with cost control is justification of the project to management. The easiest projects to approve are non-investment procedural measures. Some common-sense steps will help remove basic waste and improve efficiency.

Standard operating procedures are helpful for addressing efficiency improvements. Each operator performs tasks differently, some better than others. If the most efficient practice is desired, standard operating procedures can ensure that each operator follows the best practice. Standard operating procedures include describing in detail what actions to take and when to take them. They also include visual aids so operators can understand the steps with ease. A case discussed at the end of this chapter illustrates an industrial production project cost control process.

Cost control requires a basic understanding of cost elements, time value of money, and cash flow equivalents. The following sections present the basic techniques of cash flow analysis useful for developing effective project cost control practices. Other topics covered are cost concepts, cost estimation, cost monitoring, budgeting allocation, and inflation.

Cost Concepts and Definitions

Cost management in a project environment encompasses the functions required to maintain effective financial control of a project throughout its life cycle. Several cost concepts influence the economic aspects of managing projects. Specific analyses may require consideration of various combinations of cost aspects that are discussed below:

Actual cost of work performed—Costs of accomplishing the work performed within a given time period.

Applied direct cost—Costs over a given period for labor, materials, and other direct resources without regard to the date of commitment or date of payment. These amounts are to be charged to work in process (WIP) when the resources are actually consumed, are withdrawn from inventory for use, or are received and scheduled for use within 60 days.

Budgeted cost for work performed—Sum of budget amounts for completed work plus appropriate portions of the budgets for level of effort and apportioned effort. Apportioned effort alone is not readily divisible into short-span work packages but relates in direct proportion to measured effort.

Budgeted cost for work scheduled—Sum of budget amounts for all work packages and planning packages scheduled to be accomplished (including work in process) plus the amount of effort and apportioned effort scheduled to be accomplished within a given period.

Direct cost—Cost directly associated with operations of a project. Examples are direct material costs and direct labor costs. Direct costs can be reasonably measured and allocated to specific components of a project.

Economies of scale—Reduction of the relative weight of a fixed cost in total cost by increasing output quantity. This helps to reduce the final unit cost of a product. Economies of scale may also be called savings from mass production.

Estimated cost at completion—Actual direct cost plus indirect costs that can be allocated to the contract and the estimate of direct and indirect costs for remaining authorized work.

First cost—Total initial investment required to initiate a project or total initial cost of the equipment needed to start a project.

Fixed cost—Cost incurred regardless of project level. Fixed costs do not vary in proportion to output. Examples of fixed costs are administrative expenses, certain types of taxes, insurance premiums, depreciation, and debt service.

Incremental cost—Additional costs for changing production output from one level to another. Incremental costs are normally variable.

Indirect cost—Cost indirectly associated with project operations. Indirect costs are difficult to assign to specific components of a project. An example of an indirect cost is the purchase of computer hardware and software needed to manage project operations. Indirect costs are usually calculated as a percentage of direct costs. For example, the direct costs may be computed as 10% of direct labor costs.

Life cycle cost—Sum of all recurring and nonrecurring costs associated with a project during its entire life cycle.

Maintenance cost—Intermittent or periodic cost incurred to keep project equipment in good operating condition.

Marginal cost—Additional cost of increasing production output by one additional unit. The marginal cost is equal to the slope of the total cost curve or line at the current operating level.

Operating cost—Recurring cost needed to keep a project in operation during its life cycle. Examples are labor, material, and energy costs.

Opportunity cost—Cost of forgoing the opportunity to invest in a venture that would have produced an economic advantage. Opportunity costs are usually incurred when limited resources make it impossible to take advantage of investment opportunities. It is often defined as the cost of the best rejected opportunity. Opportunity costs can arise from missed opportunities rather than intentional rejections. In many cases, opportunity costs are hidden or implied because they typically relate to future events that cannot be accurately predicted.

Overhead cost—Cost incurred for activities performed to support operations of a project. The activities that generate overhead costs support the project efforts rather contributing directly to the project goal. The handling of overhead costs varies widely among companies. Typical overhead items are electricity, insurance premiums, security, and inventory carrying costs.

Standard cost—The normal or expected cost of a unit of output. Standard costs are established in advance. They are developed as composites of several component costs such as direct labor cost per unit, material cost per unit, and allowable overhead charge per unit.

Sunk cost—Past cost that cannot be recovered under the present analysis. Sunk costs should have no bearing on the prevailing economic analysis and project decisions. Ignoring sunk costs is always difficult for analysts. For example, if $950,000 was spent 4 years ago to buy equipment for a technology-based project, a decision on whether to replace the equipment now should not consider the initial cost. However, uncompromising analysts may find it difficult to ignore that amount. Similarly, an individual who decides to sell a personal automobile would typically try to relate the asking price to what the automobile cost when acquired. This is wrong under the strict concept of sunk costs.

Total cost—Sum of all the variable and fixed costs associated with a project.

Variable cost—Cost that varies in direct proportion to the level of operation or quantity of output. For example, the costs of material and labor required to make an item are classified as variable because they vary with changes in output levels.

Project Cash Flow Analysis

The reason for performing economic analysis is to choose between mutually exclusive projects that compete for limited resources. The cost performance of each project depends on its timing and costs. The techniques of computing cash flow equivalence permit us to put competing project cash flows on a common basis for comparison. The common basis depends on the prevailing interest rate. Two cash flows that are equivalent at a given interest rate will not be equivalent at different interest rates. The basic techniques for converting cash flows from one point in time to another are presented in the next section.

Time Value of Money Calculation

Cash flow conversion involves the transfer of project funds from one point in time to another. The following notations indicate the variables involved in the conversion process:

i = interest rate per period
n = number of interest periods
P = present sum of money
F = future sum of money
A = uniform end-of-period cash receipt or disbursement
G = uniform arithmetic gradient increase in period-by-period payments or disbursements.

In many cases, the interest rate used in performing economic analysis is set equal to the minimum attractive rate of return (MARR) of the decision maker. The MARR may also be called the *hurdle rate*, *required internal rate of return* (IRR), *return on investment* (ROI), or *discount rate*. The value of MARR is chosen with the objective of maximizing the economic performance of a project.

Compound amount factor—This procedure for the single payment compound amount factor calculates a future sum of money F equivalent to a present sum of money P at a specified interest rate i after n periods. It is calculated as $F = P(1 + i)^n$. Figure 5.1 shows the relationship of P and F.

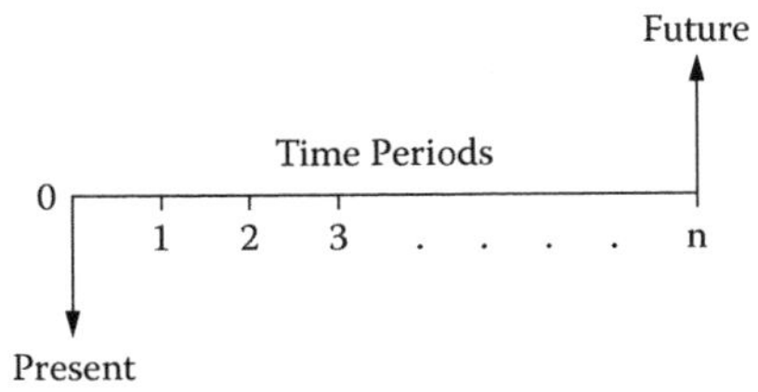

FIGURE 5.1
Single payment compound amount cash flow.

Example

\$5,000 is deposited in a project account and left to earn interest for 15 years. If the interest rate per year is 12%, the compound amount is F = \$5,000(1 + 0.12)^{15} = \$27,367.85.

Present worth factor—This factor computes P when F is given. It is determined by solving for P in the equation for the compound amount factor, that is, $P = F(1 + i)^{-n}$.

Example

Assume \$15,000 is needed to complete a project 5 years from now. How much should be deposited in a special project fund now so that the fund would accrue to the required \$15,000 in exactly 5 years? If the special project fund pays interest at 9.2% per year, the required deposit would be P = \$15,000(1 + 0.092)$^{-5}$ = \$9660.03.

Uniform series present worth factor—We use this factor to calculate the present worth equivalent P of a series of equal end-of-period amounts A. Figure 5.2 shows the uniform series cash flow. The derivation of the formula uses the finite sum of the present worth of the individual amounts in the uniform series cash flow below. The appendix at the end of this book presents some formulas for series and summation operations.

$$P = \sum_{t=1}^{n} A(1+i)^{-1}$$

$$= A\left[\frac{(1+i)^n - 1}{i(1+i)^n}\right]$$

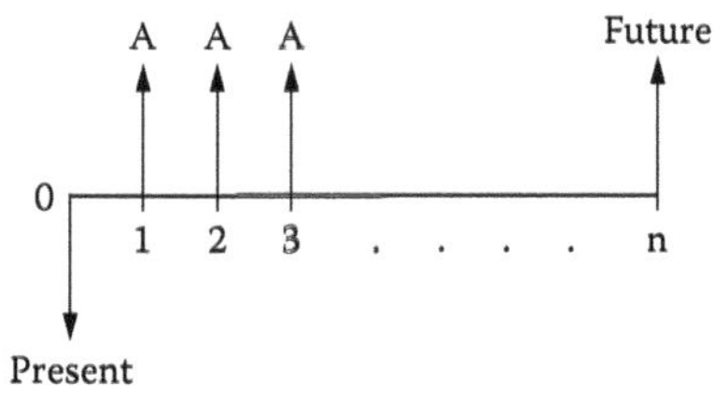

FIGURE 5.2
Uniform series cash flow.

Example

Suppose $12,000 must be withdrawn from an account to meet the annual operating expenses of a multiyear project. The project account pays interest at 7.5% per year compounded annually. If the project is expected to last 10 years, how much must be deposited in the project account now to ensure that operating expenses of $12,000 can be withdrawn at the end of every year for 10 years? The project fund is expected to be depleted to zero by the end of the last year of the project. The first withdrawal will be made 1 year after the project account is opened, and no additional deposits will be made during the project life cycle. The required deposit is calculated as:

$$P = \$12{,}000\left[\frac{(1+0.075)^{10}-1}{0.075(1+0.075)^{10}}\right]$$

$$= \$82{,}368.92.$$

Uniform series capital recovery factor—This formula enables us to calculate a uniform series of equal end-of-period payments A that are equivalent to a given present amount P. It is the converse of the uniform series present amount factor. The equation is obtained by solving for A in the uniform series present amount factor, that is,

$$A = P\left[\frac{i(1+i)^n}{(1+i)^n-1}\right].$$

Example

Suppose a device needed to launch a project must be purchased at a cost of $50,000. The entire cost is to be financed at 13.5% per year and repaid on a monthly installment schedule over 4 years so we must calculate the monthly loan payments. Assume that the first loan payment will be made exactly 1 month after the equipment is financed. If the interest rate of 13.5% per year is compounded monthly, the interest rate per month will be 13.5%/12 = 1.125%. The number of interest periods over which the loan will be repaid is 4(12) = 48 months. Thus, the monthly loan payments are calculated:

$$A = \$50{,}000\left[\frac{0.01125(1+0.01125)^{48}}{(1+0.01125)^{48}-1}\right]$$

$$= \$1{,}353.82.$$

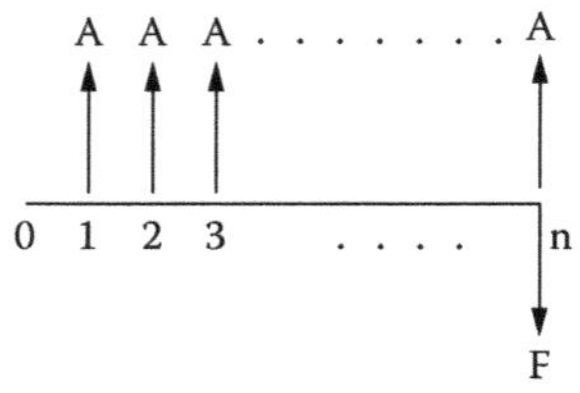

FIGURE 5.3
Uniform series compound amount cash flow.

Uniform series compound amount factor—This factor is used to calculate a single future amount that is equivalent to a uniform series of equal end-of-period payments. Figure 5.3 shows the cash flow. Note that the future amount occurs at the same time point as the last amount in the uniform series of payments. The factor is derived as:

$$F = \sum_{t=1}^{n} A(1+i)^{n-1}$$

$$= A\left[\frac{(1+i)^n - 1}{i}\right].$$

Example

If equal end-of-year deposits of $5,000 are made to a project fund paying 8% per year for 10 years, how much can be expected to be available for withdrawal from the account for capital expenditure immediately after the last deposit is made?

$$F = \$5000\left[\frac{(1+0.08)^{10} - 1}{0.08}\right]$$

$$= \$72{,}432.50.$$

Uniform series sinking fund factor—We use this factor to calculate a uniform series of equal end-of-period amounts A that are equivalent to a single future amount F. It is the reverse of the uniform series compound amount factor. The formula is obtained by solving for A in the formula for the uniform series compound amount factor, that is,

$$A = F\left[\frac{i}{(1+i)^n - 1}\right].$$

Example

How large are the end-of-year equal amounts that must be deposited into a project account to ensure that $75,000 will be available for withdrawal immediately after the 12th annual deposit is made? The initial balance in the account is zero at the beginning of the first year. The account pays 10% interest per year. Using the formula for the sinking fund factor, the required annual deposits are:

$$A = \$75{,}000\left[\frac{0.10}{(1+0.10)^{12}-1}\right]$$

$$= \$3{,}507.25.$$

Capitalized cost formula—*Capitalized* cost is the present value of a single amount equivalent to a perpetual series of equal end-of-period payments. It is calculated as an extension of the series present worth factor with an infinitely large number of periods (Figure 5.4). Using the limit theorem from calculus as n approaches infinity, the series present worth factor reduces to the following formula for capitalized cost:

$$P = \lim_{n\to\infty} A\left[\frac{(1+i)^n - 1}{i(1+i)^n}\right]$$

$$= A\left\{\lim_{n\to\infty}\left[\frac{(1+i)^n - 1}{i(1+i)^n}\right]\right\}$$

$$= A\left(\frac{1}{i}\right).$$

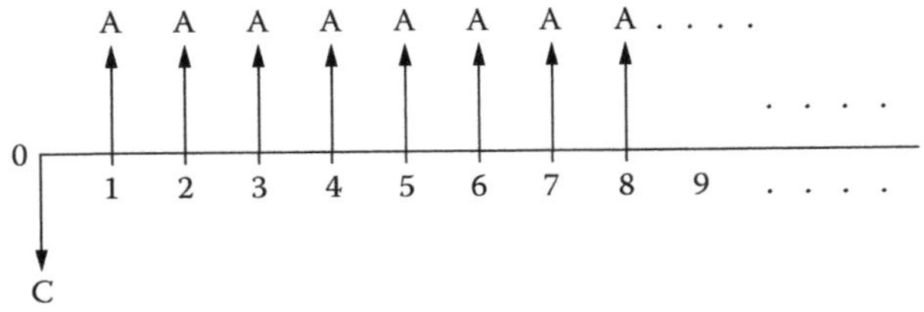

FIGURE 5.4
Capitalized cost cash flow.

Example

How much should be deposited in a general fund to service a recurring public service project at \$6,500 per year forever if the fund yields an annual interest rate of 11%? Using the capitalized cost formula, the required one-time deposit to the general fund is:

$$P = \frac{\$6500}{0.11}$$

$$= \$59{,}090.91.$$

The formulas presented above represent the basic cash flow conversion factors. Tables of the factors are widely available in books about engineering economy. Several variations and extensions of the factors are available, for example, the arithmetic gradient series factor and the geometric series factor. Variations in cash flow profiles include situations where payments are made at the beginning of each period rather than at the end or may require a series of unequal payments. Conversion formulas can be derived mathematically for special cases by using the basic factors presented above.

Arithmetic gradient series—Gradient series cash flow involves an increase of a fixed amount of cash flow at the end of each period. Thus, the amount at any time point is greater than the amount during the preceding period by a constant amount G. Figure 5.5 shows the basic gradient series in which the base amount at the end of the first period is zero. The size of the cash flow in the gradient series at the end of period t is calculated as:

$$A_t = (t - 1)G, \qquad t = 1, 2, \ldots, n.$$

The total present value of the gradient series is calculated by using the present amount factor to convert each individual amount from time t to time 0 at an interest rate of i% per period and summing the resulting present values. The finite summation reduces to a closed form:

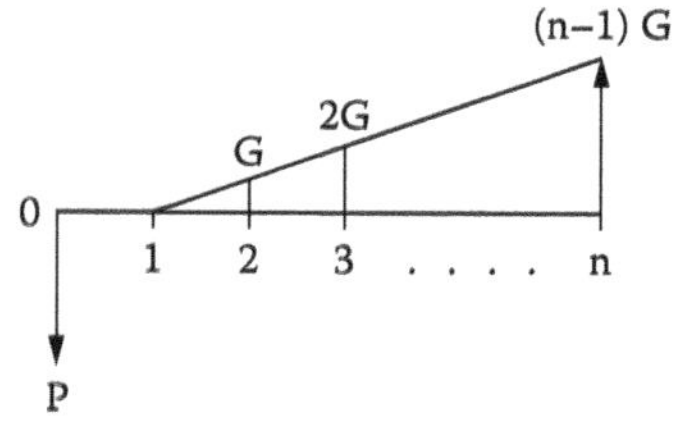

FIGURE 5.5
Arithmetic gradient cash flow with zero base amount.

$$P = \sum_{t=1}^{n} A_t(1+i)^{-t}$$

$$= \sum_{t=1}^{n} (t-1)G(1+i)^{-t}$$

$$= G\sum_{t=1}^{n} (t-1)(1+i)^{-t}$$

$$= G\left[\frac{(1+i)^n - (1+ni)}{i^2(1+i)^n}\right].$$

Example

The cost of supplies for a 10-year period increases by $1,500 every year starting at the end of year 2. There is no supplies cost at the end of the first year. If interest rate is 8% per year, determine the present amount to be set aside at time zero to cover all the future supplies expenditures. $G = 1500$, $i = 0.08$, and $n = 10$. Using the arithmetic gradient formula, we obtain:

$$P = 1500\left[\frac{1-(1+10(0.08))(1+0.08)^{-10}}{(0.08)^2}\right]$$

$$= \$1500(25.9768)$$

$$= \$38{,}965.20.$$

In many cases, an arithmetic gradient starts with some base amount at the end of the first period and increases by a constant amount thereafter. The nonzero base amount is denoted as A_1. Figure 5.6 shows this type of cash flow.

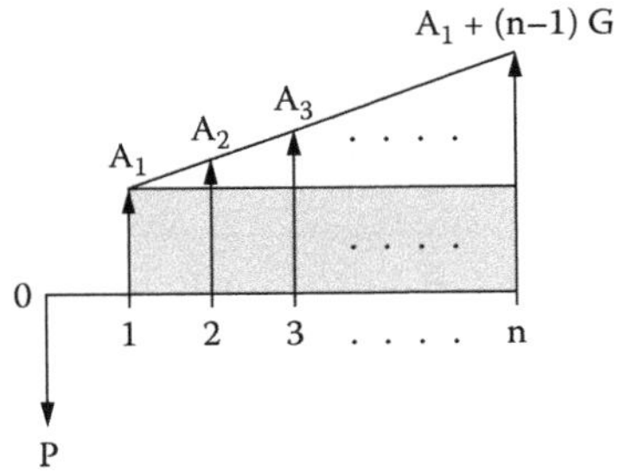

FIGURE 5.6
Arithmetic gradient cash flow with nonzero base amount.

The calculation of the present amount for such cash flows requires breaking the cash flow into a uniform series cash flow of amount A_1 and an arithmetic gradient cash flow with a zero base. The uniform series present worth formula is used to calculate the present value of the uniform series portion and the basic gradient series formula is used to calculate the gradient portion. The overall present worth is then calculated as:

$$P = P_{\text{uniform series}} + P_{\text{gradient series}}$$

$$= A_1\left[\frac{(1+i)^n - 1}{i(1+i)^n}\right] + G\left[\frac{(1+i)^n - (1+ni)}{i^2(1+i)^n}\right].$$

Increasing geometric series cash flow—The cash flow increases by a constant percentage from period to period. A positive base amount A_1 exists at the end of period 1. Figure 5.7 shows an increasing geometric series. The amount at time t is denoted as:

$$A_t = A_{t-1}(1+j), \qquad t = 2, 3, \ldots, n,$$

where j is the percentage increase in cash flow from period to period. Via a series of back substitutions, we can represent A_t in terms of A_1 instead of A_{t-1} as shown below:

$$A_2 = A_1(1+j)$$

$$A_3 = A_2(1+j) = A_1(1+j)(1+j)$$

$$\ldots$$

$$A_t = A_1(1+j)^{t-1}, \qquad t = 1, 2, 3, \ldots, n.$$

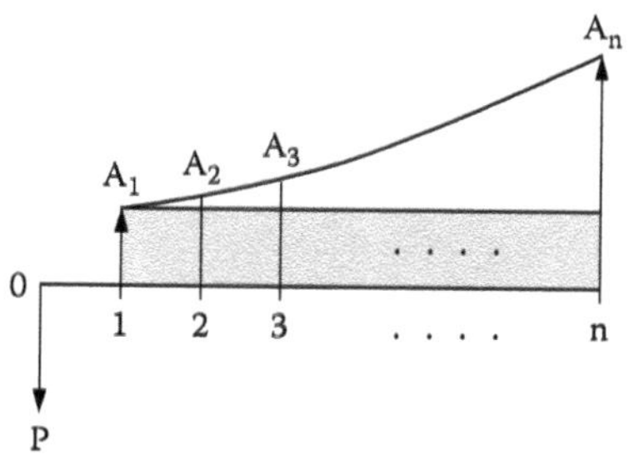

FIGURE 5.7
Increasing geometric series cash flow.

The formula for calculating the present worth of the increasing geometric series cash flow is derived by summing the present values of the individual cash flow amounts:

$$P = \sum_{t=1}^{n} A_t(1+i)^{-t}$$

$$= \sum_{t=1}^{n} \left[A_1(1+j)^{t-1} \right] (1+i)^{-t}$$

$$= \frac{A_1}{(1+j)} \sum_{t=1}^{n} \left(\frac{1+j}{1+i} \right)^t$$

$$= A_1 \left[\frac{1-(1+j)^n (1+i)^{-n}}{i-j} \right], \qquad i \neq j.$$

If $i = j$, the formula above reduces to the limit as $i \to j$:

$$P = \frac{nA_1}{1+i}, \qquad i = j.$$

Example

Suppose funding for a 5-year project is to increase by 6% every year with an initial funding of \$20,000 at the end of the first year. Determine how much must be deposited into a budget account at time zero to cover the anticipated funding levels if the budget account pays 10% interest per year. $j = 6\%$, $i = 10\%$, $n = 5$, $A_1 = \$20{,}000$. Therefore:

$$P = 20{,}000 \left[\frac{1-(1+0.06)^5 (1+0.10)^{-5}}{0.10-0.06} \right]$$

$$= \$20{,}000(4.2267)$$

$$= \$84{,}533.60.$$

Decreasing geometric series cash flow—The amounts of cash flow decrease by a constant percentage from period to period. The cash flow starts at a positive base amount A_1 at the end of period 1. Figure 5.8 shows a decreasing geometric series. The time t is denoted as $A_t = A_{t-1}(1-j)$, $t = 2, 3, \ldots, n$ where j is the percentage decrease in cash flow from period to period. As in the case of the increasing geometric series, we represent A_t as A_1:

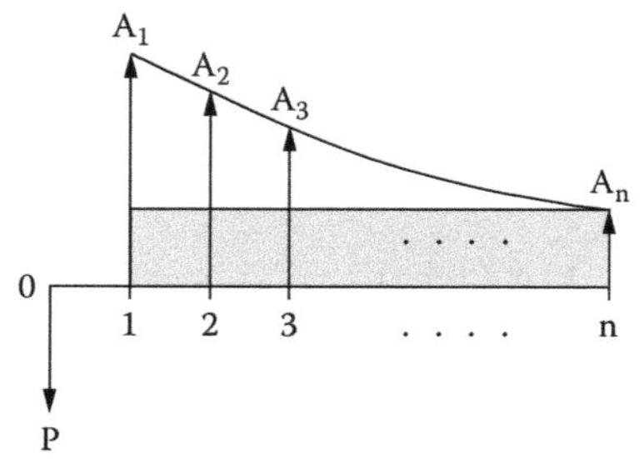

FIGURE 5.8
Decreasing geometric series cash flow.

$$A_2 = A_1(1 - j)$$

$$A_3 = A_2(1 - j) = A_1(1 - j)\,(1 - j)$$

$$\ldots$$

$$A_t = A_1(1 - j)^{t-1}, \qquad t = 1, 2, 3, \ldots, n.$$

The formula to calculate the present worth of decreasing geometric series cash flow is derived by finite summation as in the case of the increasing geometric series. The final formula is:

$$P = A_1\left[\frac{1-(1-j)^n\,(1+i)^{-n}}{i+j}\right].$$

Example

The contract amount for a 3-year project is expected to decrease by 10% every year with an initial contract of $100,000 at the end of the first year. Determine how much must be available in a reservoir fund at time zero to cover the contract amounts. The fund pays 10% interest per year. Because $j = 10\%$, $i = 10\%$, $n = 3$, and $A_1 = \$100{,}000$, we should have:

$$P = 100{,}000\left[\frac{1-(1-0.10)^3\,(1+0.10)^{-3}}{0.10+0.10}\right]$$

$$= \$100{,}000(2.2615)$$

$$= \$226{,}150.$$

Internal rate of return (IRR)—The IRR for a cash flow is defined as the interest rate that equates the future worth at time n or present worth at time 0 of the cash flow to zero. If we let i^* denote the internal rate of return, we have:

$$\text{FW}_{t=n} = \sum_{t=0}^{n} (\pm A_t)(1 + i^*)^{n-t} = 0$$

$$\text{PW}_{t=0} = \sum_{t=0}^{n} (\pm A_t)(1 + i^*)^{-t} = 0,$$

where a plus sign (+) is used in the summation for positive cash flow amounts or receipts, and a minus sign (–) is used for negative cash flow amounts or disbursements. A_t denotes cash flow amount at time t, which may be a receipt (+) or a disbursement (–). The value of i^* is also known as *discounted cash flow rate of return* or *true rate of return*. The procedure above essentially calculates the net future worth of the net present worth of the cash flow:

Net future worth = Future worth of receipts – Future worth of disbursements

$$\text{NFW} = \text{FW}_{\text{receipts}} - \text{FW}_{\text{disbursements}}.$$

Net present worth = Present worth of receipts – Present worth of disbursements

$$\text{NPW} = \text{PW}_{\text{receipts}} - \text{PW}_{\text{disbursements}}.$$

Setting the NPW or NFW equal to zero and solving for the unknown variable i determines the internal rate of return of the cash flow.

Benefit–cost ratio—The ratio of the present worth of benefits to the present worth of costs is calculated as:

$$B/C = \frac{\sum_{t=0}^{n} B_t(1+i)^{-t}}{\sum_{t=0}^{n} C_t(1+i)^{-t}}$$

$$= \frac{\text{PW}_{\text{benefits}}}{\text{PW}_{\text{costs}}},$$

where B_t is the benefit (receipt) at time t and C_t is the cost (disbursement) at time t. If the benefit–cost ratio is greater than 1, an investment is acceptable. If the ratio is less than 1, the investment is not acceptable. A ratio of 1 indicates a breakeven situation.

Simple payback period—A payback period is the length of time required to recover an initial investment. The approach does not consider the impact of the time value of money and thus is not accurate for evaluating the worth of an investment. However, it is a simple technique used widely to perform "quick-and-dirty" assessments of investment performance. Also, the technique considers only initial cost and does not include costs incurred after time zero. The payback period is defined as the smallest value of n ($n_{\min}$) that satisfies the following expression:

$$\sum_{t=1}^{n_{\min}} R_t \geq C_0,$$

where R_t is the revenue at time t and C_0 is the initial investment. The procedure calls for simple period-by-period addition of revenues until enough total is accumulated to offset the initial investment.

Example

An organization is considering installing a new computer system that will generate significant savings in material and labor requirements for order processing. The system has an initial cost of \$50,000 and is expected to save \$20,000 a year. It has an anticipated useful life of 5 years with a salvage value of \$5,000. Determine how long it will take for the system to pay for itself from the savings it is expected to generate. Because the annual savings are uniform, we can calculate the payback period by simply dividing the initial cost by the annual savings, that is,

$$n_{\min} = \frac{\$50,000}{\$20,000} = 2.5\,\text{years}.$$

Note that the salvage value of \$5,000 is not included in the above calculation because it is not realized until the end of the useful life of the asset (after 5 years). Salvage value may be considered in some cases. If it is, the amount to be offset by the annual savings will be the net cost of the asset:

$$n_{\min} = \frac{\$50,000 - \$5000}{\$20,000} = 2.25\ \text{years}.$$

Any tax liabilities associated with the annual savings must be deducted from the savings before calculating the payback period.

Discounted payback period—We introduce the *discounted payback period* approach in which revenues are reinvested at a certain interest rate. The payback period is determined when enough money has accumulated at the given interest rate to offset the initial cost and interim costs. The calculation is:

$$\sum_{t=1}^{n_{\min}} R_t(1+i)^{n_{\min}-1} \geq \sum_{t=0}^{n_{\min}} C_t .$$

Example

A new solar cell unit is to be installed in an office complex at an initial cost of $150,000. The unit is expected to generate annual cost savings of $22,500 for electricity. The unit will need to be overhauled every 5 years at a cost of $5,000 per overhaul. If the annual interest rate is 10%, find the *discounted payback period* for the unit considering the time value of money. The overhaul costs must be considered in calculating the discounted payback period.

Solution

Using the single payment compound amount factor for one period iteratively, the following solution is obtained.

Time	Cumulative Savings
1	$22,500
2	$22,500 + $22,500(1.10)^1 = $47,250
3	$22,500 + $47,250(1.10)^1 = $74,475
4	$22,500 + $74,475(1.10)^1 = $104,422.50
5	$22,500 + $104,422.50(1.10)^1 – $5000 = $132,364.75
6	$22,500 + $132,364.75(1.10)^1 = $168,101.23

The initial investment is $150,000. By the end of period 6, we have accumulated $168,101.23, more than the initial cost. Interpolating between periods 5 and 6, we obtain

$$n_{\min} = 5 + \frac{150,000 - 132,364.75}{168,101.23 - 132,364.75}(6-5)$$

$$= 5.49.$$

That is, it will take 5.49 years (5 years and 6 months) to recover the initial investment.

Investment life for multiple returns—The time required for an amount to reach a certain multiple of its initial level is often of interest in investment scenarios. The "Rule of 72" is one simple approach to calculating how long it will take an investment to double in value at a given interest rate per period. The Rule of 72 formula to estimate the doubling period is

$$n = \frac{72}{i},$$

where i is the interest rate expressed as a percentage. Referring to the single payment compound amount factor, we can set the future amount equal to twice the present amount and then solve for the number of periods n, that is, $F = 2P$. Thus, $2P = P(1 + i)^n$. Solving for n in this equation yields an expression for calculating the exact number of periods required to double P:

$$n = \frac{1n(2)}{1n(1 + i)},$$

where i is the interest rate expressed in decimals. Generally, for exact computation, the length of time required to accumulate m multiples of P is expressed as

$$n = \frac{\ln(m)}{\ln(1+i)},$$

where m is the desired multiple. For example, at an interest rate of 5% per year, the time for P to double in value ($m = 2$) is 14.21 years. This, of course, assumes that the interest rate will remain constant throughout the planning horizon. Table 5.1 tabulates the values calculated from both approaches. Figure 5.9 shows a comparison of the Rule of 72 to the exact calculation.

TABLE 5.1

Evaluation of Rule of 72

i%	*n* (Rule of 72)	*n* (Exact Value)
0.25	288.00	277.61
0.50	144.00	138.98
1.00	72.00	69.66
2.00	36.00	35.00
5.00	14.20	17.67
8.00	9.00	9.01
10.00	7.20	7.27
12.00	6.00	6.12
15.00	4.80	4.96
18.00	4.00	4.19
20.00	3.60	3.80
25.00	2.88	3.12
30.00	2.40	2.64

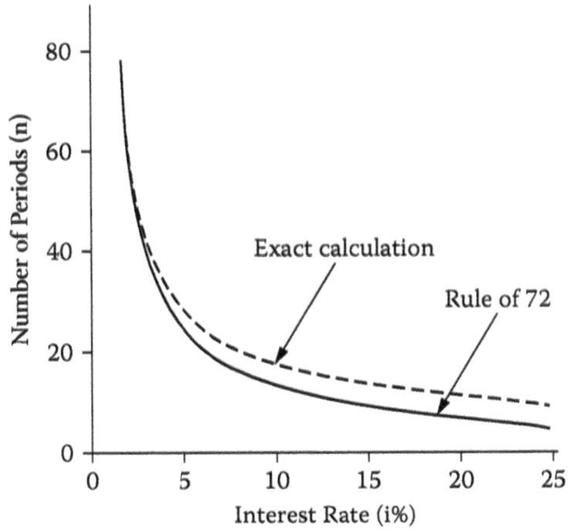

FIGURE 5.9
Evaluation of investment life for double return.

Effects of Inflation

Inflation is a major player in financial and economic analyses of projects. Multiyear projects are particularly subject to the effects of inflation. Inflation can be defined as a decline in the purchasing power of money. Some of the most common causes of inflation are:

- Increase in amount of currency in circulation
- Shortage of consumer goods
- Escalation of the cost of production
- Arbitrary increases of prices by resellers

The general effects of inflation are felt as increased prices of goods and decreases in the value of currency. In cash flow analysis, ROI for a project will be affected by the time value of money and inflation. The *real interest rate* (d) is defined as the desired rate of return in the absence of inflation. When we talk of "today's dollars" or "constant dollars," we refer to the use of real interest rates. *Combined interest rate* (i) is the rate of return combining real interest rate and inflation rate. If we denote the *inflation rate* as j, the relationship between the rates can be expressed as $1 + i = (1 + d)(1 + j)$. The combined interest rate can be expressed as $i = d + j + dj$.

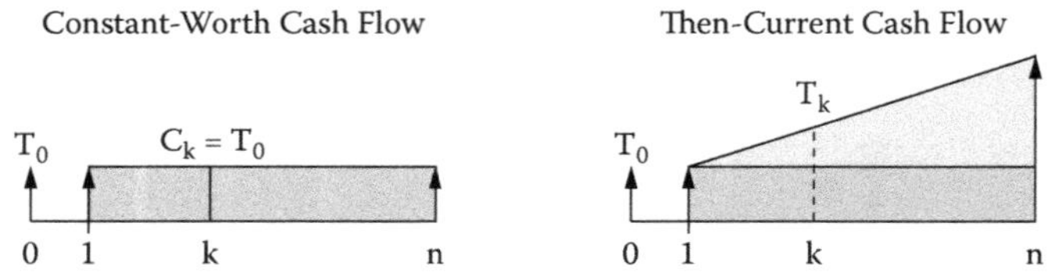

FIGURE 5.10
Cash flows showing effects of inflation processes.

If $j = 0$ (no inflation), then $i = d$. We can also define *commodity escalation rate* (g) as the rate at which individual commodity prices escalate that may be greater than or less than the overall inflation rate. In practice, several measures are used to convey inflationary effects, for example, the consumer price index, producer price index, and wholesale price index. A *market basket rate* is an estimate of inflation based on a weighted average of the annual rates of change in the costs of a wide range of representative commodities. A then-current cash flow explicitly incorporates the impact of inflation. A constant worth cash flow does not incorporate the effect of inflation. The real interest rate d is used for analyzing constant worth cash flows. Figure 5.10 shows constant worth and then-current cash flows.

The then-current cash flow in the figure is the equivalent cash flow considering the effect of inflation. C_k is the amount required to buy a certain "basket" of goods after k time periods with no inflation. T_k is the amount required to buy the same basket in k time periods if inflation is considered. For constant worth cash flow, $C_k = T_0$, $k = 1, 2, \ldots, n$. For the then-current cash flow, $T_k = T_0(1 + j)^k$, $k = 1, 2, \ldots, n$, where j is the inflation rate. If $C_k = T_0 = \$100$ under the constant worth cash flow, we have \$100 worth of buying power. If we use the commodity escalation rate g, we will have $T_k = T_0(1 + g)^k$, $k = 1, 2, \ldots, n$.

Thus, a then-current cash flow may increase based on both a regular inflation rate j and a commodity escalation rate g. We can convert a then-current cash flow to a constant worth cash flow by using $C_k = T_k(1 + j)^{-k}$, $k = 1, 2, \ldots, n$. If we substitute T_k from the commodity escalation cash flow into the expression for C_k above, we get:

$$
\begin{aligned}
C_k &= T_k(1 + j)^{-k} \\
&= T_0(1 + g)^k(1 + j)^{-k} \\
&= T_0[(1 + g)/(1 + j)]^k, \qquad k = 1, 2, \ldots, n.
\end{aligned}
$$

Note that if $g = 0$ and $j = 0$, $C_k = T_0$ or no inflationary effect. We now define effective commodity escalation rate v as $v = [(1 + g)/(1 + j)] - 1$ and express the commodity escalation rate as $g = v + j + vj$.

Inflation can exert a significant impact on the financial and economic aspects of a project. Inflation in economic terms is an increase in the amount of currency in circulation, resulting in a relatively sudden and large fall in its value. To a producer, inflation means a sudden increase in the costs of items that serve as inputs for production (equipment, labor, materials, etc). To a retailer, inflation implies higher costs of finished products. To an ordinary citizen, inflation portends an unbearable escalation of prices of consumer goods. All these views are interrelated in a project management environment.

The amount of money supply as a measure of a country's wealth is controlled by the government. With no other choice, governments often feel impelled to create more money or credit to cover old debts and pay for social programs. When money is generated at a faster rate than goods and services grow, it becomes a surplus commodity and its value (purchasing power) will fall. The result is too much money available to buy only a few goods and services. When the purchasing power of a currency falls, each individual in a product's life cycle has to dispense more currency to obtain a product. Some of the classic concepts of inflation are:

1. Increases in producer's costs are passed on to consumers. At each stage of a product's journey from producer to consumer, prices escalate disproportionately to generate good profits. The overall increase in price is directly proportional to the number of intermediaries it encounters en route to consumers. This is called *cost-driven (or cost-push) inflation.*
2. Excessive spending power of consumers forces an upward trend in prices. This spending power is usually achieved at the expense of savings. The law of supply and demand dictates that the more the demand, the higher the price. This is known as *demand-driven (or demand-pull) inflation.*
3. International economic forces can induce inflation in a local economy. Trade imbalances and fluctuations in currency values are notable examples of international inflationary factors.
4. Increasing base wages of workers generates more disposable income and leads to higher demands for goods and services. The high demands then create pulls on prices. Employers pass on the additional wage costs to consumers through higher prices. This type of inflation may be the most difficult to solve because wages set by union contracts and prices set by producers almost never fall—at least not permanently. This is known as *wage-driven (or wage-push) inflation.*
5. Easy availability of credit leads consumers to buy now and pay later, and this creates another loophole for inflation. This type of inflation is dangerous because the credit pushes prices up and leaves consumers with less money later to pay for the credit. Eventually,

many credits become uncollectible debts that may drive an economy into recession.

6. Deficit spending increases money supply and decreases the space for money to move around. "A dollar does not go far anymore" describes inflation in lay terms. The four levels of inflation are discussed below.

Mild inflation—An economy prospers when inflation is mild (2 to 4%). Producers strive to produce at full capacity to take advantage of high prices for consumers. Private investments tend to be brisk, and more jobs become available. However, the good fortune may only be temporary. Prompted by success, employers are tempted to seek larger profits, and workers ask for higher wages. They cite prosperous business as a reason to bargain for bigger shares of profits. The result is a vicious cycle in which producers seek higher prices, the unions want higher wages, and inflation starts an upward trend.

Moderate inflation—This inflation occurs when prices increase 5 to 9%. Consumers start purchasing more as a hedge against inflation. They would rather spend their money now than watch its purchasing power decline further. The increased market activity fuels further inflation.

Severe inflation—Price escalations of 10% or more are characteristic. Double-digit inflation implies that prices rise far faster than wages. Debtors benefit from this level of inflation because they repay debts with money that is less valuable then the money they borrowed.

Hyperinflation—Each price increase signals increases in wages and costs that again increase prices and the economy suffers malignant galloping inflation or hyperinflation. Rapid and uncontrollable inflation destroys an economy. Currency becomes useless because the government generates more of it to pay its obligations.

Inflation affects raw material procurement costs, salaries and wages, and other factors involved in every project. Some effects are immediate and easily observable; others are subtle and pervasive. Whatever form it takes, inflation must be considered when planning and controlling long-term projects. Large projects may be adversely affected by the effects of inflation through cost overruns and poor resource utilization. The level of inflation will determine the severity of impacts.

Breakeven Analysis

Breakeven analysis is the determination of the balanced performance level at which project income equals project expenditures. The total cost of an operation is expressed as the sum of the fixed and variable costs with respect to output quantity, that is, $TC(x) = FC + VC(x)$ where x is the number of units produced, $TC(x)$ is the total cost of producing x units, FC is the total fixed cost, and $VC(x)$ is the total variable cost to produce x units. The total revenue resulting from the sale of x units is defined as $TR(x) = px$, where p is the price per unit. The profit from producing and selling x units of product is calculated as $P(x) = TR(x) - TC(x)$.

The breakeven point of an operation is defined as the value of a given parameter that will produce neither profit nor loss. The parameter of interest may be the number of units produced, the number of hours of operation, the number of units of a resource allocated, or any other measure. At the breakeven point, we have $TR(x) = TC(x)$ or $P(x) = 0$.

Some cases may involve a known mathematical relationship between cost and the parameter of interest, for example, a linear cost relationship between the total cost of a project and the number of units produced. The cost expressions facilitate straightforward breakeven analysis. Figure 5.11 shows a breakeven point for a single project. Figure 5.12 shows multiple breakeven points when multiple projects are compared. When two project alternatives are compared, the breakeven point is the point of indifference between the two alternatives. In Figure 5.12, x_1 represents the point at which projects A and B are equally desirable, x_2 shows where A and C are equally desirable, and x_3 represents where B and C are equally desirable. Based on the figure, if we operate below a production level of x_2 units, project C is preferred. At a production level above x_2 units, A is the best choice.

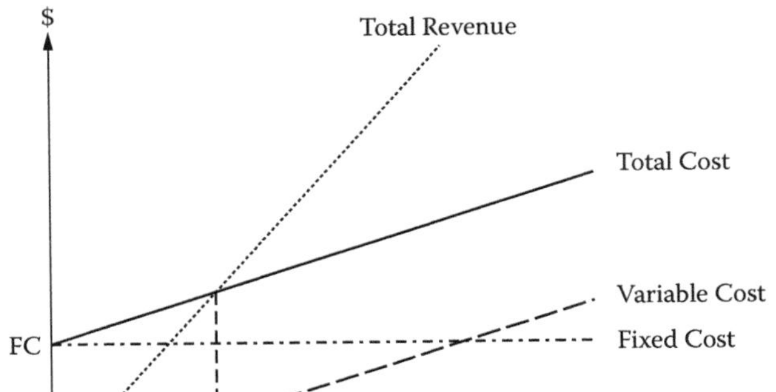

FIGURE 5.11
Breakeven point for single project.

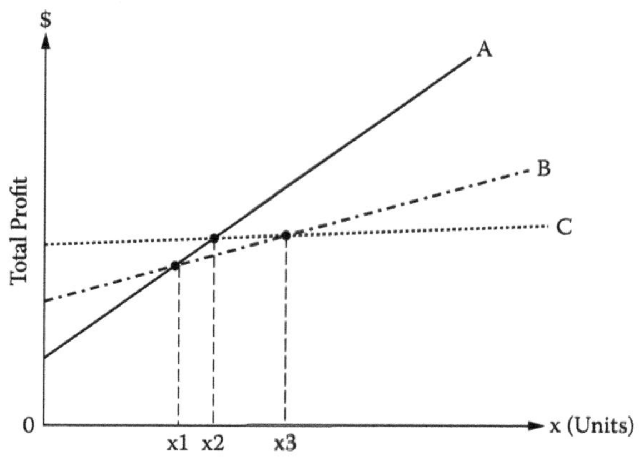

FIGURE 5.12
Breakeven points for multiple projects.

Example

Three project alternatives for a new product are under consideration. We must determine which alternative should be selected on the basis of how many units are produced per year. Past records reveal a relationship between the number of units produced per year x and the net annual profit $P(x)$ from each alternative. The level of production is projected between 0 and 250 units per year. The net annual profits (thousands of dollars) for each alternative are:

Project A: $P(x) = 3x - 200$

Project B: $P(x) = x$

Project C: $P(x) = (1/50)x^2 - 300$

This problem can be solved mathematically by finding the intersection points of the profit functions and evaluating the respective profits over the given range of product units. It can also be solved graphically. Figure 5.13 shows a plot of the profit functions called a *breakeven chart*. It indicates that Project B should be selected if 0 to 100 units are to be produced. Project A should be selected if 100 to 178* units are to be produced. Project C should be selected if more than 178 units are to be produced. If fewer than 66 units are produced, Project A will generate a net loss rather than a net profit. Similarly, Project C will generate losses if fewer than 122 units are produced.

* Decimals are rounded down to whole numbers.

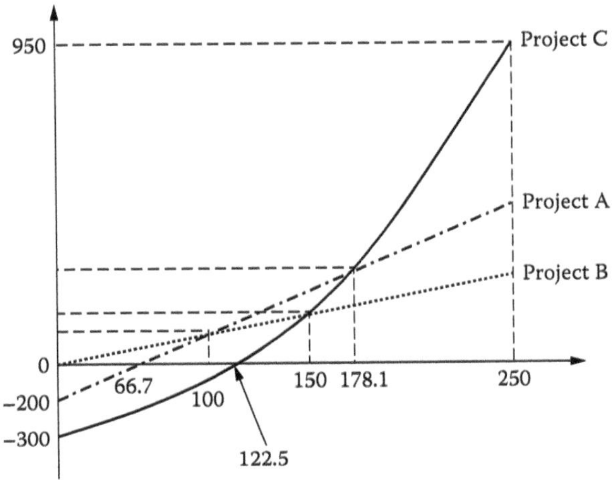

FIGURE 5.13
Plot of profit functions.

Profit Ratio Analysis

Breakeven charts allow several different types of analyses. Along with breakeven points, other measures of worth or criteria may be derived. A *profit ratio* measure (Badiru, 1996) is presented here as a further method for comparing competing projects. Profit ratio is the ratio of the profit area to the sum of the profit and loss areas in a breakeven chart:

$$\text{Profit ratio} = \frac{\text{Area of profit region}}{\text{Area of profit region} + \text{Area of loss region}}$$

Assume the expected revenue and expected total cost for a project are represented as $R(x) = 100 + 10x$ and $TC(x) = 2.5x + 250$ where x is the number of units produced and sold by the project. Figure 5.14 shows the breakeven chart. The breakeven point is 20 units. Net profits are realized if more than 20 units are produced, and net losses are realized if fewer than 20 are produced. The revenue function in Figure 5.14 represents an unusual case in which a revenue of $100 is realized when zero units are produced.

Suppose we want to calculate the profit ratio for this project if the number of units that can be produced is limited to 0 to 100 units. From Figure 5.14, the surface area of the profit region and the area of the loss region can be calculated via the standard formula for finding the area of a triangle: area = ½(base)(height). Using this formula:

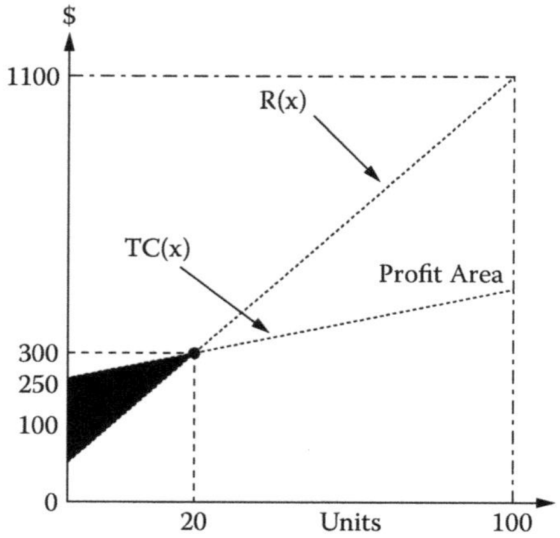

FIGURE 5.14
Area of profit versus area of loss.

Area of profit region = ½(base)(height) = ½(1100 – 500)(100 – 20) = 24,000 units.

Area of loss region = ½(base)(height) = ½(250 – 100)(20) = 1500 units.

Thus, the profit ratio is computed as

$$\frac{24,000}{24,000 + 1500} = 0.9411 = 94.11\%.$$

The profit ratio may be used as a criterion for selecting among project alternatives. In that case, the profit ratios for all the alternatives must be calculated over the same values of the independent variable. The project with the highest profit ratio will be the most suitable. Figure 5.15 presents a breakeven chart for an alternate project (Project II). Both the revenue and cost functions for the project are nonlinear. The revenue and cost are defined as $R(x) = 160x - x^2$ and $TC(x) = 500 + x^2$.

If the cost and/or revenue functions for a project are not linear, the areas bounded by the functions may not be easily determined. For those cases, it may be necessary to use techniques such as definite integrals to find the areas. Figure 5.15 indicates a loss if fewer than 3 or more than 76 units are produced. The respective profit and loss areas on the chart are calculated as:

$$\text{Area 1 (loss)} = \int_0^{3.3} \left[\left(500 + x^2\right) - \left(160x - x^2\right)\right] dx$$

$$= 802.8 \text{ unit-dollars.}$$

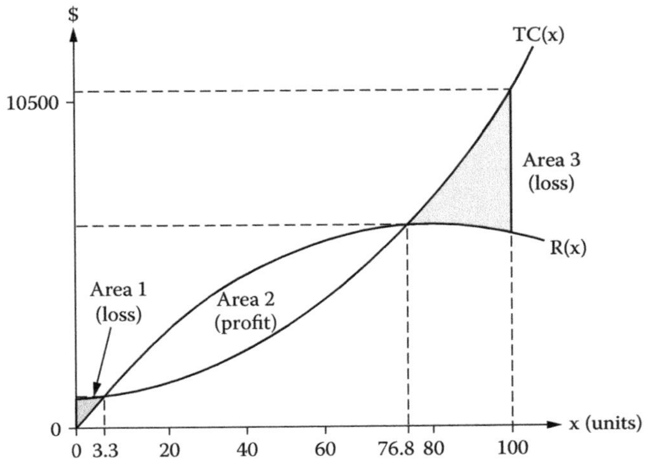

FIGURE 5.15
Breakeven chart for revenue and cost functions.

$$\text{Area 2 (profit)} = \int_{3.3}^{76.8} \left[\left(160x - x^2\right) - \left(500 + x^2\right) \right] dx$$

$$= 132{,}272.08 \text{ unit-dollars.}$$

$$\text{Area 3 (loss)} = \int_{76.8}^{100} \left[\left(500 + x^2\right) - \left(160x - x^2\right) \right] dx$$

$$= 48{,}135.98 \text{ unit-dollars.}$$

Consequently, the profit ratio for Project II is computed as:

$$\text{Profit ratio} = \frac{\text{Total area of profit region}}{\text{Total area of profit region} + \text{Total area of loss region}}$$

$$= \frac{132{,}272.08}{802.76 + 132{,}272.08 + 48{,}135.98}$$

$$= 0.7299$$

$$= 72.99\%.$$

The profit ratio approach evaluates the performance of each alternative over a specified range of operating levels. Most existing evaluation methods use single-point analysis with the assumption that the operating condition is fixed at a given production level. The profit ratio measure allows an analyst to evaluate the net yield of an alternative where the production level may shift from one level to another. An alternative, for example, may operate at a loss for most of its early life and later generate large incomes to offset the earlier losses. Conventional methods cannot easily capture transitions from one performance level to another. In addition to comparing alternate projects, the profit ratio may also be used to evaluate the economic feasibility of a single project. In such a case, a decision rule may be developed, for example, if a profit ratio is greater than 75%, accept the project; if a profit ratio is less than or equal to 75%, reject the project.

Amortization of Capital

Many capital investment projects are financed with external funds. A careful analysis is needed to ensure that the amortization schedule can be handled. A computer program such as GAMPS (graphic evaluation of amortization payments) may serve this purpose (Badiru, 1996). The program analyzes the installment payments, unpaid balance, principal amounts paid per period, total installment payment, and current cumulative equity. It also calculates the equity breakeven point for the debt analyzed. The equity breakeven point indicates the time when the unpaid balance on a loan equals the cumulative equity on the loan. With the output of this program, the basic cost of servicing project debt can be evaluated quickly. One output of the program presents the percentage of the installment payment going into equity and interest charges. The computation for analyzing project debt follows 10 steps:

1. Given a principal amount P, a periodic interest rate i (in decimals), and a discrete time span of n periods, the uniform series of equal end-of-period payments needed to amortize P is:

$$A = \frac{P[i(1+i)^n]}{(1+i)^n - 1}.$$

   It is assumed that the loan is to be repaid in equal monthly payments. Thus, $A(t) = A$ for each period t throughout the life of the loan.

2. The unpaid balance after making t installment payments is given by:

$$U(t) = \frac{A\left[1-(1+i)^{1-n}\right]}{i}.$$

3. The amount of equity or principal paid with installment payment number t is given by:

$$E(t) = A(1+i)^{t-n-1}.$$

4. The interest charge contained in installment payment number t is derived to be

$$I(t) = A\left[1-(1+i)^{t-n-1}\right],$$

where $A = E(t) + I(t)$.

5. The cumulative total payment made after t periods is denoted by:

$$C(t) = \sum_{k=1}^{t} A(k)$$

$$= \sum_{k=1}^{t} A$$

$$= (A)(t).$$

6. The cumulative interest payment after t periods is given by

$$Q(t) = \sum_{x=1}^{t} I(x).$$

7. The cumulative principal payment after t periods is computed as:

$$S(t) = \sum_{k=1}^{t} E(k)$$

$$= A\sum_{k=1}^{t} (1+i)^{-(n-k+1)}$$

$$= A\left[\frac{(1+i)^{t}-1}{i(1+i)^{n}}\right],$$

where

$$\sum_{n=1}^{t} x^{n} = \frac{x^{t+1}-x}{x-1}.$$

8. The percentage of interest charge contained in installment payment number t is

$$f(t) = \frac{I(t)}{A}(100\%).$$

9. The percentage of cumulative interest charge contained in the cumulative total payment up to and including payment number t is

$$F(t) = \frac{Q(t)}{C(t)}(100\%).$$

10. The percentage of cumulative principal payment contained in the cumulative total payment up to and including payment number t is:

$$\begin{aligned} H(t) &= \frac{S(t)}{C(t)} \\ &= \frac{C(t) - Q(t)}{C(t)} \\ &= 1 - \frac{Q(t)}{C(t)} \\ &= 1 - F(t). \end{aligned}$$

Example

A manufacturing productivity improvement project is to be financed by borrowing \$500,000 from a bank. The annual nominal interest rate for the loan is 10%. The loan is to be repaid in equal monthly installments over 15 years. The first payment on the loan is to be made exactly 1 month after financing is approved. We need a detailed analysis of the loan schedule. Table 5.2 presents a partial loan repayment schedule and shows a monthly payment of \$5373.04. Considering time t = 10 months, one can see the following results:

$U(10)$ = \$487,475.13 (unpaid balance)
$A(10)$ = \$5373.04 (monthly payment)
$E(10)$ = \$1299.91 (equity portion of the tenth payment)
$I(10)$ = \$4073.13 (interest charge contained in the tenth payment)
$C(10)$ = \$53,730.40 (total payment to date)
$S(10)$ = \$12,526.21 (total equity to date)

$f(10)$ = 75.81% (percentage of the tenth payment going into interest charge)
$F(10)$ = 76.69% (percentage of the total payment going into interest charge)

TABLE 5.2
Amortization Schedule for Financed Project

t	$U(t)$	$A(t)$	$E(t)$	$I(t)$	$C(t)$	$S(t)$	$f(t)$	$F(t)$
1	498794.98	5373.04	1206.36	4166.68	5373.04	1206.36	77.6	77.6
2	497578.56	5373.04	1216.42	4156.62	10746.08	2422.78	77.4	77.5
3	496352.01	5373.04	1226.55	4146.49	16119.12	3649.33	77.2	77.4
4	495115.24	5373.04	1236.77	4136.27	21492.16	4886.10	76.9	77.3
5	493868.16	5373.04	1247.08	4125.96	26865.20	6133.18	76.8	77.2
6	492610.69	5373.04	1257.47	4115.57	32238.24	7390.65	76.6	77.1
7	491342.74	5373.04	1267.95	4105.09	37611.28	8658.61	76.4	76.9
8	490064.22	5373.04	1278.52	4094.52	42984.32	9937.12	76.2	76.9
9	488775.05	5373.04	1289.17	4083.87	48357.36	11226.29	76.0	76.8
10	487475.13	5373.04	1299.91	4073.13	53730.40	12526.21	75.8	76.7
.	.	.	.	.	.	.	.	.
.	.	.	.	.	.	.	.	.
.	.	.	.	.	.	.	.	.
170	51347.67	5373.04	4904.27	468.77	913416.80	448656.40	8.7	50.9
171	46402.53	5373.04	4945.14	427.90	918789.84	453601.54	7.9	50.6
172	41416.18	5373.04	4986.35	386.69	924162.88	458587.89	7.2	50.4
173	36388.27	5373.04	5027.91	345.13	929535.92	463615.80	6.4	50.1
174	31318.47	5373.04	5069.80	303.24	934908.96	468685.60	5.6	49.9
175	26206.42	5373.04	5112.05	260.99	940282.00	473797.66	4.9	49.6
176	21051.76	5373.04	5154.65	218.39	945655.04	478952.31	4.1	49.4
177	15854.15	5373.04	5197.61	175.43	951028.08	484149.92	3.3	49.1
178	10613.23	5373.04	5240.92	132.12	956401.12	489390.84	2.5	48.8
179	5328.63	5373.04	5284.60	88.44	961774.16	494675.44	1.7	48.6
180	0.00	5373.04	5328.63	44.41	967147.20	500004.07	0.8	48.3

Thus, over 76% of the sum of the first 10 installment payments represents interest charges. By time $t = 180$, the unpaid balance will be reduced to zero, that is, $U(180) = 0.0$. The total payment on the loan is \$967,148.40. The total interest charge is \$967,148.20 – \$500,000 = \$467,148.20. Thus, 48.30% of the total payment goes to interest charges. Such information may be very useful for tax purposes. The table indicates that equity builds slowly while the unpaid balance decreases slowly; very little equity accumulates for the first 3 years of the loan schedule as shown in Figure 5.16. The effects of inflation, depreciation, property appreciation, and other economic factors are not included in the analysis above. A project analyst should include such factors if they are relevant to a loan situation.

The point at which the curves intersect is the *equity breakeven point*. It indicates when the unpaid balance is exactly equal to the accumulated equity or cumulative principal payment. For the example, the equity breakeven point is 120.9 months (more than 10 years). The importance of the equity breakeven point is that any equity accumulated after that point represents the amount

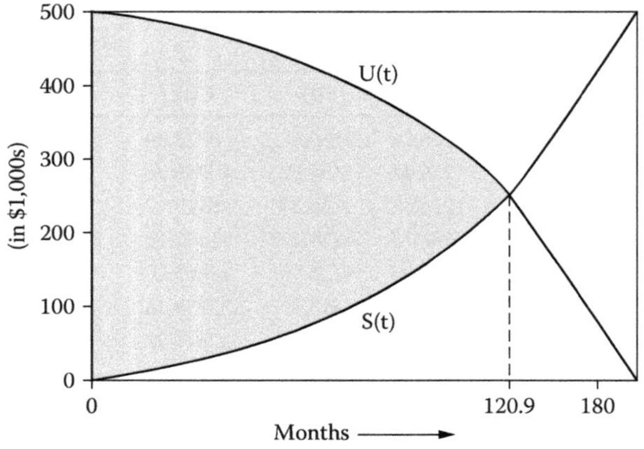

FIGURE 5.16
Plot of unpaid balance and cumulative equity.

of ownership or equity to which a debtor is entitled after the unpaid balance on the loan is settled with project collateral. This has important implications, particularly for mortgage loans. *Mortgage* is of French origin and means *death pledge*—perhaps a sarcastic reference to the burdensome nature of a mortgage loan. The equity breakeven point can be calculated directly. Let the equity breakeven point x be defined as the point where $U(x) = S(x)$, that is,

$$A\left[\frac{1-(1+i)^{-(n-x)}}{i}\right] = A\left[\frac{(1+i)^x - 1}{i(1+i)^n}\right].$$

Multiplying the numerator and denominator on the left side of the above expression by $(1 + i)^n$ and simplifying yields:

$$\frac{(1+i)^n - (1+i)^x}{i(1+i)^n},$$

on the left side. Consequently, we have:

$$(1+i)^n - (1+i)^x = (1+i)^x - 1$$

$$(1+i)^x = \frac{(1+i)^n + 1}{2},$$

which yields the equity breakeven expression:

$$x = \frac{1\text{n}[0.5(1+i)^n + 0.5]}{1\text{n}(1+i)},$$

where ln is the natural log function; n is the number of periods in the life of the loan; and i is the interest rate per period.

Figure 5.17 plots the total loan payment and the cumulative equity over time. The total payment starts from \$0.0 at time 0 and increases to \$967,147.20 by the end of the last month of installment payments. Only \$500,000 was borrowed. The total interest payment on the loan is \$967,147.20 – \$500,00 = \$467,147.20. The cumulative principal payment starts at \$0.0 at time 0 and slowly builds to \$500,001.34—the original loan amount. The extra \$1.34 is due to round-offs in calculations.

Figure 5.18 plots the percentage of interest charge in the monthly payments and in the total payment. The percentage of interest charge in the monthly

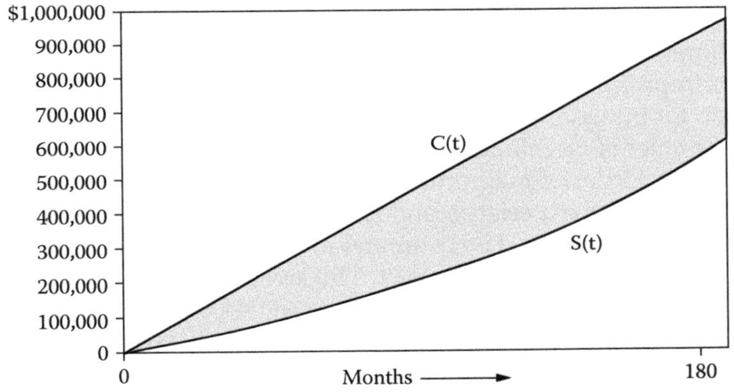

FIGURE 5.17
Plot of total loan payment and total equity.

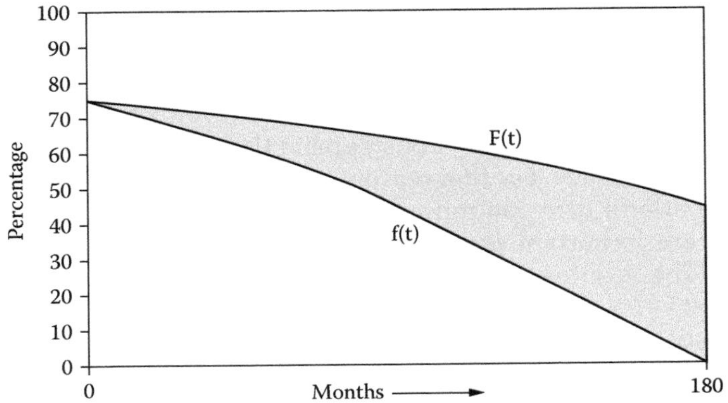

FIGURE 5.18
Plot of percentage of interest charge.

payments starts at 77.55% for the first month and decreases to 0.83% for the last month. By comparison, the percentage of interest in the total payment starts also at 77.55% for the first month and slowly decreases to 48.30% by the time the last payment is made at time 180. Table 5.2 and Figure 5.18 show that an increasing proportion of monthly payments is applied to principal as time goes on. If the interest charges are tax deductible, the decreasing values of $f(t)$ mean decreasing tax benefits from interest charges in the later months of the loan.

Project Cost Estimation

Cost estimation and budgeting help establish a strategy for allocating resources in project planning and control. The three major categories of cost estimation for budgeting are based on the desired level of accuracy. The categories are *order-of-magnitude estimates, preliminary cost estimates,* and *detailed cost estimates.* Order-of-magnitude cost estimates are usually gross estimates based on the experience and judgment of an estimator—sometimes called "ballpark" figures. The estimates are typically made without formal evaluations of the details of a project. The level of accuracy associated with order-of-magnitude estimates can range from −50% to +50% of the actual cost. These estimates provide quick cost information early in a project: 50% (actual cost) ≤ order-of-magnitude estimate ≤ 150% (actual cost).

Preliminary cost estimates are also gross but provide a better level of accuracy. In developing preliminary cost estimates, more attention is paid to certain project details. An example is an estimation of labor cost. Preliminary estimates are useful for evaluating project alternatives before final commitments are made. The level of accuracy associated with preliminary estimates can range from −20% to +20% of the actual cost: 80% (actual cost) ≤ preliminary estimate ≤ 120% (actual cost).

Detailed cost estimates are developed after careful consideration of all the major details of a project. Considerable time is required to generate a detailed cost estimate. For that reason, detailed cost estimates are usually developed after a firm commitment for a project is issued. Detailed cost estimates are important for evaluating actual cost performance during a project. The level of accuracy for a detailed estimate normally ranges from −5% to +5% of actual cost: 95% (actual cost) ≤ detailed cost ≤ 105% (actual cost).

One of two basic approaches is used to generate a cost estimate. The first is a variant approach that bases cost estimates on variations of previous cost records. The other approach is the generative cost estimation in which cost estimates are developed "from scratch" without considering previous cost records.

Optimistic and pessimistic cost estimates—Using an adaptation of the PERT formula, we can combine optimistic and pessimistic cost estimates. Let O = optimistic cost estimate; M = most likely cost estimate; and P = pessimistic cost estimate. Costs can be estimated as:

$$E[C] = \frac{O + 4M + P}{6},$$

and the cost variance can be estimated as:

$$V[C] = \left[\frac{P - O}{6}\right]^2.$$

Budgeting and Capital Allocation

Budgeting involves sharing limited resources among several project groups or functions in a project environment. Budget analysis can:

- Plan resources expenditures
- Select project criteria
- Protect project policy
- Serve as a basis for project control
- Measure performance
- Standardize resource allocation
- Provide incentive for improvement

Top-down budgeting—Data are collected from upper-level sources such as top and middle managers. The figures they supply may be based on personal judgment, past experience, or data generated by similar projects. The cost estimates are passed to lower-level managers who then break the estimates into specific work components of a project. These estimates may, in turn, be given to line managers, supervisors, and lead workers to continue the process until individual activity costs are calculated. Top management generates a global budget. The functional level worker determine specific budget requirements for project items.

Bottom-up budgeting—Basic activities, their schedules, descriptions, and labor skill requirements are used to construct detailed budget requests. Line workers familiar with specific activities provide cost estimates for each activity based on labor time, materials, and machine time. The estimates are then converted to an appropriate cost basis and merged into composite budgets at each successive level up the budgeting hierarchy. Estimate discrepancies can

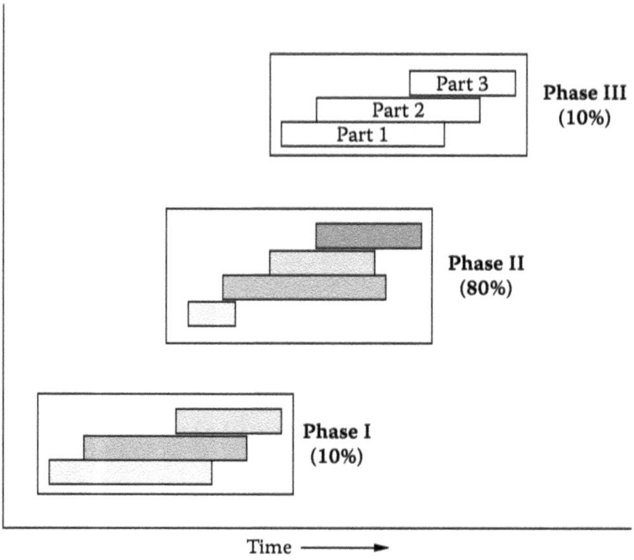

FIGURE 5.19
Budgeting by project phases.

be resolved by interventions of senior and middle management, functional managers, project managers, accountants, or standard cost consultants. Figure 5.19 shows the breaking of a project into phases and parts to facilitate bottom-up budgeting and improve schedule and cost control.

Basic budgets may be developed by evaluating the time progress of each part of a project. When all the individual estimates are gathered, the result is a composite budget estimate. Figure 5.20 shows the various components of an overall budget. The bar chart appended to a segment of the pie chart indicates the individual cost components within that specific segment. Analytical tools such as learning curve analysis, work sampling, and statistical estimation may be employed during cost estimation and budgeting.

Mathematical capital allocation—Capital rationing is selecting a combination of projects that will optimize ROI. A mathematical formula for a capital budgeting problem is:

$$\text{Maximize } z = \sum_{i=1}^{n} v_i x_i$$

$$\text{Subject to } \sum_{i=1}^{n} c_i x_i \le B$$

$$x_i = 0, 1; i = 1, \ldots, n,$$

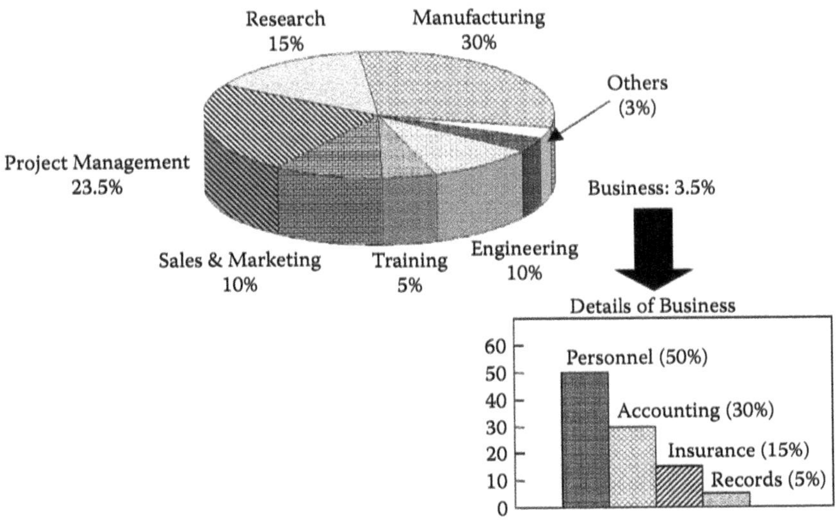

FIGURE 5.20
Budget breakdown and distribution.

where n = number of projects; v_i = measure of performance for project i (present value); c_i = cost of project I; x_i = indicator variable for project I; and B = budget availability level. A solution of this model will indicate which projects should be selected for a combination.

Example capital rationing problem—Planning of a portfolio of projects is essential for resource-limited projects. Suppose a project analyst must review N projects, $X_1, X_2, X_3, \ldots, X_N$, to determine the level of investment in each project required to maximize total investment return subject to a specified budget limit. The projects are not mutually exclusive. The investment in each project starts at a base level b_i ($i = 1, 2, \ldots, N$) and increases by variable increments k_{ij} ($j = 1, 2, 3, \ldots, K_i$), where K_i is the number of increments used for project i. Consequently, the level of investment in project X_i is defined as

$$x_i = b_i + \sum_{j=1}^{K_i} k_{ij},$$

where $x_i \geq 0 \quad \forall i$. For most cases, the base investment will be zero. In those cases, $b_i = 0$. In the modeling procedure used for this problem:

$$X_i = \begin{cases} 1 & \text{if the investment in project } i \text{ is greater than zero} \\ 0 & \text{otherwise} \end{cases}$$

and

$$Y_{ij} = \begin{Bmatrix} 1 & \text{if } j\text{th increment of alternative } i \text{ is used} \\ 0 & \text{otherwise} \end{Bmatrix}.$$

The variable x_i is the actual level of investment in project i, while X_i is a variable indicating whether project i is one of those selected for investment. Similarly, k_{ij} is the magnitude of the jth increment while Y_{ij} is a variable that indicates whether the jth increment is used for project i. The maximum possible investment in each project is defined as M_i such that $b_i \le x_i \le M_i$. There is a specified limit B on the total budget available to invest such that

$$\sum_i x_i \le B.$$

There is a known relationship between the level of investment x_i in each project and the expected return $R(x_i)$. This relationship is called the *utility function* $f(\cdot)$ for the project and may be developed through historical data, regression analysis, and forecasting models. The utility function is used to determine the expected return $R(x_i)$ for a specified level of investment in a specific project, that is,

$$\begin{aligned} R(x_i) &= f(x_i) \\ &= \sum_{j=1}^{K_i} r_{ij} Y_{ij}, \end{aligned}$$

where r_{ij} is the incremental return obtained when the investment in project i is increased by k_{ij}. If the incremental return decreases as the level of investment increases, the utility function will be *concave* as shown by $r_{ij} \ge r_{ij+1}$ or $r_{if} - r_{ij+1} \ge 0$. Thus, $Y_{ij} \ge Y_{ij+1}$ or $Y_{ij} - Y_{ij+1} \ge 0$ so that only the first n increments ($j = 1, 2, \ldots, n$) that produce the highest returns are used for project i.

Figure 5.21 shows an example of a concave investment utility function. If the incremental returns do not define a concave function $f(x_i)$, we must introduce the inequality constraints presented above into the optimization model. Otherwise, the inequality constraints may be omitted since the first inequality $Y_{ij} \ge Y_{ij+1}$ is always implicitly satisfied for concave functions. Our objective is to maximize the total return, that is, maximize

$$Z = \sum_i \sum_j r_{ij} Y_{ij},$$

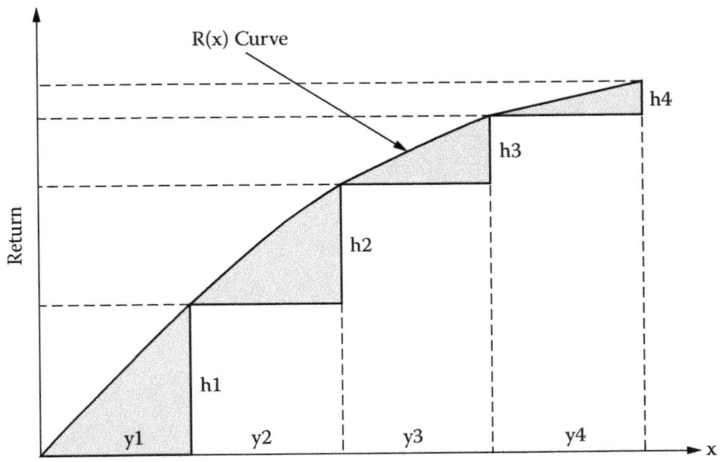

FIGURE 5.21
Utility curve for investment yield.

subject to the following constraints:

$$x_i = b_i + \sum_j k_{ij} Y_{ij} \qquad \forall i$$

$$b_i \leq x_i \leq M_i \qquad \forall i$$

$$Y_{ij} \geq Y_{ij+1} \qquad \forall i, j$$

$$\sum_i x_i \leq B$$

$$x_i \geq 0 \qquad \forall i$$

$$Y_{ij} = 0 \text{ or } 1 \qquad \forall i, j.$$

Assume we are assigned four projects ($N = 4$) with a budget limit of \$10 million. Tables 5.3 through 5.6 illustrate the respective investments and returns. All values are in millions of dollars. For example, in Table 5.3, if an incremental investment of \$0.20 million from stage 2 to stage 3 is made in project 1, the expected incremental return from the project will be \$0.30 million. Thus, a total investment of \$1.20 million in project 1 will yield a total return of \$1.90 million.

The optimization model is intended to determine how many investment increments should be used for each project, i.e., when should we stop

TABLE 5.3

Project 1: Investment Data for Capital Rationing

Stage (j)	y_{1j} Incremental Investment	x_1 Level of Investment	r_{1j} Incremental Return	$R(x_1)$ Total Return
0	—	0	—	0
1	0.80	0.80	1.40	1.40
2	0.20	1.00	0.20	1.60
3	0.20	1.20	0.30	1.90
4	0.20	1.40	0.10	2.00
5	0.20	1.60	0.10	2.10

TABLE 5.4

Project 2: Investment Data for Capital Rationing

Stage (j)	y_{2j} Incremental Investment	x_2 Level of Investment	r_{2j} Incremental Return	$R(x_2)$ Total Return
0	—	0	—	0
1	3.20	3.20	6.00	6.00
2	0.20	3.40	0.30	6.30
3	0.20	3.60	0.30	6.60
4	0.20	3.80	0.20	6.80
5	0.20	4.00	0.10	6.90
6	0.20	4.20	0.05	6.95
7	0.20	4.40	0.05	7.00

TABLE 5.5

Project 3: Investment Data for Capital Rationing

Stage (j)	y_{3j} Incremental Investment	x_3 Level of Investment	r_{3j} Incremental Return	$R(x_3)$ Total Return
0	—	0	—	0
1	2.00	2.00	4.90	4.90
2	0.20	2.20	0.30	5.20
3	0.20	2.40	0.40	5.60
4	0.20	2.60	0.30	5.90
5	0.20	2.80	0.20	6.10
6	0.20	3.00	0.10	6.20
7	0.20	3.20	0.10	6.30
8	0.20	3.40	0.10	6.40

TABLE 5.6

Project 4: Investment Data for Capital Rationing

Stage (j)	y_{4j} Incremental Investment	x_4 Level of Investment	r_{4j} Incremental Return	$R(x_4)$ Total Return
0	—	0	—	0
1	1.95	1.95	3.00	3.00
2	0.20	2.15	0.50	3.50
3	0.20	2.35	0.20	3.70
4	0.20	2.55	0.10	3.80
5	0.20	2.75	0.05	3.85
6	0.20	2.95	0.15	4.00
7	0.20	3.15	0.00	4.00

increasing the investments in a given project? Obviously, for a single project, we would continue to invest as long as the incremental returns exceeded the incremental investments. However, for multiple projects, investment interactions complicate the decision; investment in one project cannot be independent of investments in the others. The LP model of the capital rationing example was solved with LINDO software. The solution indicates the following values for Y_{ij}.

Project 1: Y11 = 1, Y12 = 1, Y13 = 1, Y14 = 0, Y15 = 0. Thus, the investment is X_1 = \$1.20 million. The corresponding return is \$1.90 million.

Project 2: Y21 = 1, Y22 = 1, Y23 = 1, Y24 = 1, Y25 = 0, Y26 = 0, Y27 = 0. Thus, the investment is X_2 = \$3.80 million. The corresponding return is \$6.80 million.

Project 3: Y31 = 1, Y32 = 1, Y33 = 1, Y34 = 1, Y35 = 0, Y36 = 0, Y37 = 0. Thus, the investment is X_3 = \$2.60 million. The corresponding return is \$5.90 million.

Project 4: Y41 = 1, Y42 = 1, Y43 = 1. Thus, the investment is X_4 = \$2.35 million. The corresponding return is \$3.70 million.

The total investment in all four projects is \$9,950,000. Thus, the optimal solution indicates that not all the \$10,000,000 available should be invested. The expected return from the total investment is \$18,300,000 or 83.92% ROI. Figure 5.22 presents histograms of the investments and returns for the four projects. The individual ROIs are shown in Figure 5.23.

The optimal solution indicates an unusually large ROI. In a practical setting, expectations may need to be scaled down to fit the realities of a project environment. Not all optimization results will be directly applicable to real situations. Possible extensions of the above model of capital rationing include the incorporations of risk and time value of money into the solution. Risk analysis is relevant, particularly if the levels of returns for the various levels of investment are not known with certainty. The incorporation of time

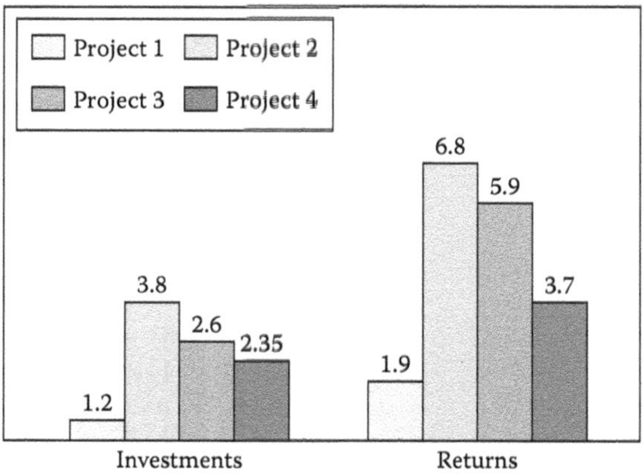

FIGURE 5.22
Histogram of capital rationing example.

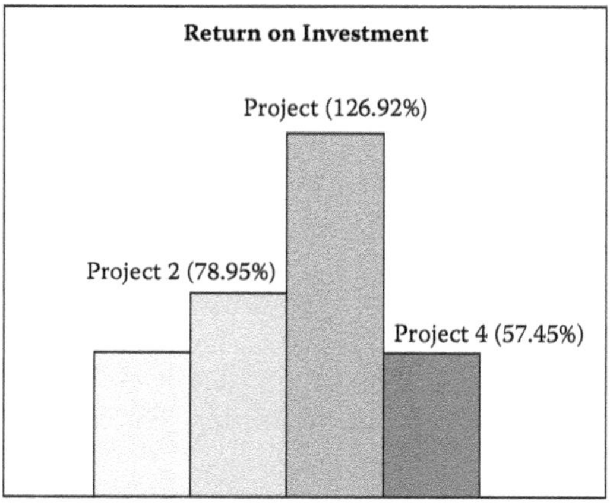

FIGURE 5.23
Histogram of returns on investments.

value of money is useful if the investment analysis is to be performed for a specific planning horizon, for example, to make investment decisions to cover the next 5 years rather than currently.

Cost Monitoring

As a project progresses, costs can be monitored and evaluated to identify areas of unacceptable cost performance. Figure 5.24 plots projected

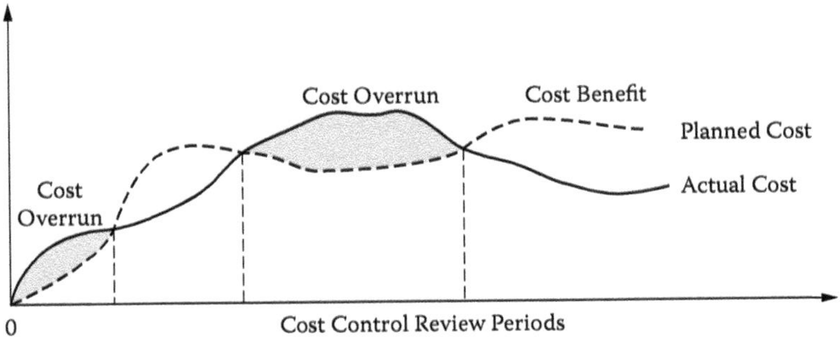

FIGURE 5.24
Evaluation of actual and projected cost.

and actual cost versus time data. The plot quickly reveals cost overruns. Similar plots may be used to evaluate costs, scheduling, and time performance. An approach similar to the profit ratio presented earlier may be used with a plot to evaluate the overall cost performance of a project over a specified planning horizon. The formula for *cost performance index* (*CPI*) is:

$$\frac{\text{Area of cost benefit}}{\text{Area of cost benefit} + \text{Area of cost overrun}}.$$

As in the case of the profit ratio, CPI may be used to evaluate the relative performances of several project alternatives or the feasibility and acceptability of a single alternative. Figure 5.25 presents another cost monitoring tool

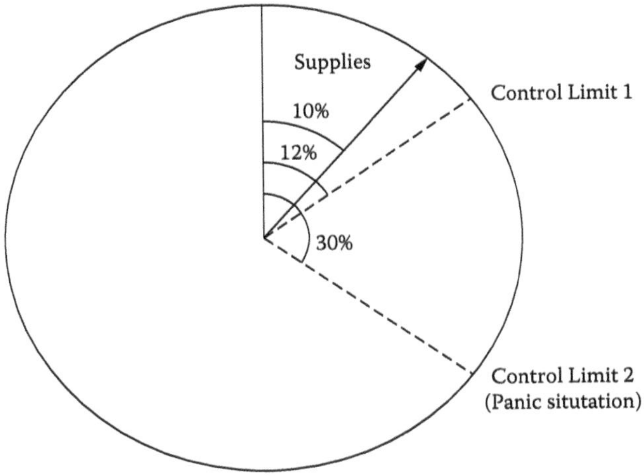

FIGURE 5.25
Cost control pie chart.

known as a *control pie chart* used to track percentages of costs going to a specific project component. Control limits can be included in a pie chart to identify out-of-control situations. Figure 5.25 shows that 10% of total cost is tied up in supplies. The control limit is 12% of total cost. Hence, the supplies expenditure is within control—at least to date..

Project Balance Technique

One other approach to monitoring cost performance is project balance. The technique helps assess the economic state of a project at a desired time point in its life cycle. It calculates the net cash flow to a specific point in time. The project balance is calculated as:

$$B(i)_t = S_t - P(1+i)^t + \sum_{k=1}^{t} \mathrm{PW}_{\mathrm{income}}(i)_k,$$

where

$B(i)_i$ = project balance at time t at an interest rate of i% per period

$\mathrm{PW}_{\mathrm{income}}(i)_t$ = present worth of net income from the project up to time t

P = initial cost of the project

S_t = salvage value at time t.

The project balanced at time t gives the net loss or net profit of the project to that time.

Activity-Based Costing

Activity-based costing (ABC) has emerged as an appealing costing technique. ABC offers an improved method to enhance operational and strategic decisions and a mechanism to allocate costs in direct proportion to activities performed. It represents an improvement over the traditional way of generically allocating costs to departments and improves the conventional approaches to allocating overhead. Using PERT/CPM, precedence diagramming, and critical resource diagramming can facilitate task decomposition to provide information for ABC. Some of the potential impacts of ABC on a production line are:

- Identification and removal of unnecessary costs
- Identification of cost impact of adding specific attributes to a product
- Indication of incremental cost of improved quality
- Identification of value-added points in a production process
- Inclusion of specific inventory carrying costs
- Providing bases for comparing production alternatives
- Assessment of "what-if" scenarios for specific tasks

ABC is only one component of activity-based management in an organization. Activity-based management involves a more global approach to planning and control of organizational endeavors based on product planning, resource allocation, productivity management, quality control, training, line balancing, value analysis, and other organizational issues. Thus, while activity-based costing is important, one must not lose sight of the universality of the environment in which it must operate. The processes involved in some functions are so intermingled that decomposition into specific components may be difficult. Major considerations for implementing ABC are:

- Resources committed to developing activity-based information and costs
- Duration and level of effort needed to achieve ABC objectives
- Level of cost accuracy that can be achieved
- Ability to track activities based on ABC requirements
- Handling volume of detailed information provided by ABC
- Sensitivity of ABC to changes in activity configuration

Case Example: Cost Control Project in Manufacturing

This section discusses a cost control project at a manufacturing company. One process yielded a low first pass quality (FPQ) result. The product was put on hold before it reached customers because it does not meet specifications. The continuous improvement team objectives are to increase FPQ to 99% and reduce internal holds by 10%. The method for reaching the objectives is:

- Determine a process that can detect holds when or before they occur.
- Implement specifications for raw ingredients.
- Determine whether specifications for upper and lower control limits are accurate.
- Train all production line personnel.
- Create SOPs and instruction manuals.
- Determine proper preventive maintenance (PM) for equipment.
- Measure initial temperatures and viscosities for initial product and compare to temperatures and viscosities after 24 hours.
- Determine possible equipment replacement and/or upgrades.

The DMAIC methodology was used. The first process was defining the key issues. The first issue was the number of holds that occurred because the product was too thick. This issue had a value of $350,000 per year. The second hold issue was a close second—$225,000 in holds per year. The source of the thickness problem was investigated. Most thick and thin holds came from the same production line. More investigation led the team to determine that 83% of holds were due to viscosity issues. The following Six Sigma tools were used during the project:

- SIPOC
- Process flow diagrams
- Cause and effect matrices
- Failure modes and effects analysis
- Brainstorming to find process defects
- Benchmarking

Because another facility manufactured the same product, benchmarking techniques were used. The FPQ, cost implications, and quantities produced were calculated for both facilities. A process map was completed to compare the differences between the facilities. The comparisons showed that different equipment, ingredients, and processes were used. Utilizing a process map, root causes for the differences were identified.

Facility A checked product against specifications at the beginning, during, and at the end of the process. Facility B checked only at the end when it was too late to make a change to the process. Facility A heated product to a temperature of 235°; Facility B did not check cooking temperatures. Facility A utilized a different measuring mechanism and a different ratio of ingredients. Facility A used a different vendor for some raw ingredients and had wider specification ranges than Facility B. Products of both facilities had different end uses. Facility B had tighter standard deviations for viscosities.

After benchmarking analysis was performed, a root cause analysis was completed to explain the differences. Both facilities were directed to use best-in-class practices. Action plans addressed each problem, required steps for resolution, potential results, and cost implications. Table 5.7 summarizes the action plans. The total upgrades were projected to cost $621,000. The savings calculations were to be completed next.

- Reduction in hog feed holds by 50%: $221,000 annually
- Downtime savings (2.92 people, 429 hours, mainly overtime): $25,000
- Reduction in rework costs (material, labor, packaging): $104,000
- Decrease of product lost by flushing (250 lb/flush; flushing frequency reduced): $57,000
- Total annual savings anticipated = $407,000

Thick Holds	Thin Holds	Totals Per Year
$163,750.00	$167,958.00	$331,708.00
$54,583.33	$55,986.00	$110,569.33
$218,333.33	$223,944.00	$442,277.33

TABLE 5.7

Outline of Action Plans for Cost Control Case Example

	Problem	Action	Required Steps	Potential Results	Cost
1	Load cells may not always be accurate	Purchase mass flow meter for bulk ingredients	Validate inaccuracy of load cells, trial mass flow meters	Mass flow meters will ensure accuracy of weights of bulk ingredients and allow accurate addition of granulated sugar	$27,000
2	Install jacketed piping from syrup mixing kettle to votators	Purchase jacketed piping	Purchase jacketed piping	Keeps product from cooling in piping, preventing crystallization and build-up	$14,000
3	Install spare votators with bypass valves	Purchase spare votators	Installation of votators with bypass valves	Installing spare votators allows production to continue while flushing out one votator at a time; Minimizes downtime, increases throughput and keeps votators flushed as required; Reduction of up to $36,000 per year (based on 100 pounds of loss per flush)	$105,000
4	Install completely new set of votators, pump and valves	Purchase and test votators	Existing equipment designed for production in 1998	Reduction in holds and variation in product	$175,000
5	Facility A checks solids of 80% of 100% mix liquid sugar, 20% of 100% mix corn syrup, then combines syrups	Implement solid readings to be taken in slurry kettles for same commodities	Purchase flow meters to take solid readings automatically	Allows solids to be manipulated at slurry kettle.	$9,000

Continued

TABLE 5.7 (continued)

Outline of Action Plans for Cost Control Case Example

	Problem	Action	Required Steps	Potential Results	Cost
6	Specifications only within one standard deviation	Make product at high and low end at two standard deviations for viscosity and show customer for acceptance	R&D and Marketing input and buy off	Increase specifications of viscosities	
7	Crystallization of sugar forms in equipment	Flush equipment every 6 batches	Training and sign off sheets implemented for flush outs	Reduces crystallization inside equipment	
8	Facility A combines syrup pumped through cooker and ensures heating to 235° to 237°; Facility B does not check temperature at this point	Check temperatures and ensure proper temperature at slurry mixing kettle	Standard operating procedure for temperature checks	Ensures consistency of temperatures and SOPs	
9	Residual water affects color and thickness of final product, making product appear dark and thin	Raise piping from syrup warming kettle to create downward pitch to 3-way valve allowing for proper water drainage during flush outs	Validate thin product step eliminated	Reduction in thin holds	$1,645
10	Existing equipment in Facility B experiences significant wear and is outdated	Replace shaft on equipment #1 with welded pins in 24" shaft; replace heat transfer tube, scraper blades, and roller bearing in equipment #2	Purchase shafts, replace tube and bearing.	Ensures product can be properly scraped off heat transfer tube wall; Reduces product build-up, required cleanouts, crystal growth, and equipment wear and damage; Requires replacement of shaft blades every 30 days along with pins, causes extensive downtime.	$36,200

TABLE 5.7 (continued)

Outline of Action Plans for Cost Control Case Example

	Problem	Action	Required Steps	Potential Results	Cost
11	Equipment blades break	Replace blades and schedule PM every 3 weeks	Add blade replacement to PM list	Reduction in broken blades due to increased PM	$201
12	Broad or no specifications for certain input variables	Proposal for input specifications	Production to ensure specifications are followed	Reduction in variation	

Savings calculations:

$221,000	Saving from decreased holds: 50%
$25,000	Downtime estimate: 50%
$103,910	Reduction of rework: 50% improvement
$57,000	Reduction in material lost by flushing: 50%
$406,910	Total saved

An additional saving arose from throughput improvements listed below and saving figures were recalculated.

3.07	Weighted average crew size
166.69	Production hour improvement
511.58	Man hour improvement
$16.50	Average rate
$8,441	Annual labor savings
$1,810	Fringe savings (21% variable fringe rate)

Savings calculations:

$221,000	New saving from decreased holds
$25,000	Downtime estimate
$103,910	Reduction of rework
$57,000	Reduction in material lost by flushing
$10,251	Productivity improvements for new throughput
$417,161	Revised savings total

References

Badiru, A.B. and O.A. Omitaomu (2007). *Computational Economic Analysis for Engineering and Industry*, Taylor and Francis, Boca Raton, FL.

Badiru, A.B. (1996). Project Management in *Manufacturing and High Technology Operations*, 2nd ed., John Wiley & Sons, New York.

6

Project Quality Control

> When Einstein was active as a professor, one of his students came to him and said: "The questions of this year's exam are the same as last year's!" "True," Einstein said, "but this year all the answers are different."
>
> **From collection of Albert Einstein's quotes**

Quality control ensures that the performance of a project conforms to specifications and meets the requirements and expectations of the project stakeholders and participants. As the quote from Einstein suggests, expectations can be designed to act as a form of control. The objective of quality control is to minimize deviation from project plans. Quality control must be performed throughout the life cycle of a project—not by a single final inspection of the product.

Quality Management: Step-By-Step Implementation

The quality management component of the project management body of knowledge consists of the elements shown in Figure 6.1. The three elements in the block diagram are carried out across the process groups presented earlier in this book. The overlay of the elements and the process groups are shown in Table 6.1. Thus, under the knowledge area of quality management, the required steps are:

Step 1: Perform Quality Planning
Step 2: Perform Quality Assurance
Step 3: Perform Quality Control

Tables 6.2 through 6.4 present the inputs, tools, techniques, and outputs of the steps. Improvement programs have the propensity to drift into anecdotal, qualitative, and subjective processes. Having a quantifiable and measurable approach helps overcome this deficiency. Figure 6.2 shows how operational efficiency transitions to effectiveness, quality, and then productivity.

Project Quality Management

1. Perform Quality Planning
2. Perform Quality Assurance
3. Perform Quality Control

FIGURE 6.1
Block diagram of project quality management.

TABLE 6.1
Implementation of Project Quality Management across Process Groups

	Initiating	Planning	Executing	Monitoring and Controlling	Closing
Project quality management		1. Perform quality planning	2. Perform quality assurance	3. Perform quality control	

TABLE 6.2
Tools and Techniques for Project Quality Management

Step 1: Perform Quality Planning		
Inputs	**Tools and Techniques**	**Outputs**
Enterprise environmental factors	Cost–benefit analysis	Quality management plan
Organizational process assets	Benchmarking	Quality metrics
Project scope statement	Experiment design	Quality check lists
Project management plan	Cost of quality (COQ) assessment	Process improvement plan
Other in-house (custom) factors of relevance and interest	Group decision techniques	Quality baseline
	Other in-house (custom) tools and techniques	Project management plans and updates
		Reports and data inferences of interest

Six Sigma and Quality Management

The Six Sigma approach originally introduced by Motorola's Government Electronics Group caught on quickly in industry. Many major companies now embrace the approach as the key to high quality industrial productivity. Six Sigma denotes six standard deviations from a statistical performance average. The Six Sigma approach allows for no more than 3.4 defects per

TABLE 6.3

Tools and Techniques for Project Quality Management

Step 2: Perform Quality Assurance		
Inputs	**Tools and Techniques**	**Outputs**
Quality management plan	Quality planning tools and techniques	Requested changes
Quality metrics	Quality audits	Recommended corrective actions
Process improvement plan	Process analysis	Organizational process assets and updates
Work performance information	Quality control tools and techniques	Project management plans and updates
Approved change requests	Other in-house (custom) tools and techniques	Other in-house outputs, reports, and data inferences of interest
Quality control measurements		
Implemented change requests		
Implemented corrective actions		
Implemented defect repair		
Implemented preventive repair		
Other in-house (custom) factors of relevance and interest		

TABLE 6.4

Tools and Techniques for Project Quality Management

Step 3: Perform Quality Control		
Inputs	**Tools and Techniques**	**Outputs**
Quality management plan	Cause and effect diagram	Quality control measurements
Quality metrics	Control chart	Validated defect repair
Quality check lists	Flowcharting	Quality baseline and updates
Organizational process assets	Histogram	Recommended corrective actions
Work performance information	Pareto chart	Recommended preventive actions
Approved change requests	Run chart	Requested changes
Deliverables	Scatter diagram	Recommended defect repairs
Other in-house (custom) factors of relevance and interest	Statistical sampling	Organization process assets and updates
	Quality inspection	Validated deliverables
	Defect repair review	Other in-house outputs, reports, and data inferences of interest
	Other in-house (custom) tools and techniques	

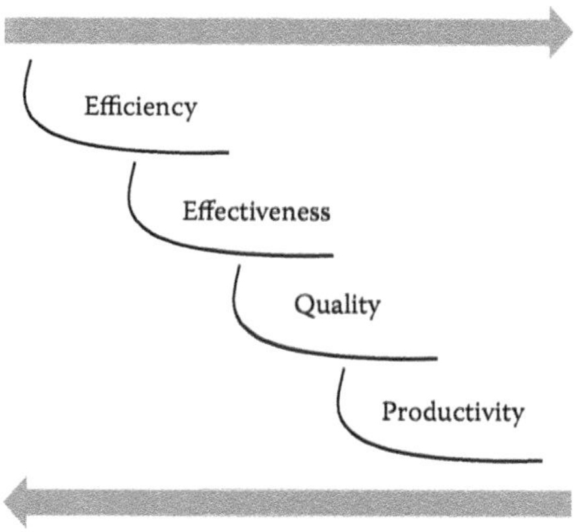

FIGURE 6.2
Foundations of STEP project quality performance success.

TABLE 6.5
Interpretation of Sigma Intervals from Mean

Process Quality Range	Percent Coverage	Interpretation of Standard
1 Sigma	68.26	Poor performance
2 Sigma	95.46	Below expectation
3 Sigma	99.73	Historical acceptable standard
4 Sigma	99.9937	Contemporary
6 Sigma	99.99999985	New competitive standard

million parts in manufactured goods or 3.4 mistakes per million activities in a service operation. To appreciate the effect of Six Sigma, consider a process that is 99% perfect and produces 10,000 defects per million parts. With Six Sigma, the process must be 99.99966% perfect to produce only 3.4 defects per million. Thus, Six Sigma pushes the limit of perfection. Table 6.5 summarizes sigma ranges and process percentage coverage levels.

Taguchi Loss Function

The Taguchi loss function defines how deviation from an intended target creates a loss in the production process. Taguchi's idea of product quality analytically models the loss to the society from the time a product is shipped to customers. Taguchi loss function measures this conjectured loss with a quadratic formula known as quality loss function (QLF):

$$L(y) = k(y - m)^2,$$

where k is a proportionality constant; m is the target value; and y is the observed value of the quality characteristic of the product in question. The quantity $(y - m)$ represents the deviation from the target. The larger the deviation, the larger the loss to the society. The constant k can be determined if $L(y)$, y, and m are known. Loss based on the QLF concept consists of several components, for example:

- Opportunity cost of not having the service of the product due to its quality deficiency. The loss of service implies that a step that should have been taken to serve the society could not be taken.
- Time lost in the search to find (or troubleshoot) the quality problem.
- Time lost (after finding the quality problem) in attempts to solve the problem. The effort to identify the problem consumes some of the time that could have been productively used to serve the society. Thus, the society incurs a loss.
- Productivity loss arising from the reduced effectiveness of the product. The decreased productivity deprives the society of a certain level of service and thus constitutes a loss.
- Actual cost of correcting the quality problem. This is, perhaps, the only direct loss that is easily recognized. The Taguchi method can help identify several subtle losses.
- Actual loss (e.g., loss of life) due to a failure of the product resulting from low quality. For example, a defective automobile tire creates potential for a traffic fatality.
- Waste (lost time and materials) due to rework and other nonproductive activities associated with low quality.

Identification and Elimination of Sources of Defects

This approach uses statistical methods to find problems that cause defects. For example, the total yield (number of nondefective units) from a process is determined by combining the performance levels of all the steps of a process. If a process consists of 20 steps and each step is 98% perfect, the performance of the overall process is $(0.98)^{20} = 0.667608$ (66.7608%). Thus, the process will produce 332,392 defects per million parts. If each step of the process is pushed to the Six Sigma limit, the process performance will be $(0.9999966)^{20} = 0.999932$ (99.9932%).

Thus, the Six Sigma process will yield only 68 defects per million parts—a significant improvement over the original process performance. In many cases, it is not realistic to expect to achieve the Six Sigma level of production, but the approach sets a high quality standard and provides a mechanism for striving to reach the goal. In effect, Six Sigma changes the ways workers perform their tasks to minimize the potential for defects.

The success of Six Sigma in industry ultimately depends on the ability to initiate and execute Six Sigma projects effectively. Thus, the project management approaches presented in this book are essential for realizing the benefits of Six Sigma. Project planning, organizing, team building, resource allocation, employee training, optimal scheduling, superior leadership, shared vision, and control are all complementary and essential to implementing Six Sigma successfully. These success factors are not mutually exclusive. In many organizations, far too much focus is directed toward the statistical training for Six Sigma at the expense of proper project management development. This explains why many organizations have not been able to achieve the much-touted benefits of Six Sigma.

The success of the Toyota production system is not due to any special properties of the approach. It arises from the consistency, persistence, and dedication of Toyota organizations in building their projects around all the essential success factors. Toyota focuses on adapting to the organizational mindset required to initiate and coordinate the success factors throughout the organization. Six Sigma requires the management of multiple projects with an identical mindset throughout an organization. The success of this requirement depends on proper application of project management tools and techniques.

Roles and Responsibilities for Six Sigma

Specific roles and responsibilities are crucial in executing Six Sigma projects. Team players fall into several categories:

Executive leadership develops and promulgates vision and direction, leads change and maintains accountability for organizational results on a full-time basis.

Employee group in which all employees support organizational vision, receive and implement Six Sigma specifications, actively support total process improvement (TPM), export the mission statement to functional tasks, and deploy improvement practices on a full-time basis.

Champion advocates improvement projects, leads business direction, and coordinates improvement projects on a full-time basis.

Project sponsor develops requirements, engages project teams, leads project scoping, and identifies resource requirements on a part-time basis.

Master belt trains and coaches black belts and green belts, leads large projects, and provides leadership on a full-time basis.

Black belt leads specific projects, facilitates troubleshooting, coordinates improvement groups, trains and coaches project team members on a full-time basis.

Green belt participates on black belt teams and leads small projects on part-time project-specific basis.

Team members provide specific operational support, facilitate inward knowledge transfers, and serve as links to functional areas on part-time basis.

Statistical Techniques for Six Sigma

Statistical process control (SPC) simply means controlling a process statistically. Early quality control researchers developed SPC using basic statistics for quality control purposes. We know that not all manufactured products are exactly alike. Some inherent variations will always appear in units of the same product. The variations in product characteristics provide a basis to use SPC for quality improvement With the help of statistical approaches, individual items can be studied and general inferences can be drawn about a process or batches of products from the process. Since 100% inspection is difficult or impractical in many operations, SPC provides a mechanism to generalize process performance. SPC uses random samples generated consecutively over time. The samples should be representative of the general process. SPC utilizes:

- Control charts ($\overline{X}$- and R-charts)
- Process capability analysis (nested design, Cp, Cp_k)
- Process control (factorial design, response surface)

Control Charts

Two of the most commonly used control charts in industry are X-bar charts and range (R-) charts. The type of chart depends on the kind of data collected: variable data or attribute data. The success of quality improvement depends on (1) the quality of data available and (2) the effectiveness of the techniques used for analyzing the data. The charts generated by both types of data are:

Variable data:

- Control charts for individual data elements (X)
- Moving range chart (MR-chart)
- Average chart ($\overline{X}$-chart)
- Range chart (R-chart)
- Median chart
- Standard deviation chart (σ-chart)
- Cumulative sum chart (CUSUM)
- Exponentially weighted moving average (EWMA)

Attribute data:

- Proportion or fraction defective chart (p-chart); subgroup sample size can vary.
- Percent defective chart (100p-chart); subgroup sample size can vary.
- Number defective chart (np-chart); subgroup sample size is constant.
- Number defective (c-chart); subgroup sample size = 1.
- Defective per inspection unit (u-chart); subgroup sample size can vary.

The statistical theory useful to generate control limits is the same for all the charts except the EWMA and CUSUM charts.

X-Bar and Range Charts

The R-chart is a time plot useful for monitoring short-term process variations. The X-bar chart monitors longer term variations where the likelihood of special causes is greater over time. Both charts utilize control lines called upper and lower control limits and central lines; both types of lines are calculated from process measurements. They are not specification limits or percentages of the specifications or other arbitrary lines based on experience. They represent what a process is capable of doing when only common cause variation exists, in which case the data will continue to fall in a random fashion within the control limits and the process is in a state of statistical control. However, if a special cause acts on a process, one or more data points will be outside the control limits and the process will no longer be in a state of statistical control.

Data Collection Strategies

One strategy for data collection requires collection from 20 to 25 subgroups. This quantity should adequately show the location and spread of a distribution in a state of statistical control. If due to sampling costs or other factors associated with the process, we are unable to utilize 20 to 25 subgroups, we can use the samples we have to generate the trial control limits and update the limits as more samples are made available. These limits will normally be wider than normal control limits and thus be less sensitive to changes in the process. Another approach is to use run charts to monitor the process until 20 to 25 subgroups become available, then control charts showing control; limits can be applied. Other data collection strategies should consider subgroup sample size and sampling frequency.

Subgroup Sample Size

The subgroup samples of size n should be taken as n consecutive readings from the process and not random samples. This is necessary to produce an accurate estimate of the process common cause variation. Each subgroup should be selected from a small time period or small region of space or product to assure homogeneous conditions within the subgroup. This is essential because the variation within the subgroup is used to generate the control limits. The subgroup sample size n can be four or five samples. This is a good size that balances the pros and cons of using large or small sample sizes for a control chart. The advantages of small subgroup sample sizes are:

- Estimates of process standard deviation based on the range are as good and accurate as those obtained by using the complex hand calculation of the standard deviation equation.
- The probability of introducing special cause variations within a subgroup is very small.
- Range chart calculation is simple and easily performed by operators on the shop floor.

The advantages of large subgroup sample sizes are:

- The central limit theorem confirms that the process average will be more normally distributed with a larger sample size.
- If a process is stable, the larger the subgroup size the better the estimates of process variability.
- A control chart based on a larger subgroup sample size will be more sensitive to process changes.

The choice of a proper subgroup is critical to the value of a control chart. If we fail to incorporate all common cause variations within subgroups, the process variation will be underestimated, leading to very tight control limits. Then the process will appear to go out of control too frequently even no special cause if present. If we incorporate special causes within subgroups, we will fail to detect special causes as frequently as expected.

Sampling Frequency

The problem of determining how frequently we sample depends on several factors, for example:

- Cost: The greater the cost of taking and testing samples, the less frequently we should sample.
- Changes in process conditions: The larger the frequency of changes to the process, the larger the sampling frequency. For example, if process conditions tend to change every 15 minutes, sample every 15 minutes. If they change every 2 hours, sample every 2 hours.
- Importance of quality characteristics: The importance of a quality characteristic to customers determines how frequently the characteristic must be sampled.
- Process control and capability: A long history of process control and capability means less frequent sampling is required.

Stable and Unstable Processes

A process is said to be in a state of statistical control if the distribution of measurement data from the process has the same shape, location, and spread over time. In other words, a process is stable when the effects of all special causes have been removed and the remaining variability is only due to common causes. Figure 6.3 shows a stable distribution. A process is said to be unstable (not in statistical control) if it changes over time because of a shifting average, shifting variability, or a combination of shifting averages and variations. Figures 6.4 through 6.6 show distributions from unstable processes.

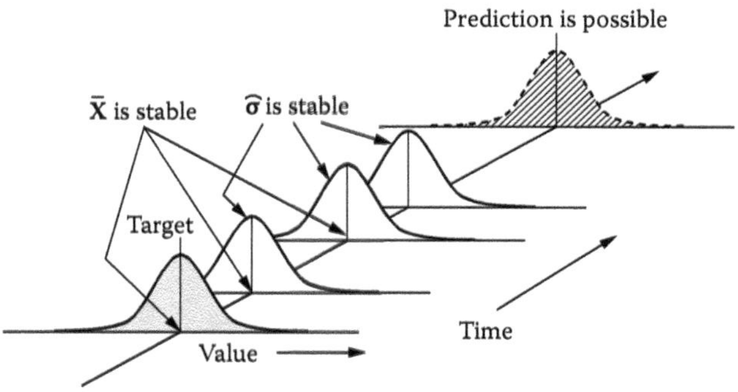

FIGURE 6.3
Stable distribution with no special causes.

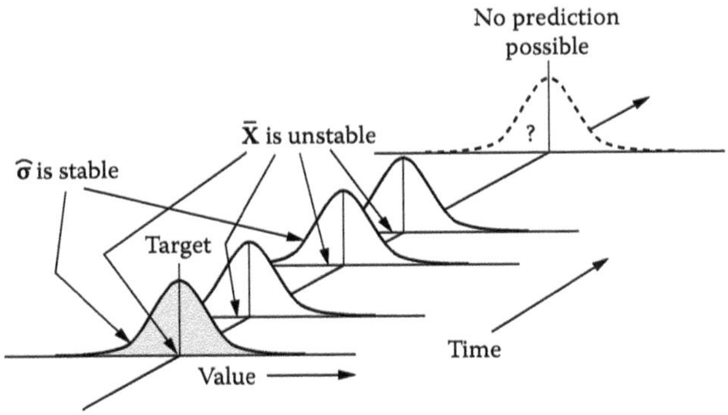

FIGURE 6.4
Unstable process average.

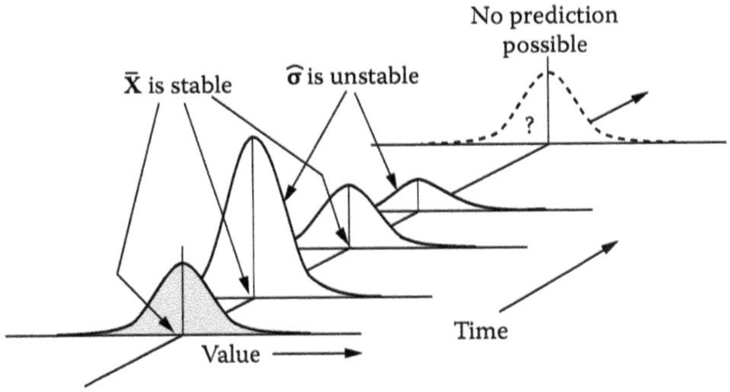

FIGURE 6.5
Unstable process variation.

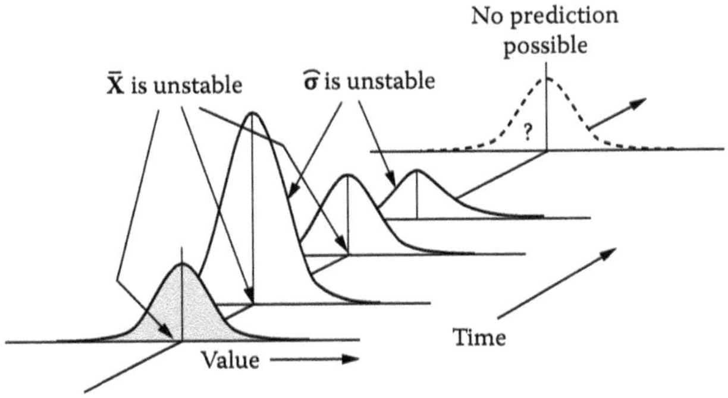

FIGURE 6.6
Unstable process average and variation.

Calculation of Control Limits

Range (R) represents the difference between the highest and lowest observations: $R = X_{highest} - X_{lowest}$.

Center lines are determined by calculating $\overline{X}$ and $\overline{R}$:

$$\overline{X} = \frac{\sum X_i}{m}$$

$$\overline{R} = \frac{\sum R_i}{m},$$

where $\overline{X}$ = overall process average; $\overline{R}$ = average range; m = total number of subgroups; and n = within-subgroup sample size.

The calculations for upper and lower control limits based on an R-chart are $UCL_R = D_4\overline{R}$ and $LCL_R = D_3\overline{R}$.

To estimate process variation:

$$\hat{\sigma} = \frac{\overline{R}}{d_2}.$$

For control limits based on an $\overline{X}$-chart, calculate the upper and lower control limits for the process average: $UCL = \overline{X} + A_2\overline{R}$ and $LCL = \overline{X} - A_2\overline{R}$.

Table 6.6 shows the values of d_2, A_2, D_3, and D_4 for different values of n. These constants are used to develop variable control charts.

TABLE 6.6

Constants for Variable Control Charts

n	d_2	A_2	D_3	D_4
2	1.128	1.880	0	3.267
3	1.693	1.023	0	2.575
4	2.059	0.729	0	2.282
5	2.326	0.577	0	2.115
6	0.534	0.483	0	2.004
7	2.704	0.419	0.076	1.924
8	2.847	0.373	0.136	1.864
9	2.970	0.337	0.184	1.816
10	3.078	0.308	0.223	1.777
11	3.173	0.285	0.256	1.744
12	3.258	0.266	0.284	1.716

Plotting Control Charts for Range and Average Charts

- Plot the range chart (R-chart) first.
- If the R-chart is in control, plot an X-bar chart.
- If the R-chart is not in control, identify and eliminate special causes, then delete points due to special causes, and recompute the control limits for the chart. If process is in control, plot an X-bar chart.
- Check whether the X-bar chart is in control; if it is not, search for special causes and eliminate them permanently.
- Perform the eight trend tests described below.

Plotting Control Charts for Moving Range and Individual Control Charts

- Plot the moving range chart (MR-chart) first.
- If the MR-chart is in control, plot an individual chart (X).
- If the MR-chart is not in control, identify and eliminate special causes, delete the special cause points, and recompute the control limits. If the MR-chart is in control, plot the individual chart.
- Check whether the individual chart is in control; if it is not, search for special causes from out-of-control points.
- Perform the eight trend tests.

Case Example: Plotting Control Chart

An industrial engineer at a manufacturing company wanted to study a machining process for producing a smooth surface on a torque converter clutch. The quality characteristic of interest is the surface smoothness of the clutch. The engineer collected four clutches every hour for 30 hours and recorded the smoothness measurements in micro inches. Acceptable values of smoothness lie between 0 (perfectly smooth) and 45 micro inches. Table 6.7 shows the data collected by the engineer.

Histograms of the individual and average measurements are presented in Figure 6.7. The hourly smoothness average ranges from 27 to 32 micro inches—much narrower than the histogram of hourly individual smoothness that ranges from 24 to 37 micro inches—averages show less variability than individual measurements. Therefore, whenever we plot subgroup averages on an X-bar chart, some individual measurements will plot outside the control limits. Figure 6.8 shows dotplots of surface smoothness for individual and average measurements. The descriptive statistics for individual smoothness are:

N = 120
MEAN = 29.367
MEDIAN = 29.00
TRMEAN = 29.287

TABLE 6.7

Data for Control Chart Example: Smoothness (micro inches)

Subgroup	I	II	III	IV	Average	Range
1	34	33	24	28	29.75	10
2	33	33	33	29	32.00	4
3	32	31	25	28	29.00	7
4	33	28	27	36	31.00	9
5	26	34	29	29	29.50	8
6	30	31	32	28	30.25	4
7	25	30	27	29	27.75	5
8	32	28	32	29	30.25	4
9	29	29	28	28	28.50	1
10	31	31	27	29	29.50	4
11	27	36	28	29	30.00	9
12	28	27	31	31	29.25	4
13	29	31	32	29	30.25	3
14	30	31	31	34	31.50	4
15	30	33	28	31	30.50	5
16	27	28	30	29	28.50	3
17	28	30	33	26	29.25	7
18	31	32	28	26	29.25	6
19	28	28	37	27	30.00	10
20	30	29	34	26	29.75	8
21	28	32	30	24	28.50	8
22	29	28	28	29	28.50	1
23	27	35	30	30	30.50	8
24	31	27	28	29	28.75	4
25	32	36	26	35	32.25	10
26	27	31	28	29	28.75	4
27	27	29	24	28	27.00	5
28	28	25	26	28	26.75	3
29	25	25	32	27	27.25	7
30	31	25	24	28	27.00	7
Total					881.00	172

STDEV = 2.822
SEMEAN = 0.258

The descriptive statistics for average smoothness are presented below:

N = 30
MEAN = 29.367
MEDIAN = 29.375
TRMEAN = 29.246
STDEV = 1.409
SEMEAN = 0.257

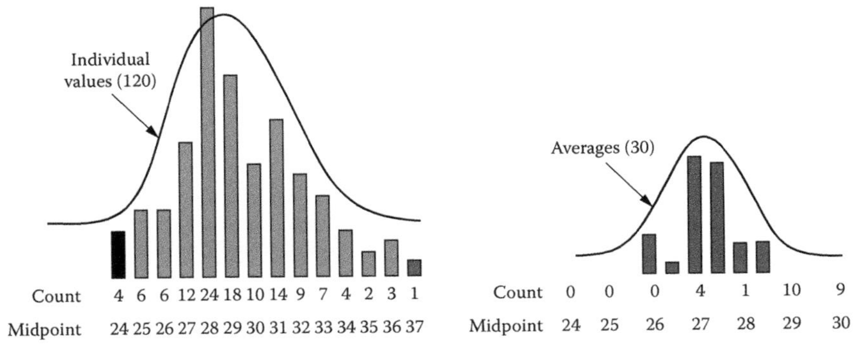

FIGURE 6.7
Histograms of individual measurements and averages for clutch smoothness.

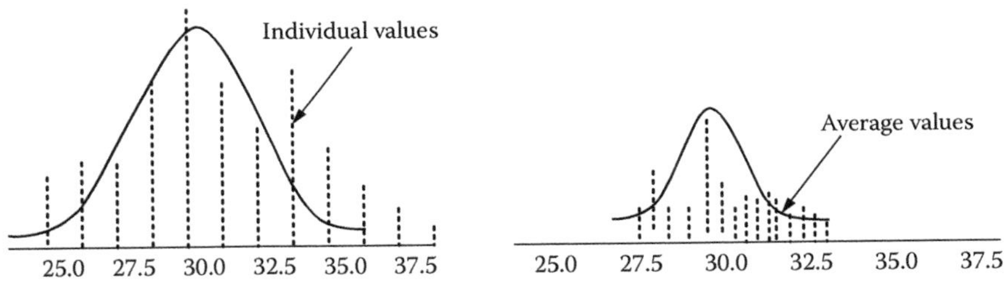

FIGURE 6.8
Dotplots of individual measurements and averages for clutch smoothness.

Calculations

Natural limit of the process = $\overline{X} \pm 3s$ (based on empirical rule):

s = estimated standard deviation of all individual samples.
Standard deviation (special and common), $s = 2.822$.
Process average, $\overline{X} = 29.367$.
Natural process limit = $29.367 \pm 3\ (2.822) = 29.367 \pm 8.466$.
The natural limit of the process is between 20.90 and 37.83.

Inherent (common cause) process variability, $\hat{\sigma} = \overline{R}/d_2$:

$\overline{R}$ from range chart = 5.83.
d_2 (for n=4) = 2.059 (from Table 6.6).
$\hat{\sigma} = \overline{R}/d_2 = 5.83/2.059 = 2.83$.

Thus, the total process variation s is about the same as the inherent process variability because the process is in control. If the process is out of control, the total standard deviation of all the numbers will be larger than $\overline{R}/d_2$.

Control limits for the range chart:

Obtain constants D_3, D_4 from Table 6.6 for $n = 4$.
$D_3 = 0$.
$D_4 = 2.282$.
$\overline{R} = 172/30 = 5.73$.
$UCL = D_4 \times \overline{R} = 2.282(5.73) = 16.16$.
$LCL = D_3 \times \overline{R} = 0(5.73) = 0.0$.

Control limits for the averages:

Obtain constants A_2 from Table 6.6 for $n = 4$.
$A_2 = 0.729$.
$UCL = \overline{X} + A_2(\overline{R}) = 29.367 + 0.729(5.73) = 33.54$.
$LCL = \overline{X} - A_2(\overline{R}) = 29.367 - 0.729(5.73) = 25.19$.

Natural limit of process $= \overline{X} \pm 3(\overline{R})/d_2 = 29.367 \pm 3(2.83) = 29.367 \pm 8.49$. The natural limit of the process is between 20.88 and 37.86 which is slightly different from ± 3s calculated earlier based on the empirical rule. This is because $\overline{R}/d_2$ was used rather than the standard deviation of all the values. Again, if a process is out of control, the standard deviation of all the values will be greater than $\overline{R}/d_2$. The correct procedure is always to use $\overline{R}/d_2$ from a process in statistical control.

Comparison with specification. Since the specification for clutch surface smoothness is between 0 (perfectly smooth) and 45 micro inches and the natural limit of the process is between 20.88 and 37.86, the process is capable of producing within the specification limits. Figure 6.9 presents the R and X-bar charts for clutch smoothness.

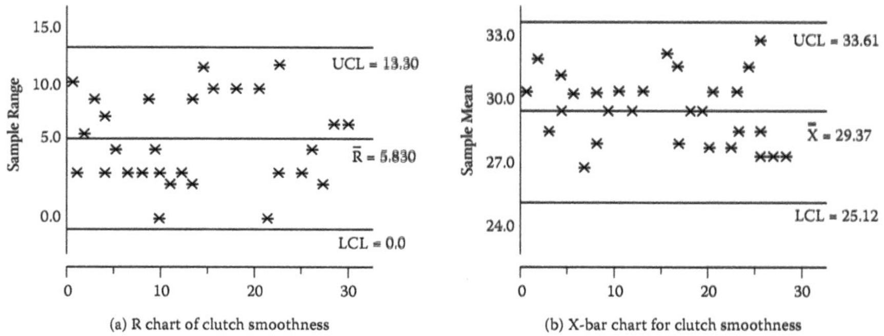

FIGURE 6.9
R- and X-bar charts for clutch smoothness.

For this case example, the industrial engineer examined the above charts and concluded that the process is in a state of statistical control. The engineer also recognized a process improvement opportunity. If the smoothness of the clutch could be held below 15 micro inches, clutch performance would improve significantly. The engineer can select key control factors to study in a two-level factorial or fractional factorial design to analyze the possibility.

Trend Analysis

After a process is out of control, zone control charting technique is a logical approach to find the sources of the variation problems by performing eight tests using MINITAB or other statistical software tools. The zone control chart is divided into three zones: zone A between ± 3σ, zone B between ± 2σ, and zone C between ± 1σ.

Test 1

Pattern: One or more points falling outside the control limits on either side of the average (Figure 6.10). **Problem source:** A sporadic change in the process due to special causes such as (a) equipment breakdown; (b) new operator; (c) drastic change in raw material quality; or (d) change in method, machine, or process setting. **Check:** Review records to determine what might have been done differently before the out-of-control point signals.

Test 2

Pattern: A run of nine points on one side of the average (Figure 6.11). **Problem source:** This may be due to a small change in the level of process average; change may be permanent at new level. **Check:** Review the beginning of the run to determine what was done differently then or earlier.

Test 3

Pattern: A trend of six points in a row, increasing or decreasing, as shown in Figure 6.12. **Problem source:** May arise from (a) gradual tool wear; (b) change in a characteristic such as gradual deterioration in the mixing or concentration

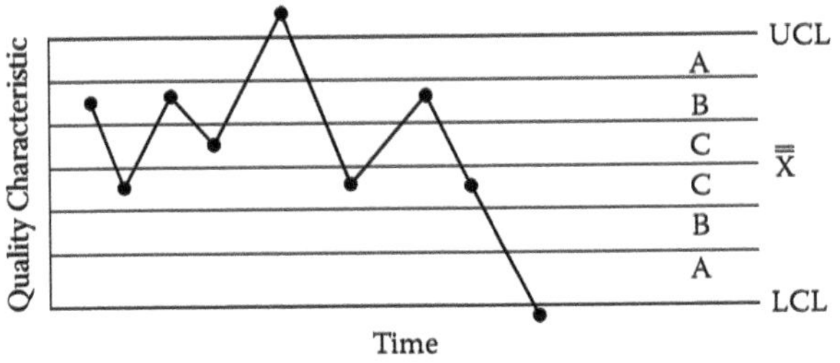

FIGURE 6.10
Test 1 for trend analysis.

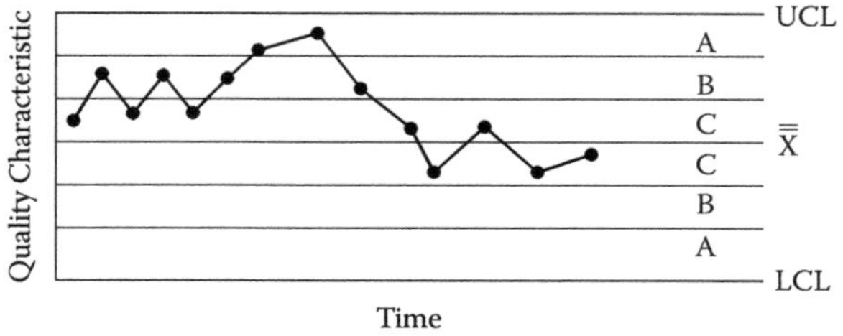

FIGURE 6.11
Test 2 for trend analysis.

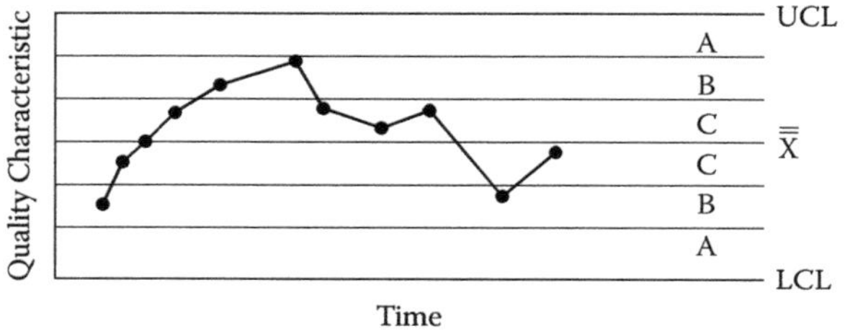

FIGURE 6.12
Test 3 for trend analysis.

of a chemical; (c) deterioration of plating or etching solution in electronics or chemical industries. **Check:** Search for the source at the beginning of the run.

These three tests are useful for achieving good control of a process. However, advanced tests based on the zone control chart can be used to detect out-of-control patterns.

Test 4

Pattern: Fourteen points in a row alternating up and down within or outside the control limits (Figure 6.13). **Problem source:** Sampling variation from two different sources such as sampling systematically from high and low

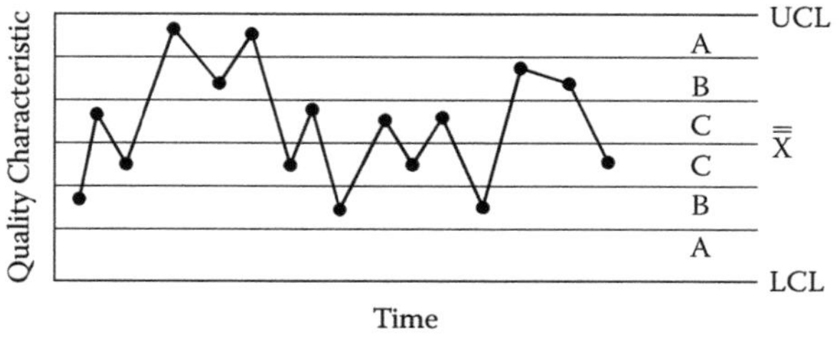

FIGURE 6.13
Test 4 for trend analysis.

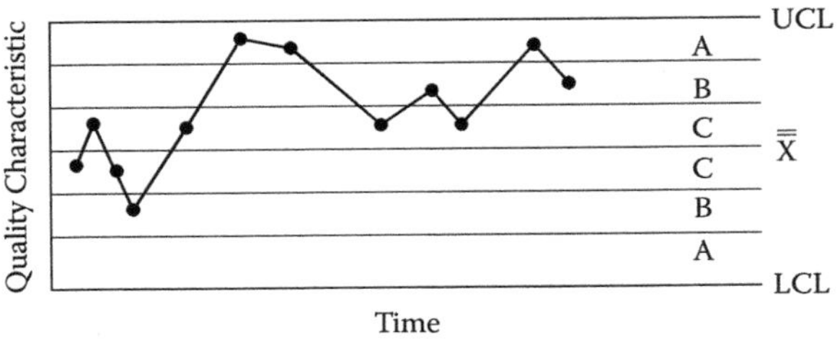

FIGURE 6.14
Test 5 for trend analysis.

temperatures or using lots with different averages may be the cause. This pattern can also be caused by constant adjustment (over-control). **Check:** Look for cycles in parameters such as humidity or temperature cycles or operator over-control.

Test 5

Pattern: Two of three points in a row on one side of the average in zone A or beyond (Figure 6.14). **Problem source:** A large, dramatic shift in the process level may be the cause. This test sometimes provides early warning, particularly if the special cause is not sporadic as in the case of Test 1. **Check:** Go back one or more points in time and determine what might have caused the large shift in the level of the process.

Test 6

Pattern: Four of five points in a row on one side of the average in zone B or beyond Figure 6.15). **Problem source:** The cause may be a moderate shift in the process. **Check:** Go back three or four points in time.

Test 7

Pattern: Fifteen points in a row on either side of the average in zone C (Figure 6.16). **Problem source:** One possibility is unnatural small fluctuations or absence of points near the control limits. At first glance, this may not seem

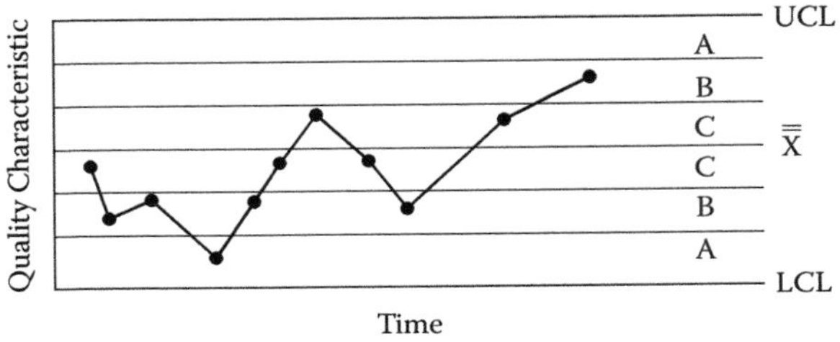

FIGURE 6.15
Test 6 for trend analysis.

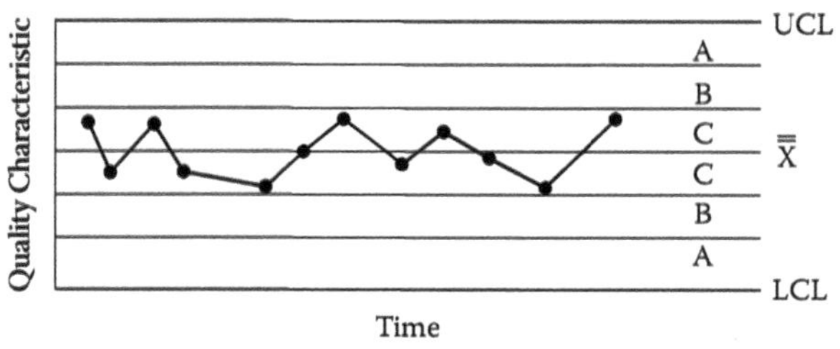

FIGURE 6.16
Test 7 for trend analysis.

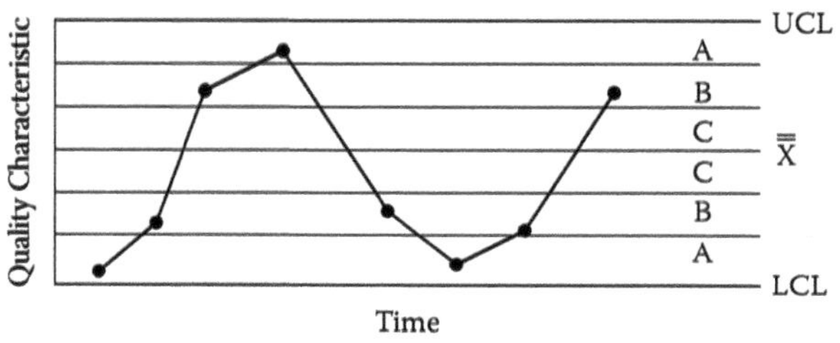

FIGURE 6.17
Test 8 for trend analysis.

negative, but it indicates poor control. Other possibilities are (a) incorrect selection of subgroups, (b) sampling from various subpopulations and combining them into a single subgroup for charting; or (c) incorrect calculation of control limits. **Check:** Look very closely at the beginning of the pattern.

Test 8

Pattern: Eight points in a row on both sides of the center line with none in zone C (Figure 6.17).

Problem source: Insufficient resolution of measurement system (see section on measurement systems). **Check:** Check whether range chart is in control.

Process Capability Analysis for Six Sigma

Industrial process capability analysis is an important aspect of managing industrial projects. The capability of a process is the spread that contains most of the values of the process distribution. It is very important to note that capability is defined in terms of distribution. Therefore, capability can only

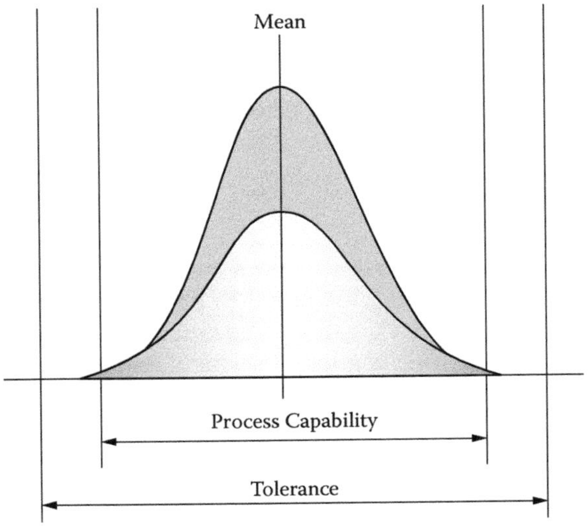

FIGURE 6.18
Process capability distribution.

be defined for a process that is stable (has distribution) with common cause variation (inherent variability). It cannot be defined for an out-of-control process (no distribution) with variation arising from specific causes (total variability). Figure 6.18 shows a process capability distribution.

Capable Process (C_p)

A process is capable ($C_p \geq 1$) if its natural tolerance lies within the engineering tolerance or specifications. The measure of process capability of a stable process is $6\hat{\sigma}$, where $\hat{\sigma}$ is the inherent variability estimated from the process. A minimum C_p value of 1.33 is generally used for ongoing processes. This ensures a very low reject rate of 0.007% and constitutes an effective strategy for prevention of nonconforming items. C_p *is* defined mathematically as:

$$C_p = \frac{USL - LSL}{6\hat{\sigma}}$$

$$= \frac{\text{allowable process spread}}{\text{actual process spread}},$$

where USL = upper specification limit and LSL = lower specification limit. C_p measures the effect of the inherent variability only. An analyst should use R-bar/d_2 to estimate $\hat{\sigma}$ from an R-chart in a state of statistical control, where R-bar is the average of the subgroup ranges and d_2 is a normalizing factor tabulated for different subgroup sizes (n). We do not have to verify control before performing a capability study. We can perform the study, then verify control afterward by using control charts. If a process is in control during the study, our estimates of capabilities are correct and valid. However, if the process is not in control, we have gained useful information and insights about corrective actions to pursue.

Capability Index (C_{pk})

Process centering can be assessed when a two-sided specification is available. If the capability index (C_{pk}) is equal to or greater than 1.33, a process may be adequately centered. C_{pk} can also be employed with one-sided specification. For a two-sided specification, it is calculated as:

$$C_{pk} = \text{Minimum}\left\{\frac{USL-\overline{X}}{3\hat{\sigma}}, \frac{\overline{X}-LSL}{3\hat{\sigma}}\right\},$$

where $\overline{X}$ = overall process average. However, for a one-sided specification, the actual C_{pk} obtained is reported and can be used to determine the percentage of observations out of specification. The overall long-term objective is to make C_p and C_{pk} as large as possible by continuously improving or reducing process variability $\hat{\sigma}$ for every iteration so that a greater percentage of the product is near the key quality characteristic's target value. The ideal is to center the process with zero variability.

If a process is centered but not capable, one or several courses of action may be necessary. One action may be integrating designed experiments to gain additional knowledge of the process and then designing control strategies. If excessive variability is demonstrated, a nested design with the objective of estimating the various sources of variability may be implemented. The sources of variability can then be evaluated to determine strategies for reducing or permanently eliminating them. Another action may be changing the specifications or continuing production and then sorting the items. A process exhibits one of three capability characteristics (Figures 6.19 through 6.21). A process may be (a) centered and capable; (b) capable but not centered; or (c) centered but not capable.

Process Capability Example

Step 1: Using data for the specific process, determine whether the process is capable. Assume that the analyst determined that the process is in a state of statistical control. For this example, the specification limits are set at 0 (lower) and 45 (upper). The inherent process variability as determined from the control chart is $\hat{\sigma} = \overline{R}/d_2 = 5.83/2.059 = 2.83$. The capability of this process to produce within the specifications can be determined as:

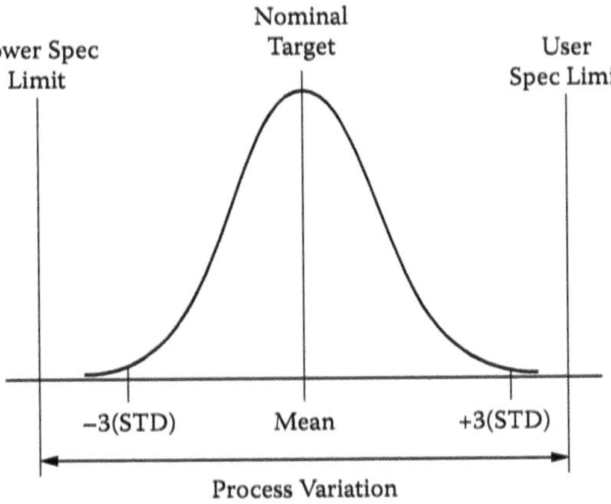

FIGURE 6.19
Process that is centered and capable.

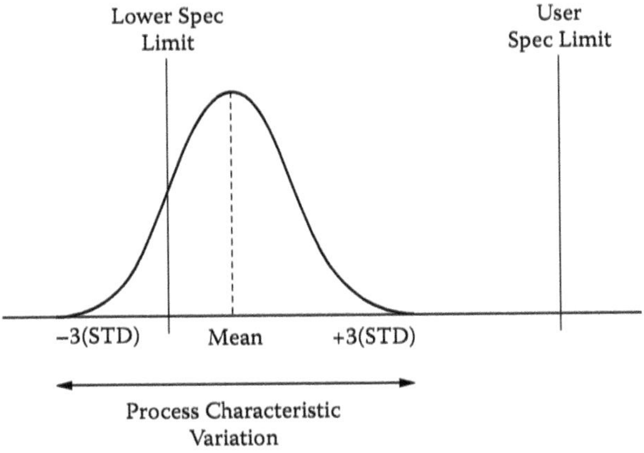

FIGURE 6.20
Process that is capable but not centered.

$$C_p = \frac{USL - LSL}{6\hat{\sigma}} = \frac{45 - 0}{6(2.83)} = 2.650\,.$$

The $C_p = 2.65 > 1.0$ indicates that the process is capable of producing clutches that will meet the specification of 0 to 45. The process average is 29.367.

Step 2: Determine whether the process can be adequately centered. C_{pk} = minimum [C_l and C_u] can be used to determine whether a process can be centered.

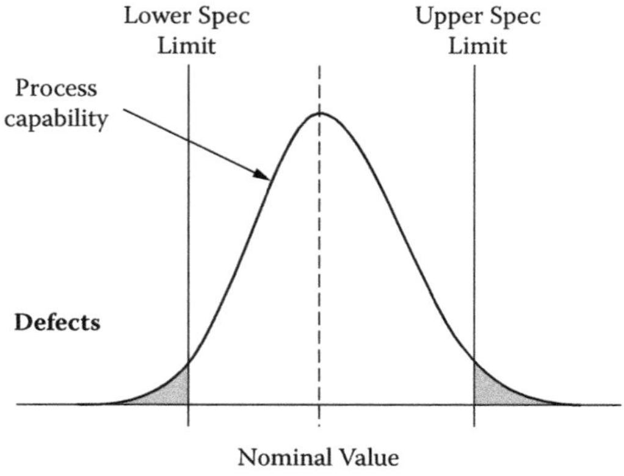

FIGURE 6.21
Process that is centered but not capable.

$$C_u = \frac{USL - \bar{X}}{3\hat{\sigma}} = \frac{45 - 29.367}{3(2.83)} = 1.84$$

$$C_l = \frac{\bar{X} - LSL}{3\hat{\sigma}} = \frac{29.367 - 0}{3(2.83)} = 3.46.$$

Therefore, the capability index C_{pk} for this process is 1.84. Since $C_{pk} = 1.84$ is greater than 1.33, the process can be adequately centered.

Possible Applications of Process Capability Index

The process capability index can be applied to:

Communications—Industry uses C_p and C_{pk} to establish a dimensionless common language useful for assessing the performance of production processes. Engineering, quality, manufacturing, and other areas can communicate and understand processes with high capabilities.

Continuous improvement—The indices can be used to monitor continuous improvement by observing changes in the distribution of process capabilities. For example, if 20% of the processes achieved capabilities between 1 and 1.67 in a month and some of these improved to between 1.33 and 2.0 the next month, improvement has occurred.

Audits—Many types of audits in use today can assess the performances of quality systems. A comparison of in-process capabilities with capabilities determined from audits can help establish problem areas.

Prioritizing improvements—A complete printout of all processes with unacceptable C_p or C_{pk} values can be extremely helpful for establishing priorities for process improvements.

Elimination of nonconforming products—For process qualification, it is reasonable to establish a benchmark capability of $C_{pk} = 1.33$ which will make nonconforming products unlikely in most cases.

Potential Abuse of C_p and C_{pk}

In spite of its many possible applications, a process capability index has potential for abuse:

Problems and drawbacks—C_{pk} can increase without process improvement even though repeated testing reduces test variability. The wider the specifications, the larger the C_p or C_{pk}, but the action does not improve the process. Furthermore, analysts tend to focus on numbers rather than on processes.

Process control aspects—Analysts tend to determine process capability before statistical control has been established. Most people are not aware that capability determination is based on process commonalities. The appearance of special causes of variation makes prediction impossible and capability indices unclear.

Non-normality—Some processes result in non-normal distributions for some characteristics. Since capability indices are very sensitive to departures from normality, data transformation may be used to achieve approximate normality.

Computation—Most computer-based tools do not use $\overline{R}/d_2$ to calculate σ. When analytical and statistical tools are coupled with sound managerial approaches, an organization can benefit from a robust implementation of improvement strategies.

Cp and C_{pk} are capability analyses that can only be performed with normal data. It is very easy to use any data for capability analyses especially on software systems that calculate automatically. The first step in a capability analysis is to obtain continuous data and check for normality. If the state of data is not normal, the special cause variation must be found. Outlier data points may be taken out only if the reason for the outlier (temperature change, shift change, etc.) is known. Once a reason is found for an outlier, the outlier can be removed and the data can be checked again for normality. If no root cause is found for the outlier, more data must be taken, but capability analyses should not be done until normality is proven.

Lean Principles and Applications

The "lean" approach that emerged as a sound managerial principle has been successfully applied to many industrial operations. Lean is defined as the identification and elimination of sources of *waste* in operations. Recall that Six Sigma involves the identification and elimination of sources of *defects*.

When Lean and Six Sigma are coupled, an organization can derive a double benefit from reducing wastes and defects. The combination is known as Lean Six Sigma. It enables an organization to achieve higher product quality, improved employee morale, better satisfaction of customer requirements, and more effective utilization of limited resources. The basic principle of lean is to take a close look at the elemental compositions of a process so that non-value-adding elements can be located and eliminated.

Kaizen

By applying the Japanese concept of kaizen ("take apart and make better"), an organization can redesign its processes to be lean (devoid of excesses). In a mechanical design sense, this can be likened to finite element analysis that identifies how the component parts of a mechanical system fit together. Identifying these basic elements allows improvement opportunities to be recognized easily and quickly. It should be recalled that the project management task of breaking down work structures facilitates the identification of task-level components of an endeavor. Consequently, using a project management approach facilitates the achievement of lean objectives. In the context of quality management, Figure 6.22 shows a process decomposition hierarchy that may help identify elemental characteristics that may harbor waste, inefficiency, and quality impedance. The functional relationships (f) are:

Task = f(activity)

Subprocess = f(task)

Process = f(sub-process)

Quality system = f(process)

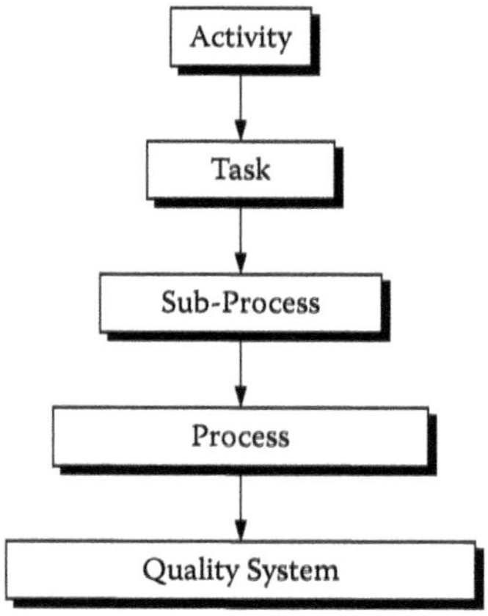

FIGURE 6.22
Hierarchy of process components.

Thus, quality improvement can be achieved by hierarchically improving a process and all its elements. Industry fads come and go. Over the years, we witnessed the introduction and demise of many techniques that were hailed as panaceas for industry's ailments. Some of the techniques survived the test of time because they do indeed hold promise. Lean is such a technique if it is viewed as an open-ended and focused application of the many improvement tools that emerged over the years.

The adoption of lean principles by the U.S. Air Force has given more credence to its application. The Air Force embarked on a massive endeavor to achieve widespread improvement in operational processes. The endeavor is called AFSO21 (Air Force Smart Operations for the 21st century or simply Air Force Smart Ops 21) and it requires the implementation of appropriate project management practices at all levels. AFSO21 is a coordinated effort to improve Air Force operations throughout the rank and file levels. It is an integrative process of using lean principles, the theory of constraints, Six Sigma, BPI, MBO, TQM, 6s, project management, and other classical management tools. However, the implementation of lean principles constitutes about 80% of AFSO21 efforts. One tool among lean practices and procedures is the lean task value rating system that compares and ranks elements of a process for retention, rescoping, scaling, or elimination purposes.

Lean Task Value Rating System

To identify value-adding elements of a lean project, the component tasks must be ranked and assessed comparatively. The method below applies relative ratings to tasks and is based on the distribution of a total point system. The total points available to a composite process or project are allocated across individual tasks. The steps are:

1. Let T represent the total points available to tasks.
2. $T = 100(n)$, where n = number of raters on the rating team.
3. Rate the value of each task on the basis of specified output (or quality) criteria on a scale of 0 to 100.
4. Let x_{ij} be the rating for task i by rater j.
5. Let m = number of tasks to be rated.
6. Organize the ratings by rater j:

   Rating for Task 1: x_{ij}

   Rating for Task 2: x_{2j}

   ...

   Rating for Task m: x_{mj}

   Total Rating Points 100.

TABLE 6.8

Lean Task Rating Matrix

	Rating by Rater $j = 1$	Rating by Rater $j = 2$	...	...	Rating by Rater n	Total Points for Task i	w_i
Rating for task $i = 1$							
Rating for ask $I = 2$							
...							
...							
Rating for task m							
Total points from Rater j	100	100	...	...	100	$100n$	

7. Tabulate the ratings by the raters as shown in Table 6.8 and calculate the overall weighted score for each Task i from:

$$w_i = \frac{1}{n}\sum_{j=1}^{n} x_{ij}.$$

The w_i values are used to rank order the tasks to determine their relative value-added contributions. Subsequently, using a preferred cut-off margin, the low or non-contributing activities can be slated for elimination. A comprehensive lean analysis can identify the important versus unimportant and urgent versus not urgent tasks to determine priorities. Waste task elements are found within the unimportant and not urgent quadrants and should be eliminated. Using the familiar Pareto distribution format, Table 6.9 illustrates task elements within a 20% waste elimination zone.

It is conjectured that activities that fall in the not important and not urgent zones run the risk of generating points of waste in any productive undertaking. That zone should be the first target of review for finding tasks that can be eliminated. Despite "sacred cow" activities that an organization must retain for political, cultural, or regulatory reasons, attempts should still be made to categorize all task elements of a project. The long-established industrial engineering time-and-motion study technique is making a comeback due to increased interest in eliminating waste via lean initiatives.

TABLE 6.9

Pareto Analysis of Unimportant Process Task Elements

	Urgent	Not Urgent
Important	20%	80%
Not important	80%	20%

Lean Six Sigma in Project Management

Lean and Six Sigma use analytical tools as bases for pursuing their goals, but the achievement of the goals is predicated on a structured approach to production activities. Proper project management practiced at the outset of an industrial endeavor will pave the way for achieving Six Sigma results and realizing lean outcomes. The key in any project endeavor is to have a structured design so that diagnostic and corrective steps can easily be pursued. If the proverbial "garbage" is allowed to creep into a project, a lean Six Sigma clean-up will require much more time, effort, and cost.

7

Lean Principles for Project Control

"Haste makes waste. Care preempts the need for haste."

In any enterprise, maintaining excess inventory is strongly discouraged. Excess inventory consumes space, cost, and valuable attention that could be better directed elsewhere. Excess inventory represents latent value that is not actualized and thus does not contribute to organizational goals while it remains in inventory status. If excess inventory is seen as undesirable, why then are we willing to accept or tolerate excess steps in project execution? The basic principle of lean is to identify and eliminate waste from work processes. Non-value-adding elements in a project create additional risks for it to fall out of control. To achieve and sustain project control, every aspect of waste in a project must be eliminated. Unlike in a production environment, waste in a project structure can be subtle and seemingly innocuous. A project team must be vigilant and continually seek opportunities to apply lean principles throughout a project value stream.

Lean and Six Sigma

Lean is a continual pursuit of the identification and elimination of waste from every facet of an operation with the ultimate goal of providing exceptional quality, on-time delivery, and value-adding service at the lowest possible cost. Six Sigma is a continuous striving for statistical standardization of processes with the end result of increasing quality and reducing cost within acceptable control limits.

Both lean and six-sigma efforts are predicated on data and facts rather than sentiments and emotions. Facts identify where the boundary of "fat" and "muscle" resides to ensure that only fat and not muscle is cut when implementing lean methods. Data facilitates the identification of process flow relationships and product characteristics so that variations can be reduced in the pursuit of eliminating defects.

Lean manufacturing focuses on eliminating waste in manufacturing by reducing wait times and inventory, maximizing scheduling production, reducing batch sizes, line balancing, and reducing process time. The technique also eliminates waste at every phase of production, streamlines work

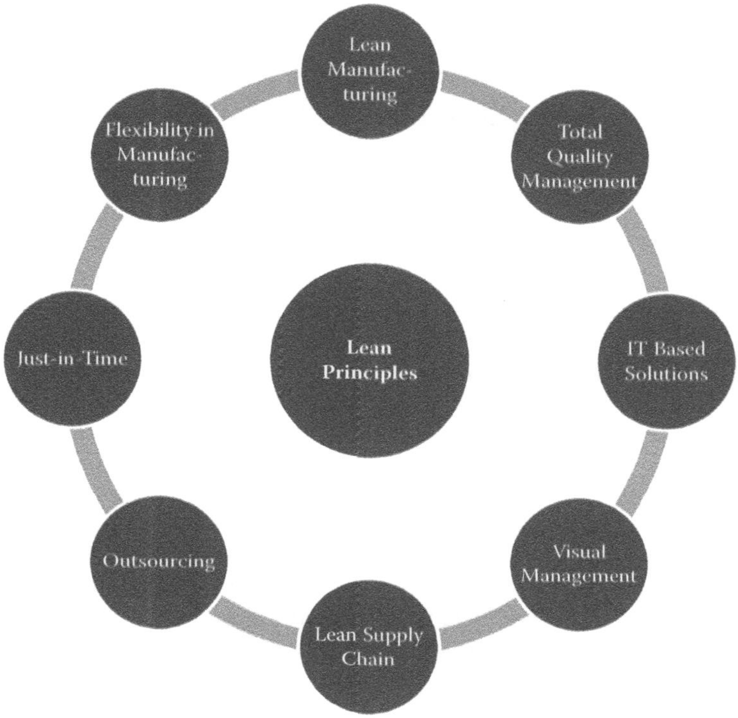

FIGURE 7.1
Cycle of lean principles.

flows to achieve improved efficiencies, increases flexibility for key products or processes, introduces new products easily via benchmarking, reduces work in process (WIP), meets customer demands, produces higher quality products at lower costs and higher margins, and finally consumes less labor, space, and inventory and uses them more efficiently. Figure 7.1 illustrates the basic lean principles.

By clearly defining value-added tasks to benefit the end customer, a manufacturing operation can eliminate wastes. Assigning monetary values to non-value-added tasks is a required step in an effort to reduce waste. Identifying all steps followed through the design, order, and production processes will also highlight non-value-added activities.

Muda is a Japanese term for waste—any activity that consumes resources but creates no value for customers. Muda consists of scrap; defects, errors, and reworks; excessive transportation, handling and movement; and faster-than-desired production that creates excess inventory. After non-value-added activities are determined, value-added processes can be continually checked by capability analyses, total productive maintenance

(TPM), and adequate capacity management techniques. Manufacturing "just in time" by utilizing kanbans or long-term inventory technology is another key to lean. Finally, the 5s principles ("a place for everything and everything in its place") are valuable for organizing work places. The 5s levels are:

Sort—Identify and eliminate necessary items and dispose of unneeded materials that do not belong in an area. This reduces waste, creates a safer work area, opens space, and helps visualize processes. It is important to sort through the entire area. The removal of items should be discussed with all personnel involved. Items that cannot be removed immediately should be tagged for subsequent removal.

Sweep—Clean the area so that it looks like new and clean it continuously. Sweeping prevents an area from getting dirty in the first place and eliminates further cleaning. A clean work place indicates high standards of quality and good process controls. Sweeping should eliminate dirt, build pride in work areas, and build value in equipment.

Straighten—Have a place for everything and everything in its place. Arranging all necessary items is the first step. It shows what items are required and what items are not in place. Straightening aids efficiency; items can be found more quickly and employees travel shorter distances. Items that are used together should be kept together. Labels, floor markings, signs, tape, and shadowed outlines can be used to identify materials. Shared items can be kept at a central location to eliminate purchasing more than needed.

Schedule—Assign responsibilities and due dates to actions. Scheduling guides sorting, sweeping, and straightening and prevents regressing to unclean or disorganized conditions. Items are returned where they belong and routine cleaning eliminates the need for special cleaning projects. Scheduling requires checklists and schedules to maintain and improve neatness.

Sustain—Establish ways to ensure maintenance of manufacturing or process improvements.

Sustaining maintains discipline. Utilizing proper processes will eventually become routine. Training is key to sustaining the effort and involvement of all parties. Management must mandate the commitment to housekeeping for this process to be successful.

The benefits of 5s include (a) a cleaner and safer workplace; (b) customer satisfaction through better organization; and (c) increased quality, productivity, and effectiveness.

Kaizen events normally are key processes for starting a 5s project. *Kai* means to break apart or disassemble so that one can begin to understand. *Zen* means to improve. A kaizen process focuses on an improvement objective by breaking the process into its basic elements to understand it, identify wastes, create improvement ideas, and eliminate the wastes identified. The basic philosophy of kaizen is to manufacture products safely, when they are needed, and in proper quantities needed while reducing cycle and lead times.

Kaizen increases productivity, reduces WIP, eliminates defects, enhances capacity, increases flexibility, improves layout, and establishes visual management and measures.

Kaizen increases productivity by revealing operator cycle time, eliminating waste, balancing workloads, utilizing value-added tasks, and producing to demand. Kaizen reduces WIP by determining needed and unneeded inventory. Similar production outputs are grouped to balance production. Set-up times can be reduced; batches of outputs can be transported in smaller quantities. Preventive maintenance schedules can be established to aid consistent quality. Kaizen eliminates defects by asking "why" five or more times and reducing inventory so that improper manufacturing operations are caught quickly. Work is done under stable conditions and mistake-proof techniques are utilized. Kaizen enhances capacity and increases flexibility by finding and eliminating production bottlenecks caused by humans and machines. Waste is identified and eliminated. Layout flexibility promotes efficient flow of objects, people, and machines. Environments must be safe and clean and allow regular preventive maintenance. Staffing is minimal and walking distances are shortened. Work should enter and exit an area at the same place. Staff should communicate and the work balance should be even; workers should be able to help each other if needed.

Kaizen distinguishes equipment and operation improvements. Equipment-based improvements involve capital and time, may require major modifications, and may not produce cost savings. Operational-based improvements (a) change standard operating procedures; (b) change positions of layouts, tools, and equipment; (c) simplify tools by adding chutes, knock-out devices, and levers; (d) improve equipment efficiency without drastic modification; and (e) involve little cost and focus on cost reduction

Kaizen focuses on a single piece of equipment or set of flows, synchronized movements, shortened transfer distances, movement of inventory into designated or finished states, and maintaining buffers between flows to keep flows from disturbing each other. Quality is always an issue for kaizen, lean, and Six Sigma. Defects are to be reduced, flows are to be improved, and processes must be streamlined. Finally, safety and environmental issues must be considered when operations and layouts are changed. Safety takes priority over any cost savings or productivity increases. Kaizen has ten basic rules:

1. Think "outside the box." No new idea should be considered a bad idea.
2. Determine how a task can be done, not how it cannot be done.
3. No excuses. Question current practices.
4. Perfection may not come immediately; improvements are required.
5. Mistakes should be corrected as soon as possible.
6. Ideas should be quick and simple, and not involve great amounts of money.
7. Find value in other people's ideas.

8. Ask "why" at least five times to find root causes.
9. Consult more than one person to find a true solution.
10. Kaizen ideas are long term.

Visual management and measures promote successful layouts. Visual management involves bins, cards, tags, signals, lights, alarms, and other signaling mechanisms. Visual systems include:

- Indicators such as signs, maps, and displays convey passive information.
- Signals such as alarms or lights are assertive devices.
- Controls provide aggressive information by monitoring size, weight, width, or length.
- Guarantees such as sensors, guides, and locators provide assured information.

Visual management systems cover tasks such as housekeeping via 5s, standard operating procedures, detecting errors and defects, and eliminating errors by mistake proofing processes. Some of these techniques are known as methods time measurements or Maynard operations systems techniques (MOST). Once visual management techniques are completed, data evaluation can be performed as described below.

A *first pass yield* (FPY) indicates the number of good outputs from a first pass at a process or step. The formula is FPY = (# accepted)/(# processed). The formula for the first pass yield ratio is % FPY = [(# accepted)/(# processed)] × 100. If 45 outputs are satisfactory among 60 produced, the FPY = 45/60 or 0.75. The % FPY = 0.75 × 100 = 75%. This number does not include rework of the rejected product.

Rolled throughput yield (RTY) covers an entire process. If a process involves three activities with FPYs of 0.90, 0.94, and 0.97, the RTY would be 0.90 × 0.94 × 0.97 = 0.82. The %RTY = 0.82 × 100 = 82%.

Continuous flow processing (CFP) is another aspect of operational kaizen. It is characterized by continuous process flows, production paced according to Takt time (explained later in this chapter), and pulling subsequent processes. To achieve CFP, machines must be arranged in the order of the process, one piece flow of production must occur, multi-tasking must take place, easy moving must be possible, and U-cell layouts should be utilized. Pull production occurs when material is "pulled" from process to process only when needed. If a subsequent process already has the material, nothing is done.

Kanbans are communication signals that control inventory levels while ensuring even and controlled production flow. Kanbans signal times to start, times to change set-ups, and times to supply parts. Kanbans work only if monitored consistently.

Value stream mapping reveals why excessive waste is introduced. The steps for value stream mapping are:

- Map current state processes.
- Summarize current state processes.
- Map future state processes.
- Summarize future state processes.
- Develop short-term plans to move from current to future state.
- Develop long-term plans to move from current to future state.
- Implement risks, failures, and processes for transitions.
- Map key project owners, key dates, and future dates of future states.
- Continue to map new future states when future states are met.

The value stream adds value from a customer's view as shown in Figure 7.2. All steps in value stream mapping should deliver a product or service to customers. A value stream involves multiple activities that convert inputs into outputs. All processes follow sequences of starts, lead times, and ends. The first goal is to reduce problematic areas by understanding the main causes for defects. The next step is to reduce lead time.

The final step is to find the percentage of value-added time as shown in Figure 7.3. The formula is % VAT = (sum of activity times)/(lead time) × 100. When the sum of activity times equals lead time, the value-added time is 100%. For most processes, % VAT = 5 to 25%. If the sum of activity times equals the lead time, the time value is not acceptable and activity times should be reduced. Lead time is the time from process start to end or time from receipt to delivery. Activity time is the time required to complete

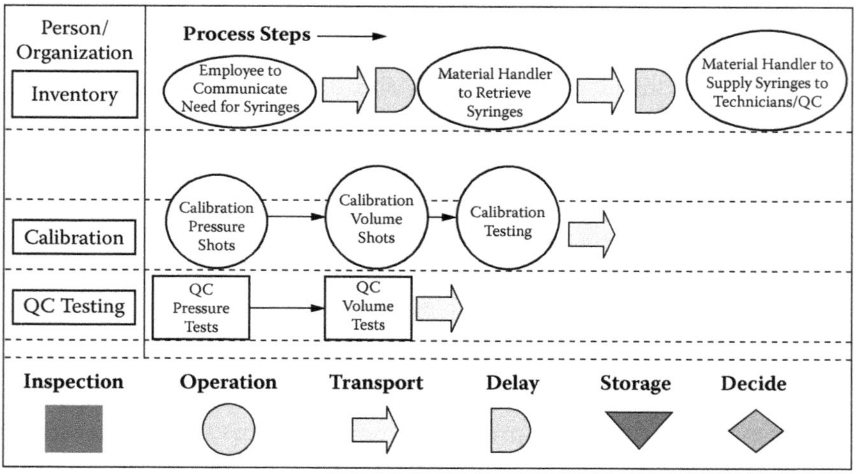

FIGURE 7.2
Value stream mapping.

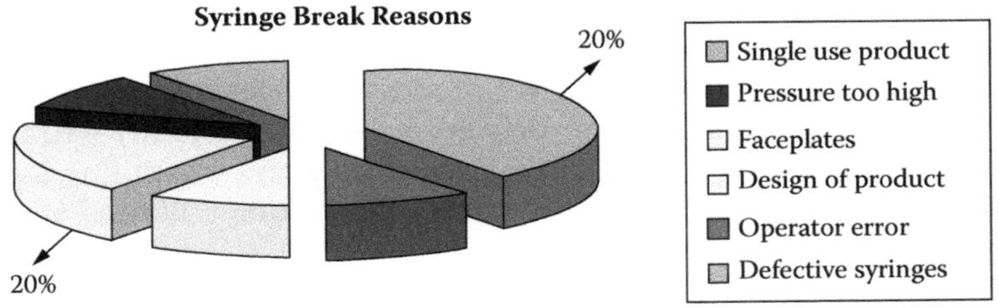

FIGURE 7.3
Identifying main causes of defects.

one output in an activity. Cycle time is the average time in which an activity yields an output calculated as the activity time divided by the number of outputs produced in one cycle of an activity. Value-added definitions and results are:

Customer value-added time (VAT)

- Benefits customers and produces competitive advantage.
- Customers will pay for the tasks.

Non-value-added time (NVAT)

- Customer wants to lower the prices by eliminating task.
- Task is eliminated or merged with upstream or downstream tasks.

Business value-added time (BVAT)

- Required for business reasons such as conforming to laws or regulations.
- Required to reduce financial risks and liabilities.
- Required due to the nature of the process.

Specific steps must be considered in determining the flow of a process: (a) design; (b) order taking; (c) production; and (d) correct location.

Design—After a design is known and a prototype is in place, the different teams will know what production, engineering, and design groups are required. In the background, the purchasing department must ensure that proper parts are ordered and in place for production. The design for the product, even when under development, follows different queues through different departments. Rework occurs and must be considered in making flow decisions.

Quality function deployment (QFD) is a lean approach that maps proper design with specifications of a product and determines proper teams and decision-making processes along the way. QFD helps find value-added processes and eliminates the need for rework. A department may have to be moved to ensure a proper design for producing a customer-focused product.

Order taking—This function ensures that orders are fully processed. Consistency is important to customers and maintains credibility of a business. Scheduling departments normally set production dates and relay shipping information to sales. If a customer delivery is late, the customer will usually call sales. The sales department will call scheduling and scheduling will contact production to ask about status. At times, production will need to change its schedule to handle urgent or back-ordered items to satisfy a customer. For these reasons, production scheduling must be consistent and timely.

Sales and scheduling departments were created during a reengineering movement in the early 1990s intended to process orders more quickly. The reengineering based on lean production flow was successful. Sales and scheduling are core members of the production team that plan the sales at the same time the product design is developed. The focus is to plan sales and production capabilities so that both orders and product flow easily from sales to delivery. Building products to order is efficient and helps reduce the need for expediting.

Takt time is a kaizen tool used in the order taking phase. *Takt* is a German word for pace. Takt time is defined as time per unit. It measures operation times to keep production on track. To calculate Takt time, the formula is time available/production required. Thus, if a required production is 100 units per day and 240 minutes are available, the Takt time = 240/100 or 2.4 minutes to keep the process on track. Individual cycle times should be balanced to the pace of Takt time. To determine the number of employees required, the formula is (labor time/unit)/Takt time. Takt in this case is time per unit. Takt requires visual controls and helps reduce accidents and injuries in the workplace. Maintaining a close connection of sales and production will in turn reduce muda (waste).

Production—Production and scheduling must work closely to prevent waste. Changeovers are costly and time consuming so it makes sense to make large batches of parts that require changeovers. Versatility is key for changeovers that are minimally invasive. Maintaining proper inventory can be tricky when dealing with a mix of high volume production and complicated parts. Ensuring the availability of correct parts for the next process at the right time starts with scheduling and then becomes a function of production. A master schedule normally utilized to ensure product quality.

In recent years, material requirements planning (MRP) systems were created to eliminate manual scheduling. MRPs can track inventory, order materials, and transmit processes or requirements for the next step of production. MRPs led to capacity planning tools that evaluate the capacity of machines to produce goods to prevent major bottlenecks and capacity constraints. One recurring issue is the ability to implement a changeover with clear communication, without confusion, and with minimal muda. Waiting and material waste normally result from poor planning of changeovers. Waiting and wasting can escalate to lead time and batch issues. Sometimes engineering or design changes in the middle of production lead to even more confusion. Rework results from faulty communication.

These problems led to the development of just-in-time (JIT) strategies. JIT is a value streaming methodology created by Taiichi Ohno at Toyota in the 1950s. Western industry embraced it in the early 1980s. JIT is a system to produce and deliver proper items at proper times in proper amounts by having upstream activities occur right before downstream activities so that single-piece flow is possible. Flow, pull, standard work, and Takt are the key elements.

JIT is only effective if machine changeovers are reduced so that upstream manufacturing operations produce small numbers of parts until the next process downstream needs the next part. This prevents accumulation of excess stock. JIT works properly only when scheduling is effective and consistent. It is important to use JIT principles for inventory and production to ensure that inventory does not greatly exceed production. JIT and TPM have increased productivity more than 200% in the last 5 years.

Flow is an important aspect of JIT because it enhances efficiency. Value-added activities help with flow, but flow may not be easy to analyze. Proper flow can decrease processes by half. The first step to defining flow is finding value-added activities. The focus should be on the main process or product, not the intermediate steps, in other words, what absolutely must happen to yield a product or service. Understanding what customers want is key.

Standardized work, also known as standard operating procedure (SOP), consists of a definitive set of work procedures with all tasks organized optimally to meet customer needs. Standardization of the process and the activities allow for consistent times and completion of entire processes.

Poka yoke (mistake-proofing) is another initiative for improving production systems. It eliminates product defects by installing processes to prevent, correct, or draw attention to human errors. Poka yoke was developed because people are prone to making mistakes. Poka yoke designs processes to prevent defects from occurring or moving to the next step. When an error occurs, a poka yoke system will alert the proper person to correct the problem.

Correct location—This technique places design and production operations in the correct order to satisfy customers. Increasing throughput of machines or outsourcing can ensure components are made at the proper time and place. Keeping the end product in mind is the key to utilizing correct location. Centralized areas for final assemblers can help but effective flow must be considered. A U-shaped flow or process in order of operation normally prevents bottlenecks. Improvements of process machinery should also be considered. Without correct location, logistics costs increase and large inventories ensue. Increased inventories of products that have shelf life limits can create large amounts of waste because the products have to be discarded after expiration.

Flows of all processes and activities must be considered. Value-added activities represent the first priorities along with creating lean environments by utilizing effective tools and processes, and reducing wastes to ensure continuous flows. While flows are important in lean manufacturing, another potential problem area is having a great technique but not producing goods wanted by customers. This leads us to the next lean topic: pull.

Pull Technique

Pull is the simple practice of not producing a product upstream until the customer downstream needs it. The philosophy sounds easy, but its management is complicated. Normally large batches of general inventory are in stock, but ensuring specialized and costly inventories for specific customers is far more difficult. Understanding how to be versatile and utilizing co-manufacturers for these special circumstances are keys to lean production utilizing a pull system. Customers for high-end or specialized products must understand the lead times required.

Push systems are not accurate because they normally involve production schedules based on projected customer demands. Changeovers should be kept to a minimum to produce versatile products instead of large quantities of a special product. Small "stores" of parts between operations are created to reduce inventory. When a process customer uses inventory, it should be replenished. If inventory is not needed, it is not replenished.

Lean production for pull requires 90% machine use or availability and limits downtime for maintenance and changeovers to 10% of machine time. Kanbans are used in lean production to ensure that flow is pulled by the next step and a visual indicator signals needs for inventory or activity. These steps enable production to respond directly to customer need without producing excess inventory or requiring further work. Quality should also soar because production is based on customer needs and not production per minute or per hour. Figure 7.4 illustrates pull flow.

Heinjunka is the leveling of production and scheduling based on volume and product mix. Instead of building products according to the flow of customer orders, this technique levels the total volume of orders over a specific time so that uniform batches of different product mixes are made daily. The result is a predictable matrix of product types and volumes. For heinjunka to succeed, changeovers must be managed easily. Changeovers must be as minimally invasive as possible to prevent time wasted because of product mix. Another key to heinjunka is making sure that products are needed by customers. A product should not be included in a mix simply to produce inventory if it is not demanded by customers. Long changeovers should be investigated to determine the reason and devise a method to shorten them. Consider a factory that requires long times for changeovers because of the

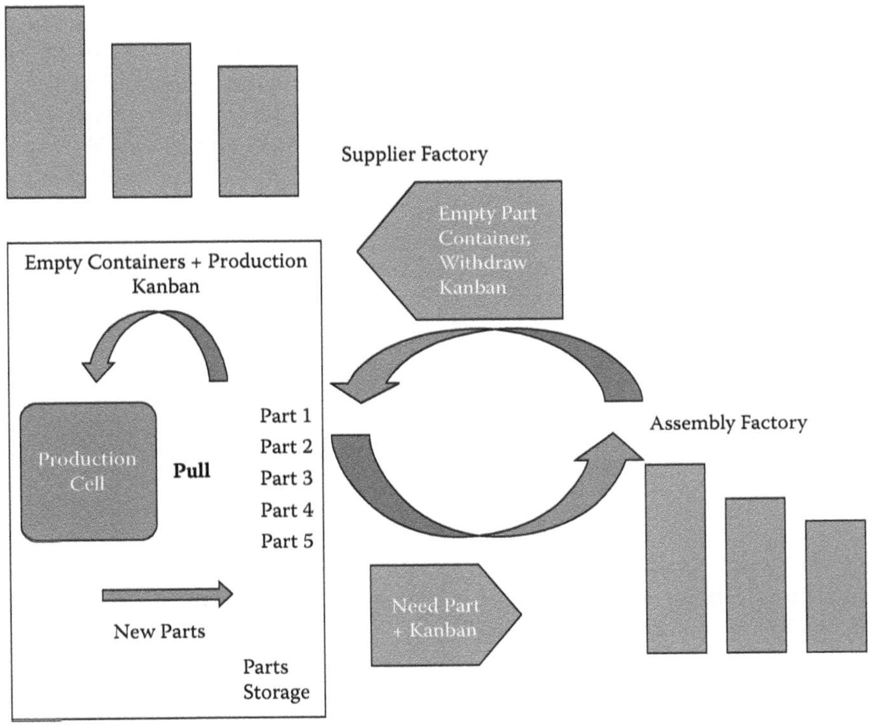

FIGURE 7.4
Pull flow system.

need to clean equipment between batches. One potential solution is having additional equipment or spare parts available to reduce cleaning time. Another option is to use different types of production equipment to assure versatility. The four goals are to (1) be flexible enough to produce what customers want when they want it, (2) reduce potential loss from unwanted or unsold goods; (3) utilize balancing techniques for labor and machinery; and even demands on upstream processes and factory suppliers.

The lean house is a visual representation of quality, JIT, customer focus, and continuous improvement. It illustrates how the components depend on each other. Continuous improvement is the core or foundation of the house.

First pass quality (FPQ) is a component of lean, TPM, Six Sigma, and other quality improvement plans. The philosophy simply means a road to perfection. Production can be stopped or delayed frequently if problems arise. If quality checks are installed in the design phase of a project, success is more likely. Quality checks should occur at each process hand-off so that quality is monitored throughout production. Toyota's principle of stopping a process to build in quality is called jidoka. It involves detecting when a process goes wrong and immediately stopping it. The process is then reset after the problem is solved; the mistake should not recur and no rework is required. The basic philosophy is to detect defects as they occur and automatically stop production to correct the problem before the defect continues downstream. Jidoka is also known as autonomation (human-aided equipment that stops moving forward if a defect is present). FPQ is also a component of error-proofing or poka yoke. FPQ is continuous. It is key to addressing quality issues at the beginning and implementing preventive measures.

Quality can be a simple initiative when based on simple principles and the involvement of all team members. The four key tools of a simple quality method are:

1. Go and see.
2. Analyze the situation.
3. Utilize one-piece flow.
4. Ask "why" five times.

Asking "why" five times is not pessimistic. It is a technique to find the root cause of a problem so that FPQ can be implemented. It also eliminates blaming others for mistakes that cause muda.

Another way of maintaining FPQ or achieving perfection is to subject processes to continuous improvement. Even if a process works well, continuous improvement activities focus on making it as error-free as possible. Reducing cycle times or potential process errors to minimize equipment replacements and process renovations will eliminate a large number of process defects. Four lean principles lead to perfection: (1) value specification; (2) value identification; (3) flow; and (4) pull. Competing *for* perfection instead of *against* competitors is the way to achieve FPQ. Continuous improvement requires standardized tasks and visual management.

Standardized tasks or SOPs—These measures are put in place so employees can follow specific instructions and complete work in a consistent manner. Standardized work is normally based on (1) Takt time; (2) task performance sequence; and (3) inventory required for employees to complete tasks. These standards stabilize the process and empower employees to find improvement opportunities.

Visual management—Visual controls prevent the masking of problems and allow errors to be seen immediately. The 5 S's (sort, sweep, straighten, schedule, and sustain; Figure 7.5) come into play with visual controls. Visual management utilizing 5s clears out rarely used items by sorting, organizing and placing them properly, straightening, cleaning work areas by sweeping, and ordering by scheduling tasks and using preventive maintenance techniques, and using discipline to sustain progress. Visual controls improve

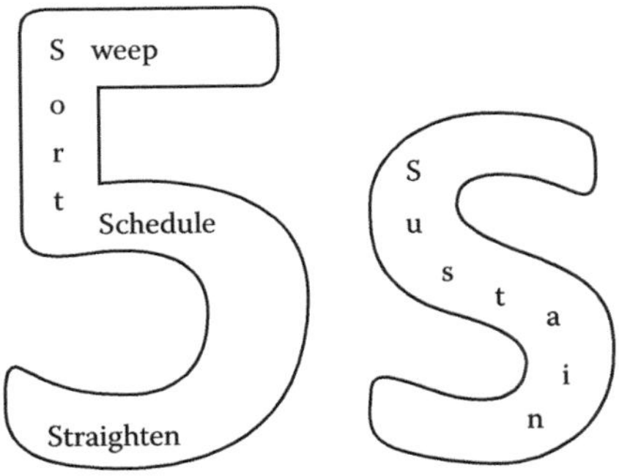

FIGURE 7.5
5s principles.

value-added flows of processes by clearly demonstrating how work should be done and revealing errors if SOPs are not followed. Well-designed process flows, graphs, charts, and other visual tools reflect improvement and motivate employees to continue improving.

Lean principles and other quality-driven initiatives are customer focused. Customer satisfaction is based on customer perceptions and expectations. If a service level exceeds customer expectation, over-processing has occurred. Customer perception has little or no correlation with expectations or actual service levels. Over time, customer expectations increase and over-production decreases. Actual service levels normally remain fairly constant but customer perception fluctuates. Figure 7.6 illustrates the lean house concept.

The foundation of the house provides stability without variation. Chaos is absent; information flow is constant, and processes are stable, consistent, and thus capable (a process may be stable but not capable). A process must be predictable from the customer's view and must reflect what the customer wants.

The two main elements of lean culture are trust and leading as a teacher. Remember that people want quality in their work, but they make mistakes. Errors only turn into defects if they continue and eventually become components of tasks. Trust requires employees to have confidence in management and the processes management dictates. Leading as a teacher means finding the sources of problems. In other words, finding a problem is a good thing, and solving the right problems is a way of life. Learning through mistakes helps implement the lean culture and empowers others. Continuous learning and development of others transform employees into successful managers and enables a future full of leaders. Lean involves a culture change and should be seen as more than a group of quality tools used to drive improvements.

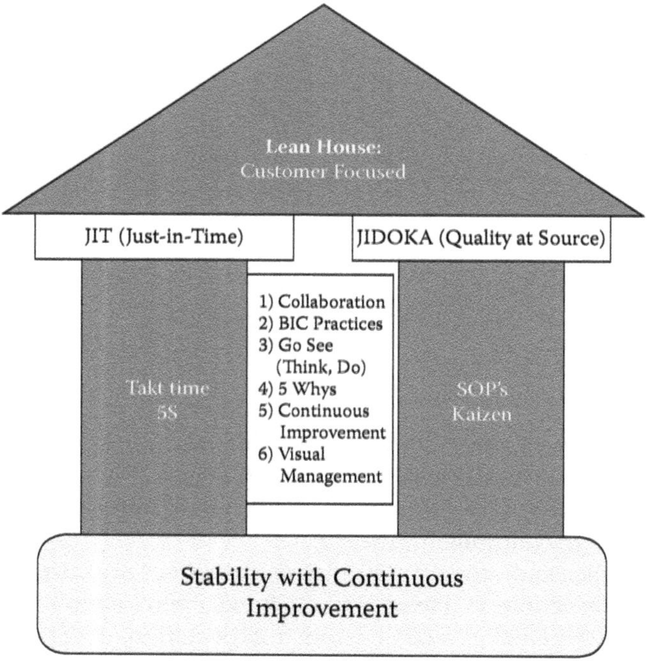

FIGURE 7.6
Lean house.

Lean Action Plan

An action plan to achieve lean requires short- and long-term strategies. The most important step of a lean effort is getting started. A change agent is required: upper management that can lead the business and change the culture. The change agent does not necessarily have to be the highest executive, but support must come from high level staff members. Gaining knowledge is the next short-term step of lean. The change agent must learn the processes, obtain the resources, and be willing to teach. After knowledge is found, value stream mapping of the process or area for improvement should be performed. Visible activity leads to improvements. If a company is using lean for the first time, it is advisable to start with a process that performs very poorly so that minimal changes can produce large improvements. The lean effort can then advance to improving more complex processes. Immediate feedback is required after lean techniques are put in place. Changes should be seen and the workplace should have a sense of what change steps worked and did not work. Lean should start small, then expand. The scope can be expanded by simple benchmarking. Finding opportunities where the lessons learned from the first success can be easily implemented into another process or project is the easiest way to move forward. After benchmarking, more complicated processes and "thinking outside the box" can be pursued.

The long-term approach is to create an organization that automatically thinks lean and includes lean activities in every step without having to be told what to do. Reorganizing a business by product family and value streams can be part of a long-term approach. Certain functions of personnel can be combined and activities should be put in units that allow easy access. Upstream and downstream processes are dependent and should complement each other.

After a flow production set-up is implemented, overstaffing and excess movements become obvious and must be handled. A growth strategy for extra people must be devised to achieve buy-in from the entire organization. Employees must be assured that they will not lose their jobs if they are to be empowered to follow lean principles. Negative reactions can be eliminated if clear responsibilities and visions are communicated. Negativity on the part of upper management must be eliminated to ensure buy-in from the rest of the business. Finally, continuous improvement should be an ongoing effort. Lean thinking should become a way of life. The change may take 6 to 12 months, and the transformation will be sustaining. Transparency via visual management and communication must be apparent for all parts of the business. After the transformation is complete, top-down leadership will be transformed to bottom-up initiatives that empower employees to find improvement opportunities to benefit themselves and the business. The time frame for a lean project is:

Getting started: 1 to 6 months

Creating a new organization or restructuring: 6 to 24 months

Implementing lean techniques and systems and pursuing continuous improvement: 2 to 4 years

Completing transformation: Up to 5 years

In conclusion, lean practices increase quality at all production levels by eliminating waste, finding value-added activities, engaging all employees, increasing throughput, giving customers what they want and expect, and continuously improving operations. Lean manufacturing increases margins and profitability, reduces non-value-added activities and waste, and leads to a successful and proactive corporation with satisfied employees. The goal is to begin with a top-down management structure (Figure 7.7) and achieve a bottom-up approach that allows employees to be engaged and proactive enough to think about change and act as change agents while focusing on continuous improvement.

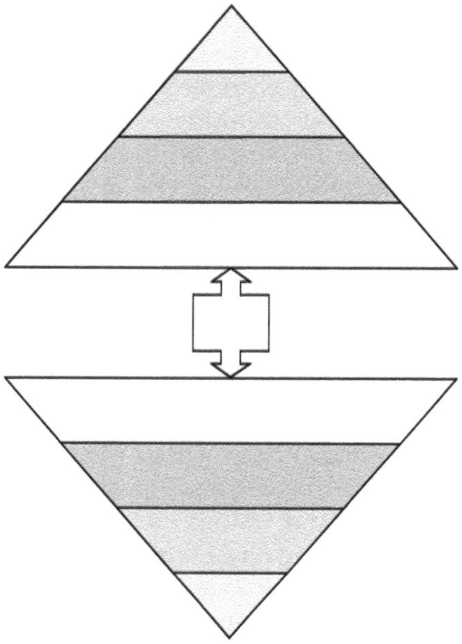

FIGURE 7.7
Top-down approach to bottom-up.

8

Six Sigma and Statistical Modeling

"A small leak can sink a great ship."

Benjamin Franklin

Without care, a leak may ultimately flood an entire process. As Benjamin Franklin notes, even a small leak in a process can mushroom into a big gap that can lead to project failure. Taking care of a process requires understanding it fully. Statistical modeling is one good way to achieve a better understanding of a process that can then lead to implementation of improvement techniques such as Six Sigma. Statistics involves tabulating, depicting, and describing data sets. It constitutes a body of formal techniques that attempt to infer the properties of a large collection of data from inspection of a sample from the collection. The basics of statistics involve measures of central tendency and variation. The tools include SIPOC (suppliers, inputs, processes, outputs, customers), key process inputs and outputs, value stream mapping, process maps, cause and effect (C&E) matrices, FMEA, continuous measurement system analysis, attribute measurement system analysis, and process capabilities. Variation measures are central tendency, shape, and spread.

Six Sigma and Statistics

Six Sigma is a quality improvement technique that strives for near perfection in a process by reducing variation in outputs. The process can be a production system or a service facility. Six Sigma is a rigorous, data-driven approach for eliminating defects. The statistical representation of Six Sigma quantitatively reveals how a process is performing. To achieve Six Sigma, a process must produce fewer than 3.4 defects per million opportunities. A Six Sigma defect is any event outside customer specifications. A Six Sigma opportunity is the total number of chances for a defect.

Central tendency in statistics is the way quantitative data clusters around a particular value. The mean, median, and mode are the most common forms of central tendency. The measures of spread are the range, variance, and standard deviations. The mean is the sum of all measurements divided by the number of observations in the data set. The median is the middle value that separates the higher half from the lower half of the data set. The mode is the most frequent value in the data set. The concept of shape relates to probability distribution, usually in the context of finding an appropriate distribution to use to model the statistical properties of a population based on a sample from the population. The shape of a distribution is measured by functions of skewness and kurtosis.

Skewness measures the asymmetry of the probability distribution of a random variable. Skewness can be positive, negative, or even undefined. Qualitatively, a negative skew indicates that the tail on the left side of the probability density function is longer than the one on the right and most of the values (including the median) lie to the right of the mean. A positive skew indicates that the tail on the right side is longer than that on the left side, and most values lie to the left of the mean. A zero value indicates a relatively even distribution on both sides of the mean, typically (but not necessarily) implying a symmetric distribution. Kurtosis (from the Greek word κυρτόσ, kyrtos or kurtos, meaning bulging) measures the "peakedness" of the probability distribution of a random variable.

Common cause and special cause are the two main types of variations. Common cause variation is characterized by a normal or predictable pattern of variation over time. Detecting patterns over time creates expectations. Special cause variation arises from changes over time that are not expected, for example, unstable data outside control limits. Special cause variation results from actions due to natural, inherent, or specific causes. The two main types of data important for statistical techniques are attribute (qualitative) and variable (quantitative). Attribute data is a yes-or-no (pass or fail; good or bad) situation, for example, it indicates whether a product is defective or not defective. Attribute data is also known as discrete measures.

Variable data is classified as discrete or continuous. Discrete data utilizes numbers, for example, the number of apples on a tree. Continuous data *continues* to refine values such as temperature, pressure, or velocity. Continuous data is a more efficient measure and paints a bigger picture. These data analysis measures can be applied to an infinitely divisible scale. The selection of a proper statistical technique depends on the type of data available. Figure 8.1 shows selection quadrants.

Selecting Proper Statistical Techniques

A chi-square test (also called the chi squared test or χ^2 test) is any statistical hypothesis in which the sampling of a test statistic reveals a chi-square distribution when the null hypothesis is true or asymptotically true—the

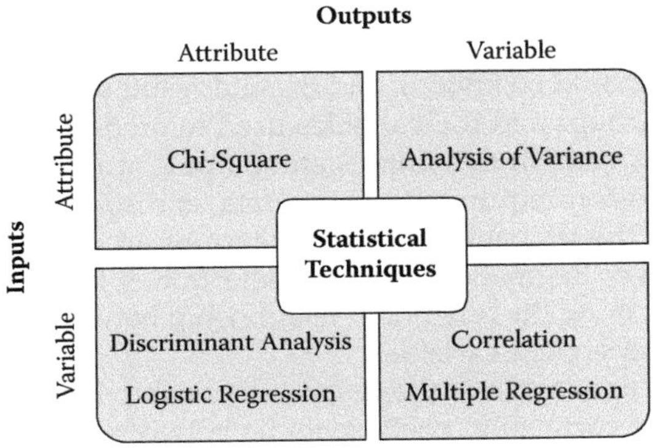

FIGURE 8.1
Selecting proper statistical techniques based on data.

sampling distribution (if the null hypothesis is true) can be made to approximate a chi-square distribution as closely as desired by making the sample size large enough. The null hypothesis can only be rejected if the probability that the findings are due to chance is 0.05 or lower. Chi-square should be used when the inputs and outputs are variable.

The chi-square is typically used to test whether the proportion of Y values meeting a certain criterion are equal across all levels of X. Stated another way, the chi-square can determine whether a Y variable is or is not dependent on a certain X variable. Each observation in a chi-square test must fall in one and only one category and must be independent of every other observation. The data for the dependent Y variable must be in count or frequency form or must be converted to whole numbers. The expected frequency for every X must be greater than 5.

Analysis of variance (ANOVA) is used when the inputs are attribute data and the outputs are variable data. ANOVA is a collection of statistics and associated models in which the observed variance of a specific variable is segregated into components of different sources of variation. ANOVA tests whether the means of several groups are equal, and then generalizes t-tests into more than two groups. ANOVA can be more beneficial than a two-sample t-test because multiple two-sample t-tests would increase the changes for Type I errors (discussed later in this chapter). ANOVA is useful for comparing two, three, or more means. ANOVA assumes all approaches take a linear model. If P is less than 5% in an ANOVA, then at least one group mean is different. The test statistic is the F test that shows the signal-TO-noise ratio. The F result is close to 1.00 when group means are similar.

Discriminant analysis is a statistical method that explains the relationship of a dependent variable and one or more independent variables. The dependent variable is explained or predicted from the values of the independent variables. Discriminant analysis is similar to regression analysis and ANOVA. The key difference between discriminant analysis and the other methods is the dependency variable. The main utility of discriminant analysis is predicting groups on linear combinations of interval variables. Both groups and interval values should be known, and the end result should predict relationships between groups and the variables used to predict them. Discriminant analysis requires variable data inputs and outputs attribute data.

Logistic regression requires the same data as discriminant analysis. It is used to predict the probability of occurrence of an event by fitting it into a logistical curve. Regression analysis determines the best mathematical expression describing the functional relationship between one response and one or more independent variables.

Correlation is a measure of the linear relationship between two data sets of variables. The correlation coefficients lie between –1 and +1. A –1 value indicates a negative or inverse dependence; 0 shows complete independence; and +1 denotes complete positive or direct dependence. Correlation coefficients between 0.8 and –0.8 are significant. Correlation coefficients between –0.8 and 0.8 are not significant. The greater the sample size, the smaller the correlation value deemed statistically significant.

Regression analysis can investigate and model the relationships of variables. Simple linear regression relates one continuous Y with one continuous X. Multiple linear regression relates one continuous Y with multiple continuous Xs. The models must also be linear. A decision tree (Figure 8.2) should be used for hypothesis testing. In general, a low P value equals a difference or correlation.

Six Sigma Methodology

Six Sigma provides tools to improve the capabilities of business processes while reducing variations. It leads to defect reduction and improved profits and quality. Six Sigma is a universal scale that compares business processes based on their limits to meet specific quality limits. The system measures defects per million opportunities (DPMOs). The Six Sigma name is based on a limit of 3.4 defects per million opportunities.

Figure 8.3 shows a normal distribution that underlies the statistical assumptions of the Six Sigma model. The Greek letter σ (sigma) marks the distance on the horizontal axis between the mean μ and the curve inflection point. The greater this distance, the greater is the spread of values encountered. The figure shows a mean of 0 and a standard deviation of 1, that is, $\mu = 0$ and $\sigma = 1$. The plot also illustrates the areas under the normal curve within different ranges around the mean. The upper and lower specification limits (USL and LSL) are $\pm 3\ \sigma$ from the mean or within a six-sigma spread. Because of the properties of the normal distribution, values lying as far away as $\pm 6\ \sigma$ from the mean are rare because most data points (99.73%) are within $\pm 3\ \sigma$ from the mean except for processes that are seriously out of control.

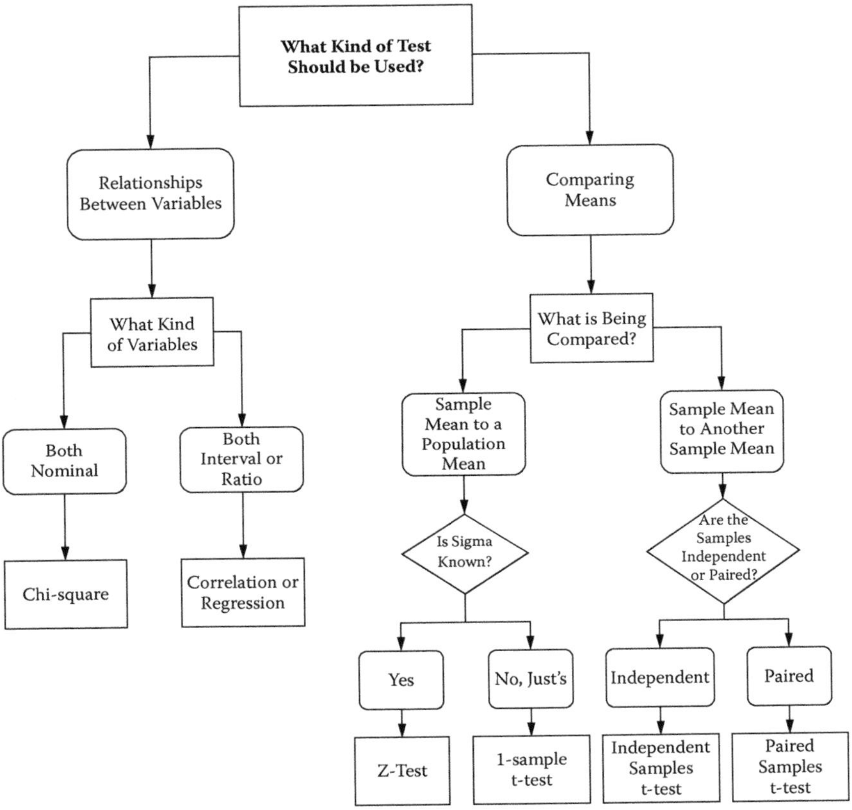

FIGURE 8.2
Decision tree for selecting proper statistical techniques.

Six Sigma Spread

Six Sigma allows no more than 3.4 defects per million parts manufactured or 3.4 errors per million activities in a service operation. To appreciate the effect of Six Sigma, consider a process that is 99% perfect (10,000 defects per million parts). Six Sigma requires the process to be 99.99966% perfect to produce only 3.4 defects per million, that is 3.4/1,000,000 = 0.0000034 = 0.00034%. That means that the area under the normal curve within ±6 σ is 99.99966% with a defect area of 0.00034%. Six Sigma pushes the limit of perfection! Table 8.1 depicts long-term DPMO values that correspond to short-term sigma levels. Six Sigma involves a five-point sequence known as DMAIC (design, measure, analyze, improve, control). The DMAIC roadmap outline is:

Define: Initiate the project, define the processes, and determine key customer requirements.

Measure: Understand the process, evaluate risks from process inputs, develop and evaluate measurement systems, and measure current process performance.

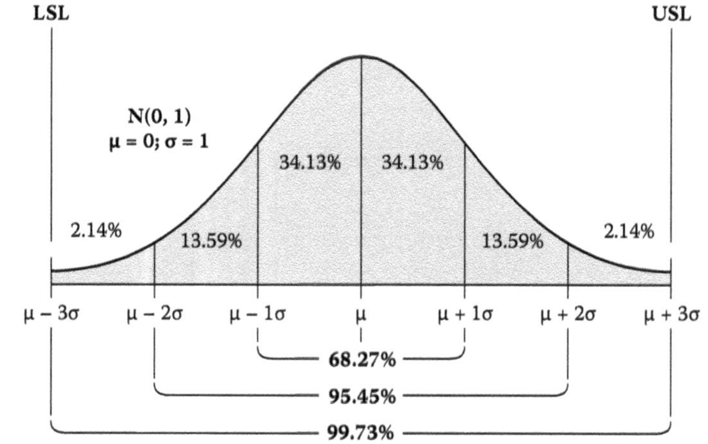

FIGURE 8.3
Areas under the normal curve.

TABLE 8.1
Defects per Million Opportunities and Sigma Levels

Sigma Spread	DPMO	Percent Defective	Percent Yield	Short-Term C_{pk}	Long-Term C_{pk}
1	691,462.00	69	31	0.33	–0.17
2	308,538.00	31	69	0.67	0.17
3	66,807.00	7	93.3	1	0.5
4	6,210.00	0.62	99.38	1.33	0.83
5	233.00	0.02	99.98	1.67	1.17
6	3.40	0	100	2	1.5

Analyze: Examine data to prioritize key input variables, identify waste from overproduction, waiting, transport, over-processing, excess inventory, movement, defects, and rework.

Improve: Verify critical inputs using planned experiments, design improvements, pilot new processes.

Control: Finalize control systems and verify long-term capability.

It is important to begin by understanding purpose, importance, focus, deliverables, measures, and resources in relation to DMAIC. A project charter *defines* the scope and project team. Selecting a team that has knowledge of the process and expertise and includes cross-functional members is vital. A process map should always be an early step to determine key inputs and outputs and review suppliers and customers to see where variations occur. Understanding what customers want and need is essential. Note that the customer is not always the end customer; a customer could be an operator on a production line.

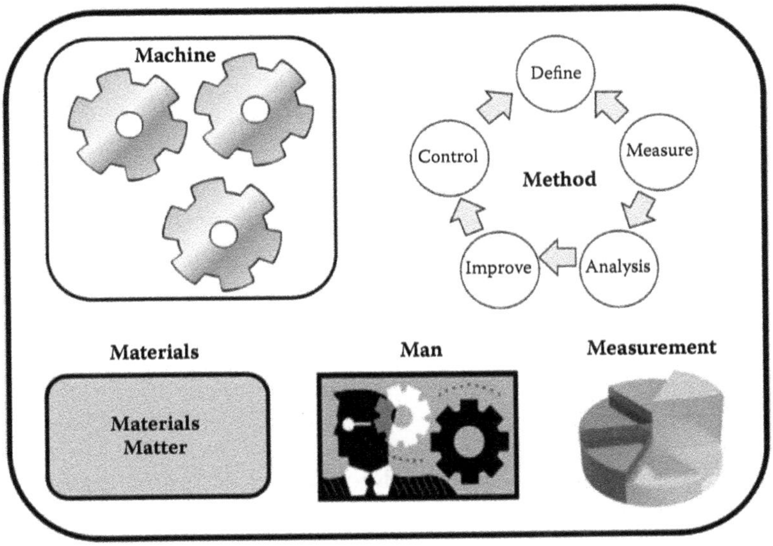

FIGURE 8.4
Five Ms (machine, method, materials, man, and measurement).

The next step is to *measure* current process performance. Developing the proper measurement systems will help the team to understand the process and find opportunities. Understanding the causes and effects of the key inputs on the outputs helps prioritize initiatives. The best next step is to perform a failure modes and effects analysis (FMEA) to identify risk factors and help formulate process controls. This is a key point for establishing action items, assigning key dates, and holding employees accountable for certain tasks. Other key areas in the measure phase are reviewing or preparing standard operating procedures (SOPs). Do they exist? Should they be updated? Are they followed? Do operators need retraining?

Understanding and *analyzing* the data measures are key steps. The *analyze* phase may lead the team to return to the *measure* phase to gather more data. Defining and analyzing machine, method, materials, man, measurement (the 5 Ms; Figure 8.4) at this point will reveal over-production, waiting, transport, over-processing, inventory, movement, defect and rework wastes (the 7 Ws; Figure 8.5).

5 Ms Lead to 7 Ws

The seven wastes in manufacturing are listed above. Improving processes can proceed after data gathering and analysis. Determining where to find "the biggest bang for the buck" is the main factor in determining improvements to be implemented. Severity is an important effect of safety and other risk failures. The frequency with which a specific cause occurs and creates

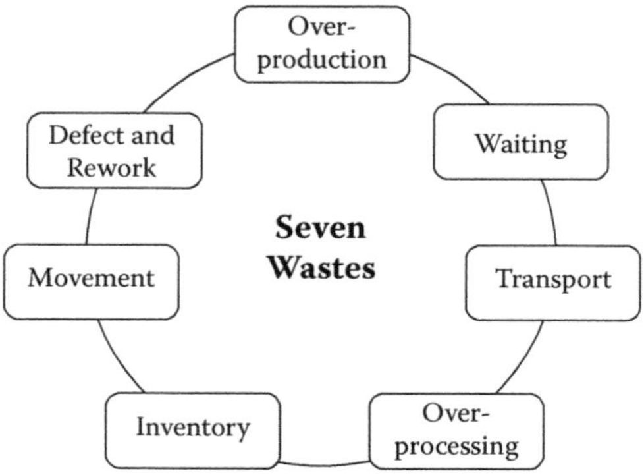

FIGURE 8.5
Seven Ws (wastes).

failure modes also should be considered when planning improvements. Finally, detection or the ability of current controls to detect or prevent failures should be noted. Poka yoke (mistake-proofing) is a simple method for preventing defects from business operations. Poka yokes are strategies used to detect or prevent defects. The goal of improving is to change processes, products, and equipment to solve the problems defined, measured, and analyzed. The project charter detailing goals should be prepared at this point.

Finally, controlling improvements is essential to sustaining Six Sigma to achieve successful results. Verifying long-term capability is important to prevent the "old way of doing things" from continuing after improvements are made. A control plan needs to cover process improvements, process steps, inputs, outputs, process specifications, P_{pk} data, measurement techniques, percent R&R (product tolerances), sample sizes, sample frequencies for continuous testing, control methods, and reaction plans. Other components of a control plan are lists of accountable personnel and key completion dates. The following tools should be used with Six Sigma and will be explained in detail below:

Histogram
Flow chart
Process mapping
SIPOC
Voice of customer
Cause and effect (C&E) diagram
FMEA
Graphical analysis
Pareto diagram
Hypothesis testing
T testing

Confidence interval
Capability analysis
Gage R&R
Control plan

Histogram—As shown in Figure 8.6, a histogram plots the frequency of values grouped together as a bar graph. Histograms are handy for determining location, spread, and shape. Outliers can easily be identified. The height equals the frequency and the width equals a range of values. A histogram with a bell-shaped curve is normal.

Flow chart—A flow chart uses arrows to indicate components of a process using arrows (Figure 8.7).

Process mapping—A process map begins with a flow chart and adds details such as process inputs (Xs) and process outputs (Ys). The technique can identify systems needing measurement studies, output variables for capability studies, missing elements, and non-value-added steps. Process mapping of each critical process should be performed at the beginning of an improvement project. It should note major tasks, subprocesses, process boundaries, and X and Y factors. Main process mapping begins with brainstorming, a listing of tasks to be completed, work instructions, and fishbone diagrams.

SIPOC—The acronym stands for suppliers, inputs, processes, outputs, and customers. It is a common tool used in process mapping (Table 8.2). The first step is defining the process in basic terms. This is followed by identifying external suppliers (inputs) and customer requirements (outputs). The list of suppliers should also name customers. Remember, the customer is not always the end customer and may be an employee such as a key operator on a production line.

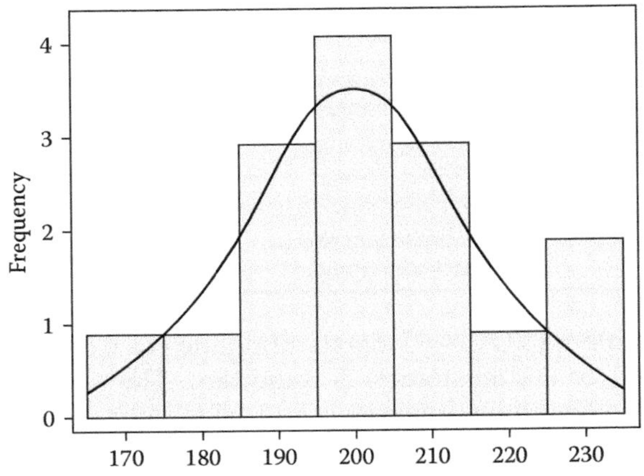

FIGURE 8.6
Histogram of data set with normal curve.

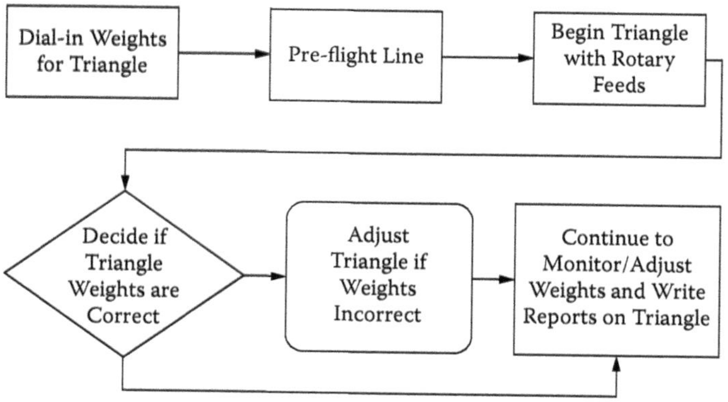

FIGURE 8.7
Process flow chart.

TABLE 8.2
SIPOC Example

Supplier	Inputs	Processes	Outputs	Customers
XYZ Chicken	Line operators	Dial-in weights for triangle	Tray with qualifying amount of product	Touch-up personnel
Ravioli suppliers	Product for line	Preflight line	Tray with missing product	Heat seal operator
Other meat suppliers	Triangle interface	Start triangle	Tray with insufficient product	Line operators
	Touch up personnel	Adjust triangle if weights are incorrect	Tray with too much product	Grocers
	Triangle operators	Monitor triangle and write reports		
		Bring pallets of meat to line		
		Open boxes and supply triangle		
		Break down empty boxes		
		Supply product to line		
		Empty frozen rejects back into triangle		

Voice of customer (VOC)—Understanding the voice of the customer (process owner and/or end customer) is important. The technique consists of setting up 30- to 45-minute interviews and asking questions. It may be beneficial to ask participants to bring key documents, flow charts, forms, lists, or procedures relevant to the interview. A detailed process map (Figure 8.8) should include SIPOC, VOC inputs and outputs, a cross functional map, and a subprocess map.

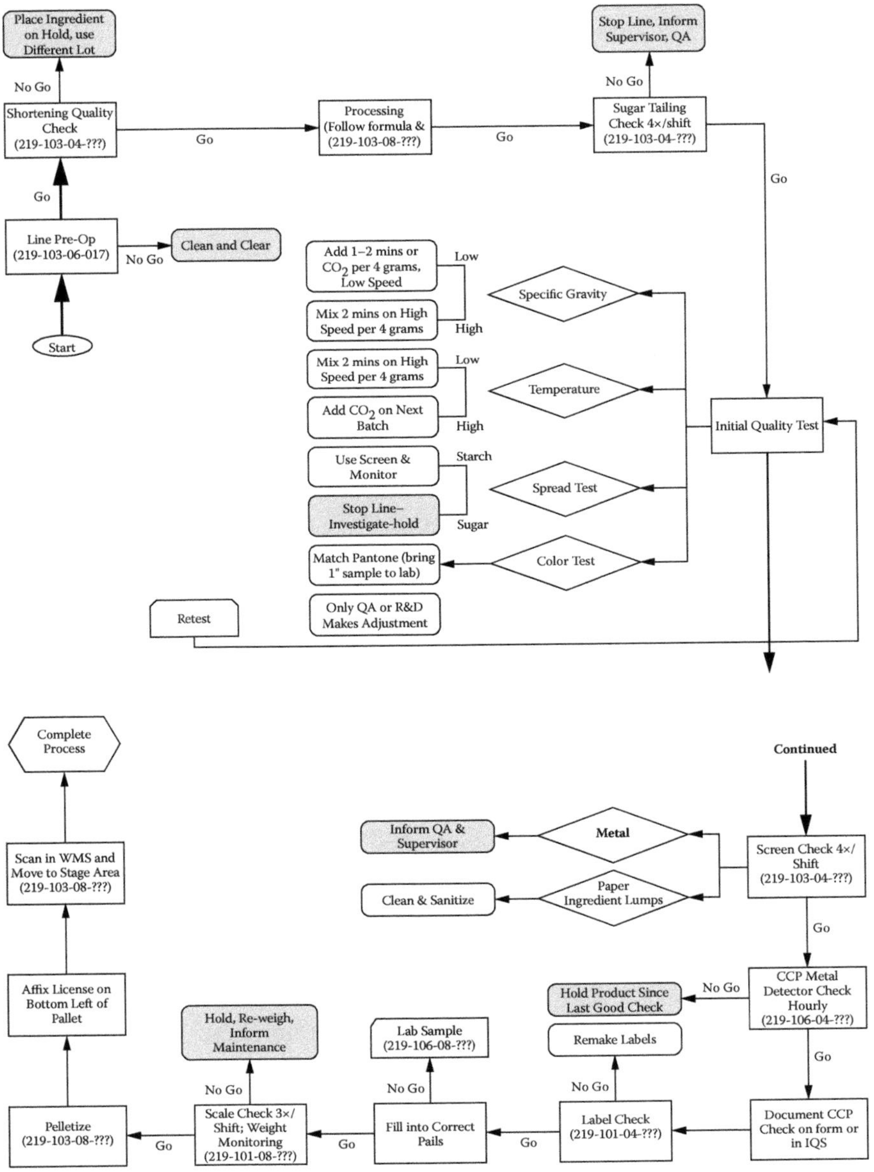

FIGURE 8.8
Process map (two parts).

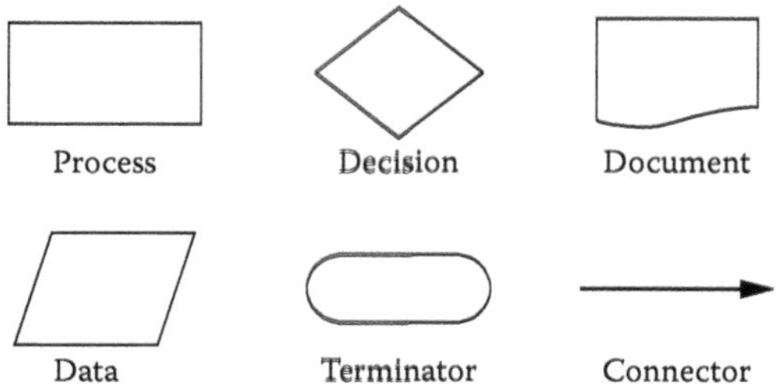

FIGURE 8.9
Process map symbols.

Value added time should be captured as a percentage of total time. A process map should include a flow chart showing process steps without major details; it should show decision points. Cross functional maps can utilize swim lanes to rearrange flows into different functions to show hand-offs and timing. Cross resource maps drill down further and illustrate resource-to-resource hand-offs. The main process map symbols are shown in Figure 8.9. Colors can help clarify tasks. Green represents movement, red indicates a queue or stop; yellow represents a delay.

A RACI matrix can be developed within a process map to indicate responsibility and authority levels. R = indicates the individual with primary *responsibility* and accountability for the task or project. A = the person or group with the *authority* to say a process is completed. C = the person who should be *consulted* about actions to be taken. I = those who must be informed about activities or actions. A process map may be modified if more information is revealed during the *measure* phase.

Cause and effect (C&E) matrix—A C&E matrix is a key tool for understanding customer requirements. It is a brainstorming tool developed to identify potential causes of a problem and inputs in the *measure* phase. A C&E matrix is sometimes called an Ishikawa or fishbone diagram (Ishikawa devised the diagram that resembles the head and bones of a fish). The Critical-to-quality measure is placed at the head of the fish. The bones represent the categories. The steps for creating the matrix are:

List the Ys across the top. See Figure 8.10 as a reference.

Clarify the team's assignment of significance ratings to the Ys and Xs.

Assign a rank for each Y using a scale of 1 to 10 where 1 denotes the lowest importance and 10 is the most significant. If all the Ys have the same rating, rate them all with the same arbitrary number.

List in the first column all the potential Xs that may affect any of the Ys. The Xs should be taken from the C&E diagrams. Ignore overlaps; each X should be listed only once. Rate the correlations of Xs to Ys on the following scale:

Cause and Effect Matrix

		Rating of Importance to Customer	10	9	9	7	7	6										
			1	2	3	4	5	6	7	8	9	10	11	12	13	14	15	
			Missed Trays	Too Little Product on Tray	Too Much Product on Tray	Product Bouncing out of Tray	Improper Placement	Scrap Caused by Machine										Total
	Process Step	**Process Input**																
	Monitor Triangle and Take Reports	Triangle Operator	9	9	9	1	9	3										340
1	Dial in Weights for Triangle	Triangle Operator	9	9	9	1	3	3										298
	Adjust Triangle if Weights are Incorrect	Triangle Operator	9	9	9	1	3	3										298
	Start Triangle	Triangle Operator	3	3	3	1	3	1										118
	Open Boxes and Supply Triangle	Supply NWF Operator	3	3	3	0	0	0										84
	Supply Line with Product	Supply Line	3	3	3	0	0	0										84
	Empty Triangle Frozen Rejects Back into Triangle	Supply NWF Operator	3	1	1	1	1	1										68
	Pre-Flight Line	Maintenance	3	0	0	0	3	0										51
2	Bring Pallets of Meat to Line	Supply Line	1	1	1	0	0	0										28
	Break Down Empty Boxes Supplied to Triangle	Supply NWF Operator	0	0	0	0	0	0										0
Total			430	342	342	35	154	66	0	0	0	0	0	0	0	0	0	
		Lower Spec																
		Target																
		Upper Spec																

FIGURE 8.10
Cause and effect matrix.

0 = no effect or correlation

1 = small effect or weak correlation

3 = medium effect or correlation

9 = strong effect or correlation

Multiply each rating by the weight and sum across the row, noting the result in the last column.

The Xs with the highest totals are the most significant. A C&E matrix ensures that the Xs and Ys of a project are in line with the benefits it should produce.

Failure mode and effects analysis (FMEA)—This technique systematically identifies and prevents product and process problems before they occur. FMEA can prevent defects, enhance safety, and increase customer satisfaction. FMEA reveals all the ways a process or product can fail. It is not limited to product problems such as user mistakes. The objectives are to identify sources of risk, create links to mapping and C&E matrices, and measure process accuracy. An FMEA is a structured approach that reveals potential failures. It prioritizes actions that should have been taken to reduce risk, formulates defect categories for performance measurement, generates a process control plan, and can help evaluate the design of a process or service.

FMEA is intended to identify sources of failure and eliminate them. It can also analyze new processes, identify deficiencies in a process control plan, establish action priorities, evaluate risks from process changes, identify potential variables in process improvement, guide development of new processes, and determine when to start improvements. Certain activities should precede FMEA: (1) process mapping; (2) preparation of C&E matrix; and (3) first pass assessment of control plans. FMEA inputs are process maps, the C&E matrix, and SOPs. The outputs are an initial list of defects to be measured, a prioritized list of actions, and an initial process control plan. Understanding where risks arise is the key to a successful FMEA. An FMEA should be conducted by a cross functional team to prevent data bias. An FMEA should answer the following questions:

What are the inputs?

What can go wrong with the inputs?

What are the effects on the inputs?

What are the effects on the outputs?

How serious are the effects on the outputs?

What are the causes?

How often do they occur?

How can they be prevented?

How well can they be prevented?

What can be done?

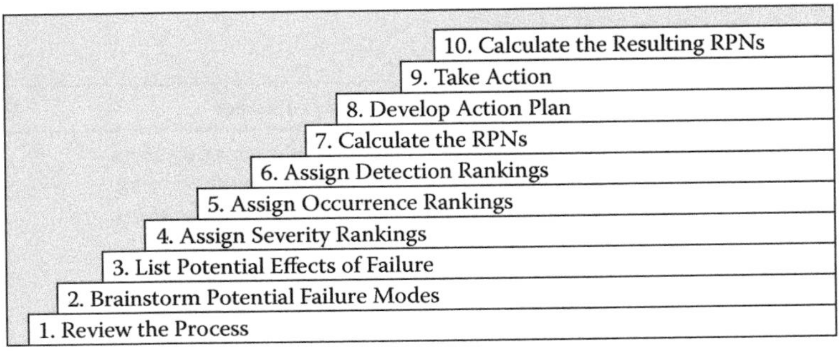

FIGURE 8.11
Steps of FMEA.

TABLE 8.3
FMEA Detection Matrix

Detection	Criterion: Likelihood a Defect Will Be Detected by Test Before Product Advances to Next Process	Ranking
Almost impossible	Test content detects < 80% of failures	10
Very remote	Test content must detect 80% of failures	9
Remote	Test content must detect 82.5% of failures	8
Very low	Test content must detect 85% of failures	7
Low	Test content must detect 87.5% of failures	6
Moderate	Test content must detect 90% of failures	5
Moderately high	Test content must detect 92.5% of failures	4
High	Test content must detect 95% of failures	3
Very high	Test content must detect 97.5% of failures	2
Almost certain	Test content must detect 99.5% of failures	1

Figure 8.11 depicts the 10 basic steps of an FMEA. Table 8.3 is an FMEA detection matrix.

Risk failures involve three factors: (1) severity—the consequence of a failure; (2) occurrence—the probability or frequency of a failure; and (3) detection—the probability that a failure will be detected before its impact is realized. A risk priority number (RPN) is calculated by multiplying severity, occurrence, and detection: RPN = severity × occurrence × detection. The failure modes with the highest RPNs should be handled first. Special attention should be given to items with high severities.

The failure mode indicates how a specific process input fails. The effect is the impact on internal and external customer requirements, but the focus should be on external customers. Severity is the impact of the effect on customer requirements and may involve safety and other risks. The failure mode scale ranges from 1 (not severe) to 10 (very severe); items

TABLE 8.4

FMEA Criteria for Severity Ranking

Effect	Criterion: Severity of Effect	Ranking
Hazardous without warning	Very high ranking when potential failure mode affects safety and involves noncompliance without warning	10
Hazardous with warning	Very high ranking when potential failure mode affects safety and involves noncompliance with warning	9
Very high	Process not operable; loses its primary function	8
High	Process operable but with reduced functionality and unhappy customer	7
Moderate	Process operable but difficult; customer is uncomfortable	6
Low	Process operable but uncomfortable; reduced level of performance; customer is dissatisfied	5
Very low	Process not in 100% compliance; most customers can notice defect	4
Minor	Process not in 100% compliance; some customers can notice defect	3
Very minor	Process not in 100% compliance; few customers can notice defect	2
None	No effect	1

TABLE 8.5

FMEA Criteria for Probability of Failure

Probability of Failure	Possible Failure Rate	Ranking
Almost inevitable	≥1 in 2	10
	1 in 3	9
High: repeated failures	1 in 8	8
	1 in 20	7
Moderate: occasional failures	1 in 80	6
	1 in 400	5
	1 in 2000	4
Low: very few failures	1 in 15,000	3
	1 in 150,000	2
Remote: failure unlikely	≥1 in 1,500,000	1

scoring 0 should not be included. Table 8.4 lists the criteria for severity in an FMEA. Occurrence is the frequency with which a cause occurs and creates a failure mode. Frequency is rated from 1 (not likely) to 10 (very likely). Table 8.5 summarizes the criteria for probability of failure in an FMEA. Detection is the ability of current control schemes to detect or prevent the causes before creating a failure mode or detecting failure modes before they cause effects.

Potential
Failure Mode and Effects Analysis
(Process FMEA)

FMEA Number
Page 1 of 1
Prepared by: __________
FMEA Date (Orig.):

Core Team:

Design Responsibility
Key Date:

Process Step / Function	Potential Failure Mode	Potential Effect(s) of Failure	SEV	CLASS	Potential Causes(s)/ Mechanism(s) Of Failure	Current Process Controls Prevention	OCCUR	Current Process Controls Detection	DETEC	R.P.N.	Recommended Action(s)	Responsibility & Target Completion Date	Action Results: Actions Taken	SEV	OCC	DET	RPN

FIGURE 8.12
FMEA template.

Reviewing current SOPs to find items that need action may be beneficial at this point. Figure 8.12 presents an FMEA template. FMEAs can help teams become more comfortable with proposals to process changes by considering what might go wrong and taking steps to prevent problems.

Graphical analysis principles—The basic graphical analysis tools that demonstrate quality are dotplots, histograms, normal plots, Pareto diagrams, stratification (second level Pareto), boxplots, scatter plots, and marginal plots. Plotting data is the first step of graphical analysis. After plotting, the data can be graphed to reveal key aspects of the process. Graphical analysis shows a "big picture" of data and more effective than manipulating numbers. It provides visual explanations of data. The tools and types of data are listed in Table 8.6.

Data can be described by the behavioral features of the plotted data with multiple data points for the variable over time, across products, or on different machinery, etc. The accumulation of the data points can be viewed as a distribution of the values. The data can be represented through dot plots, histograms, and normal curves. Figure 8.13 shows a dotplot distribution. Each dot represents one "event" of output at a given value. As the dots accumulate, the nature of the distribution can be seen as a "distribution."

Histograms and related plots—Figure 8.14 shows data grouped into intervals and plotted into a histogram. The same data can be viewed with a

TABLE 8.6
Quality Tools and Types of Data

Demonstration Means	Quality Tool	Type of Data
Tabular	Frequency distribution	
Graphical	Histogram	Continuous
	Run chart	Time series
	Boxplot	Continuous and categorical
	Pareto diagram	Categorical
	Dotplot	Continuous
	Scatter plot	Two continuous variables
Numerical	Measure of central tendency (mean)	
	Measure of spread (range, variation)	

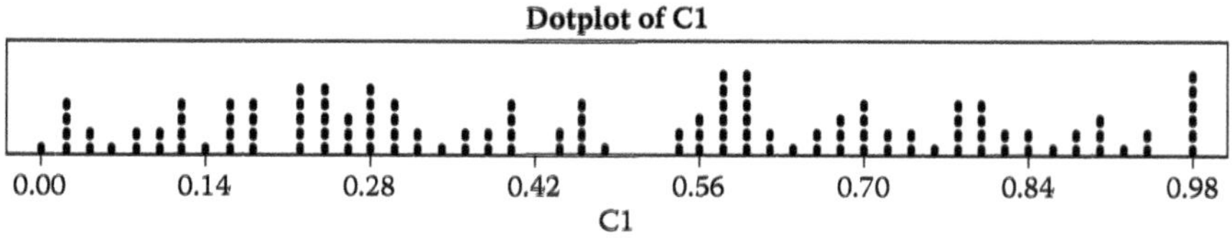

FIGURE 8.13
Dotplot.

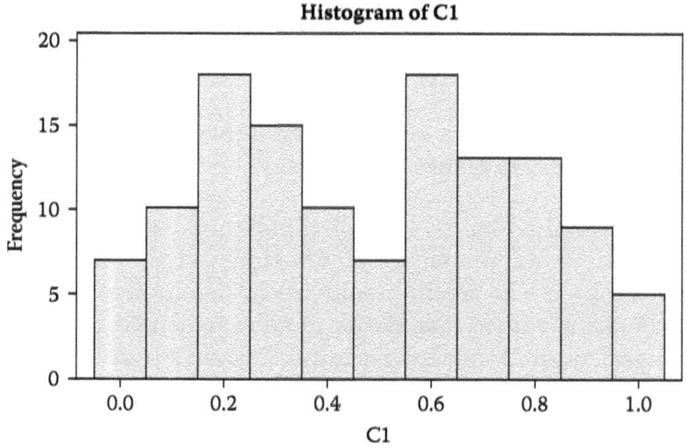

FIGURE 8.14
Histogram.

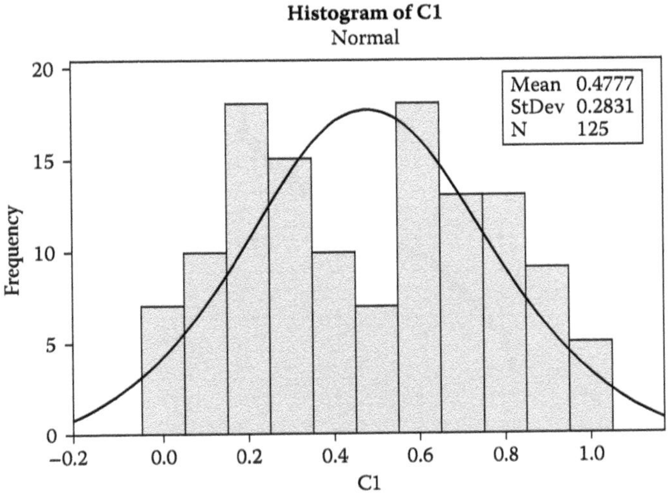

FIGURE 8.15
Histogram of data with normal curve.

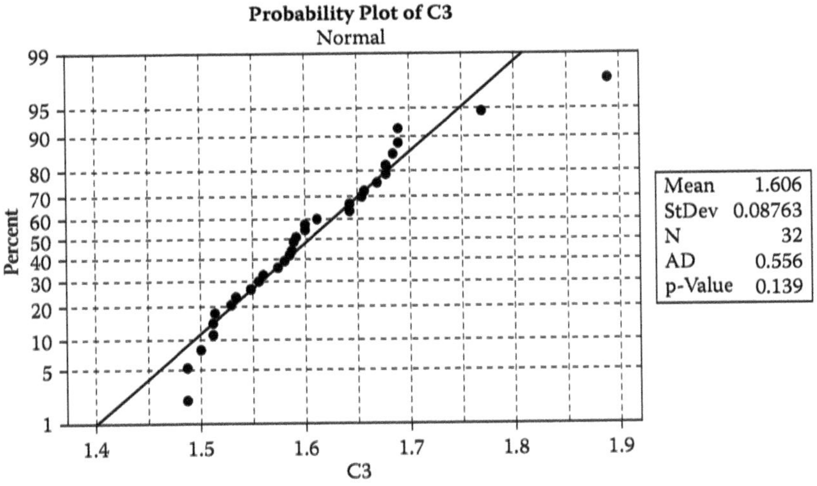

FIGURE 8.16
Probability plot of normal data.

normal distribution to see the shape (Figure 8.15). Normal probability plots can be graphed to determine the normality of the data. Normal data distribution resembles a straight line. The graph in Figure 8.16 demonstrates normal data. Figure 8.17 demonstrates non-normal data.

Data may also be plotted over time and compared to data that is not time-specific. Run charts and individual charts are plotted in this manner as

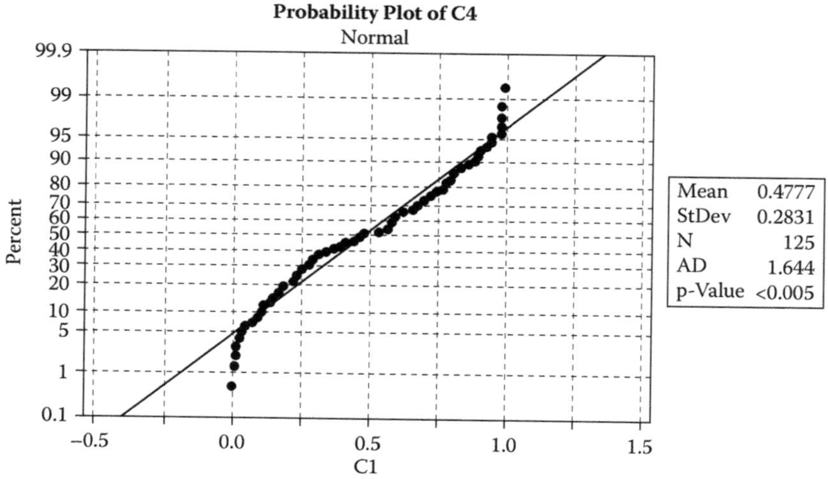

FIGURE 8.17
Probability plot of not-normal data.

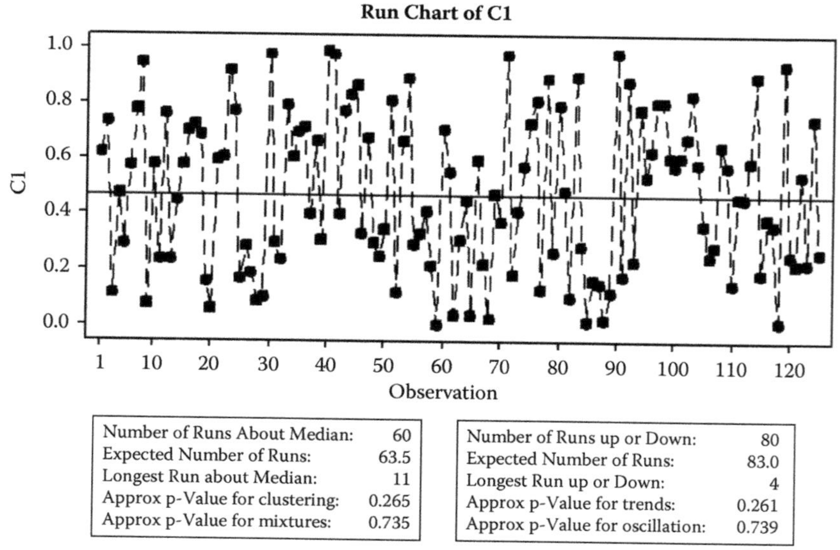

FIGURE 8.18
Run (R) chart.

shown in Figures 8.18 and 8.19. The pattern sought in a run chart is a group of consecutive points crossing the median. Any time the data crosses the median is considered a point of run. An individual chart looks like a run chart but it shows process control limits (±3s). Boxplots reveal differences in distributions, e.g., the spread and center of data (Figure 8.20).

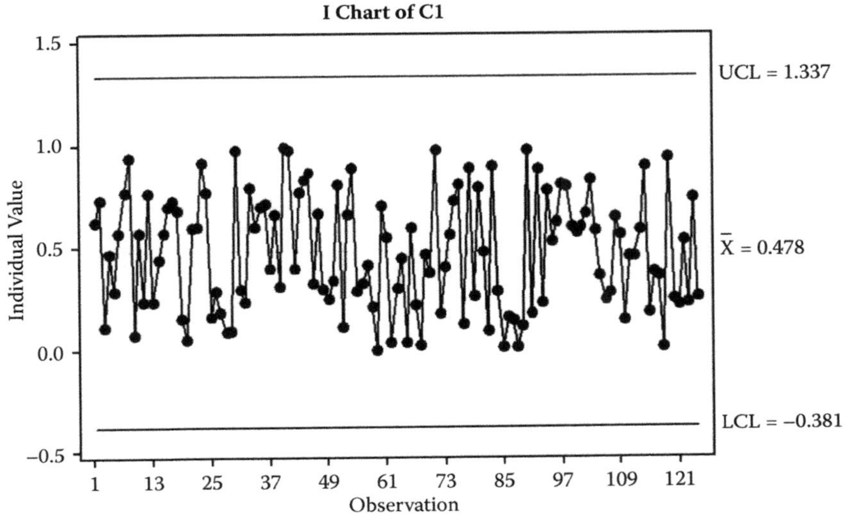

FIGURE 8.19
Individual (I) chart.

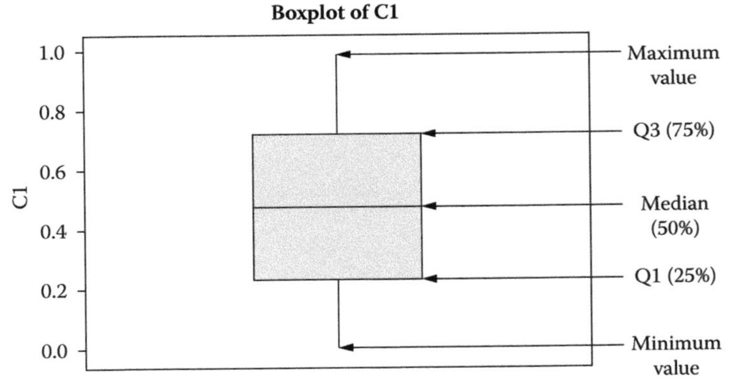

FIGURE 8.20
Boxplot.

Pareto diagram—These essential tools help prioritize improvement targets. The Pareto rule is to focus on the 20% of the problems that cause 80% of poor performance. Figure 8.21 is a Pareto plot of RPNs. Figure 8.22 is a second-level Pareto chart. The second-level charts break first-level data down to a deeper level. Note the three row tables below the bars that represent count, percent, and cumulative percentage. The cumulative percentage drives the red line spanning the graph which is based on the right axis

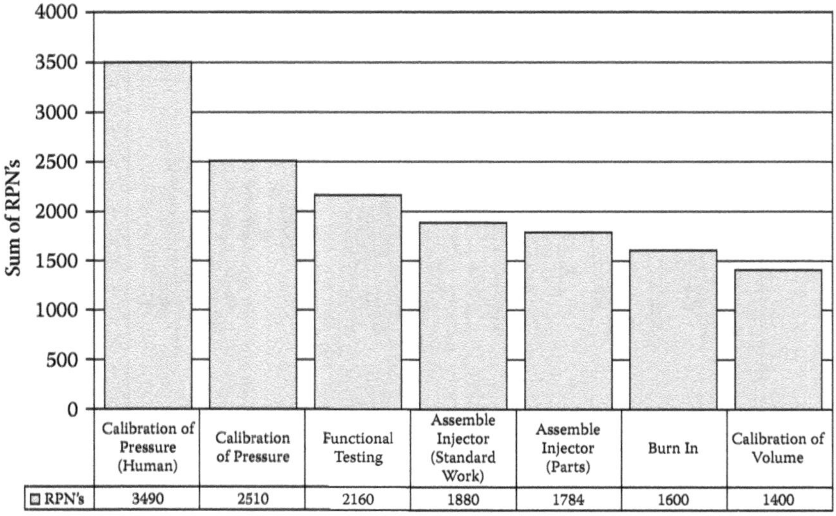

FIGURE 8.21
Pareto chart.

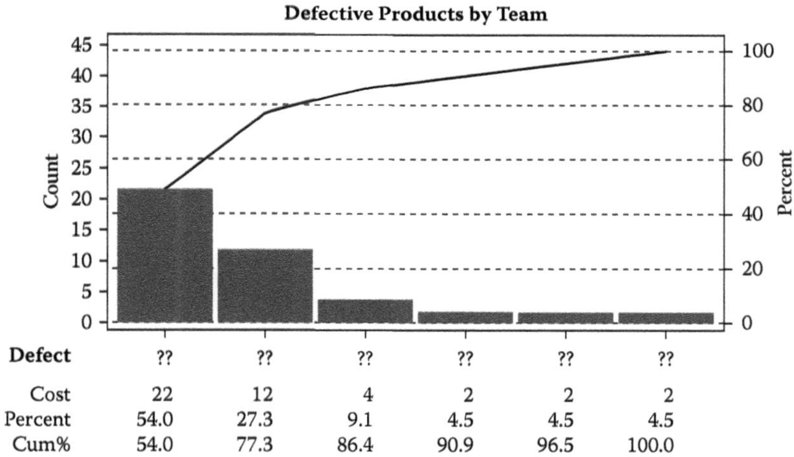

FIGURE 8.22
Second level Pareto chart.

labeled percent. The next sets of graphs reveal relationships between two variables. Scatter plots and marginal plots are two effective methods to illustrate these relationships.

Scatter plot—One variable (typically X) is placed on the horizontal axis (abscissa). The Y variable is placed on the vertical axis (ordinate). The two

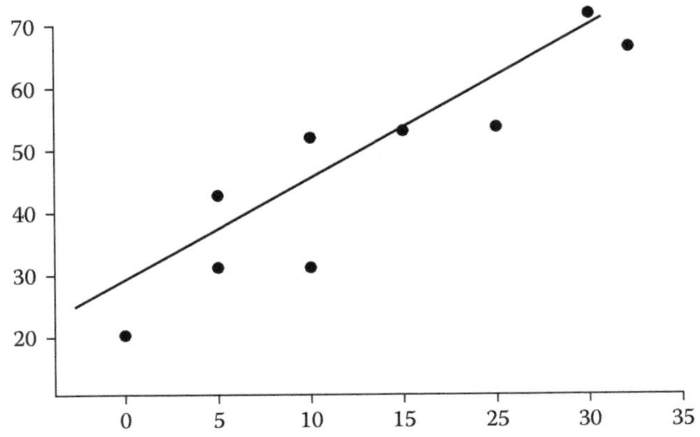

FIGURE 8.23
Scatter plot.

variables are compared to see whether a correlation is present (Figure 8.23). The line through the variable points showing an upward slope represents the direction of the relationship.

Marginal plot—Figure 8.24 shows an example. Marginal plots are combinations consisting of a scatter plot and a histogram, boxplot, or dotplot. Two situations can be covered in one graph.

In summary, data analysis involves basic steps: (1) review of data to identify abnormalities (errors or outliers); (2) analysis of data to gain a visual sense; and (3) analysis of data to gain a numerical sense. Descriptive statistics provides a way to review numerical and graphical data to understand its center, spread, and shape. The purpose of graphical analysis is to plot the data.

Hypothesis Testing

Hypothesis testing is similar to the scientific method used to determine whether a proposed explanation for an observation works in practice. Hypothesis testing uses smaller samples to estimate population parameters. Ensuring the randomness of parameters is important for hypothesis testing to ensure that inferences are correct. Hypothesis testing components are:

H_o = Null hypothesis, the general or default position, most likely stating there is no relationship between two variables.

H_1 = Alternative hypothesis, position stating that there is no difference between the two variables.

p = Probability value.

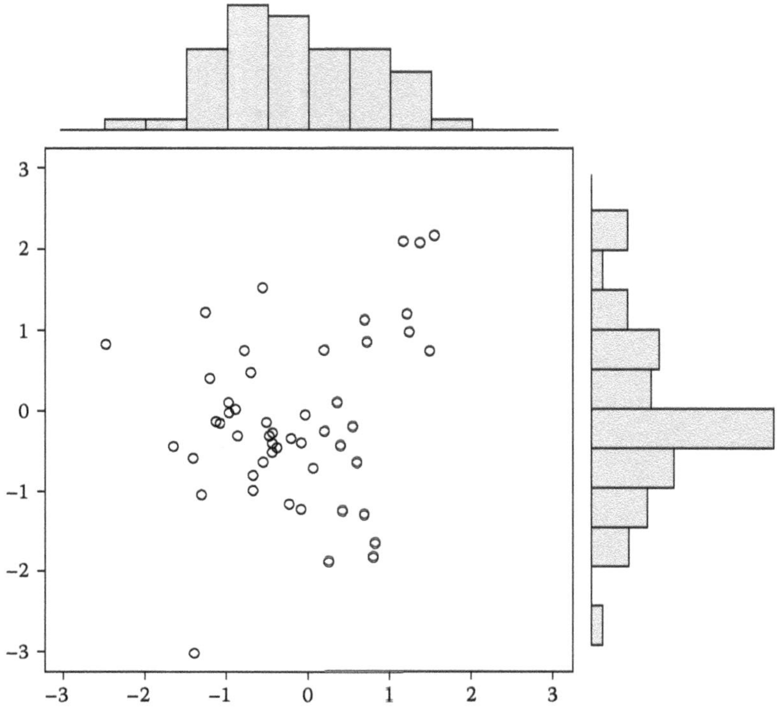

FIGURE 8.24
Marginal plot.

The following assumptions should be made for hypothesis testing:

H_o = Mean x = mean y
H_1 = Mean x ≠ mean y
H_o = Slope of the line is 0
H_1 = Slope of the line is not 0
H_o = Variance x = variance y
H_1 = Variance x ≠ variance y
H_o = Variable x is independent of variable y
H_1 = Variable x is not independent of variable y

Hypothesis testing explains an unknown It proves a hypothesis is true or false by comparing x and y (arbitrary variables). The assumption is always that the null hypothesis is true. The data proves whether the decision is accepted or rejected from the assumption. If the null hypothesis is rejected, the alternative hypothesis is accepted. A degree of confidence (confidence interval) is required to make these assumptions. Confidence

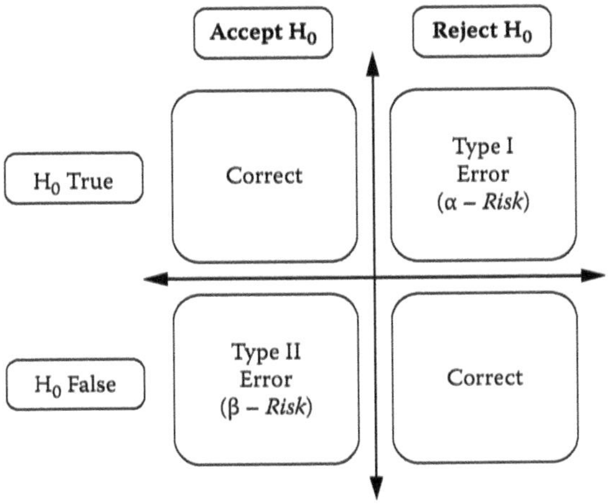

FIGURE 8.25
Key decision errors.

intervals are also known as risk levels. Key decision errors and other terms are defined below and can be seen in Figure 8.25.

Type II error—This is the biggest risk error. It indicates the change that a false negative will occur. The result will be a faulty product. A Type II error occurs when it is assumed there is no difference between the test and control factors when there is a difference.

Type I error—This error indicates the chance that a false positive will occur. It occurs when the test and control factors are assumed to be different when in fact they are the same.

Alpha risk—This is the maximum risk or probability of making a Type I error and is also called the significance level. This probability is always greater than 0, and is normally set at 0.05 or 5%. A scientist should make a risk percentage error based on positivity.

Beta risk—This is the risk or probability or making a Type II error by overlooking an effective solution to a problem. It will produce a faulty product. Beta risk must be minimized.

The *p value* is also extremely important to remember when doing hypothesis testing. If $p < 0.05$, you must reject the null hypothesis. If the null hypothesis is rejected, the data is not normal. In most cases, the confidence interval is based on 0.05 or 5%.

The *null hypothesis* (H_o) indicates no difference between x and y. It is always assumed until proven otherwise. The hypothesis is proven when test and control distributions are the same.

The *alternative hypothesis* (H_1) shows there is a difference between x and y. It is considered true if H_o is rejected or the test and control distributions are different.

Significant difference is the result of a statistical hypothesis test in which the difference is large enough to be a chance cause.

Power is the ability of a statistical test to detect something important when there is reason to expect that something important will occur.

Test statistic is a standardized value that represents the feasibility of a false positive. It is distributed in a known manner so that a probability for the observed value can be determined. The more likely the false positive, the smaller the absolute value of the test statistic, and also the greater the probability of observing the variable within its distribution.

t Testing

This type of test evaluates the importance of an effect (the difference between two means). The t values are in units of standard deviation and have associated probability values. The p value is normally set at 0.05 or 5%. If a sample size is extremely small, the p value can be set at 0.010 or 10%. The higher the t value, the more separate the two peaks of a distribution are. The t-tests normally yield three types of hypotheses: a left-tailed test, a right-tailed test, and a two-tailed test. The difference between two means is tested. The more spread between the means, the higher the t value computed by the formula below:

$$t = \frac{\bar{X}_1 - \bar{X}_2}{\sqrt{\left(\frac{(N_1 - 1)S_1^2 + (N_2 - 1)S_2^2}{N_1 + N_2 - 2}\right)\left(\frac{1}{N_1} + \frac{1}{N_2}\right)}}.$$

The following calculation applies when N_1 and N_2 are equal.

$$t = \frac{\bar{X}_1 - \bar{X}_2}{\sqrt{\left(\frac{S_1^2 + S_2^2}{N_1}\right)}}.$$

A sample t-test is used to compare data from a new test population to data from an old population of data. It is useful for comparing data populations of the same caliber. A paired t-test is used to compare paired data (e.g., products manufactured on the same machine during different shifts).

Confidence Interval

Confidence interval is an alternative to hypothesis testing. Essentially, confidence intervals are upper and lower limits about an estimate within a certain percentage of "confidence" when the true value is known. Confidence intervals are used to determine whether a null hypothesis (H_o = mean x – mean y = 0) should be accepted or rejected. The difference between the two population means is 0. If the two intervals of the two populations do not cross 0, there is confidence that there is a difference between the two. Confidence intervals are based on variability associated with estimated parameters, sample sizes, and alpha risks. They estimate the variabilities of sample statistics.

Capability Analysis

Capability analysis ensures that a process meets requirements. It can be performed with both attribute and variable types of data. The technique measures short- and long-term process capabilities. The key capability indices are C_p and C_{pk} or P_p and P_{pk} for long-term capabilities. The following formulas are used for capability indices:

$$C_p = \frac{USL - LSL}{6\hat{\sigma}}$$

$$C_pU = \frac{USL - \bar{X}}{3\hat{\sigma}}$$

$$C_pL = \frac{\bar{X} - LSL}{3\hat{\sigma}}$$

$$C_{pk} = \text{Min}(C_pU,\ C_pL).$$

The interpretations in Table 8.7 are used for capability values. The C_{pk} value should be greater than 1.5 to indicate "good" capability. C_p and C_{pk} use a pooled estimate of the standard deviation whereas P_p and P_{pk} use a long-term estimate of the standard deviation. C_{pk} denotes what a process is capable of doing if no subgroup variability is present. P_{pk} indicates actual process performance. Normally, C_{pk} is smaller than P_{pk} since P_{pk} represents both

TABLE 8.7

Interpretation of Capability Values

$\pm\sigma$	DPMO	Percent Good	C_p
3	2700	99.73	1.00
4	63	99.9937	1.33
5	0.57	99.999943	1.67
6	0.002	99.9999998	2.00

within- and between-subgroup variability. C_{pk} represents only between-subgroup variability. The steps for performing a capability study are:

Set up the process to the best parameters and identify key process input variables.

Identify subgroups

Run the product over a short time span to minimize the impact of special cause variation.

Observe the process and take notes throughout.

Measure and identify key process output variables.

Run capability analyses to review normality, statistical process control, and histograms.

Run capability analyses to determine short-term and total standard deviations.

Identify mean shifts and variations.

Estimate long-term capability.

Develop action plans based on the resulting data.

Setting short- and long-term goals based on capability analyses will result in successful action plans based on real time data. The goals of capability studies are:

Move the P_{pk} to P_p or center the process

Move the P_p to C_{pk} or reduce the variation

Move the C_{pk} to C_p or have random variation

A Six Sigma process has a C_p of 2.00 and a P_{pk} of 1.5.

Gage R&R

Gage R&R is a measurement systems analysis (MSA) technique that uses continuous data based on the principles that:

Data must be in statistical control.

Variability must be small compared to product specifications.

Discrimination should be about one-tenth of product specifications or process variations.

Possible sources of process variation are revealed by measurement systems. Repeatability and reproducibility are primary contributors to measurement errors. The total variation is equal to the real product variation plus the variation due to the measurement system. The measurement system variation is equal to the variation due to repeatability plus the variation due to reproducibility. Total (observed) variability is an additive of product (actual) variability and measurement variability.

Discrimination is the number of decimal places that can be measured by the system. Increments of measure should be about one-tenth of the width of a product specification or process variation that provides distinct categories. *Accuracy* is the average quality near to the true value. The *true value* is the theoretically correct value. *Bias* is the distance between the average value of the measurement and the true value, the amount by which the measurement instrument is consistently off target, or systematic error. *Instrument accuracy* is the difference between the observed average value of measurements and the true value.

Bias can be measured based by instruments or operators. Operator bias occurs when different operators calculate different detectable averages for the same measure. Instrument bias results when different instruments calculate different detectable averages for the same measure.

Precision encompasses total variation in a measurement system, the measure of natural variation of repeated measurements, and repeatability and reproducibility. *Repeatability* is the inherent variability of a measurement device. It occurs when repeated measurements are made of the same variable under absolutely identical condition (same operators, set-ups, test units, environmental conditions) in the short term. Repeatability is estimated by the pooled standard deviation of the distribution of repeated measurements and is always less than the total variation of the system.

Reproducibility is the variation that results when measurements are made under different conditions. The different conditions may be operators, set-ups, test units, or environmental conditions in the long term. Reproducibility is estimated by the standard deviation of the average of measurements from different measurement conditions.

The *measurement capability index* is also known as the precision-to-tolerance (P/T) ratio. The equation is $P/T = 5.15 \times \sigma_{MS})/\text{tolerance}$. The P/T ratio is usually expressed as a percent and indicates what percent of the tolerance is taken up by the measurement error. It considers both repeatability and the reproducibility. The ideal ratio is 8% or less; an acceptable ratio is 30% or less. The 5.15 standard deviation accounts for 99% of MS variation and is an industry standard.

The P/T ratio is the most common estimate of measurement system precision. It is useful for determining how well a measurement system can perform with respect to the specifications. The specifications, however, may

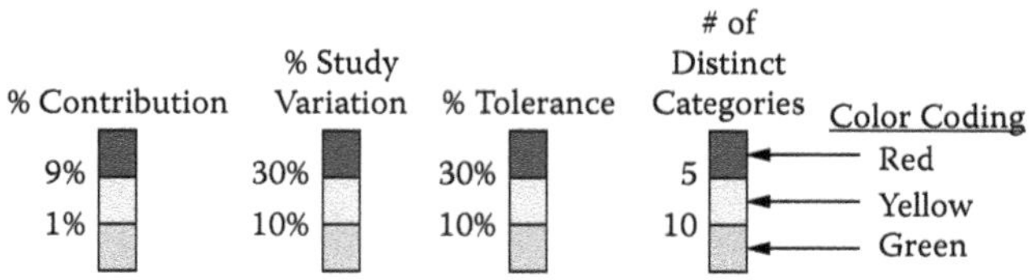

FIGURE 8.26
Gage R&R standards.

be inaccurate or need adjustment. The %R&R = $(\sigma_{MS}/\sigma_{Total}) \times 100$ formula addresses the percent of the total variation taken up by measurement error and includes both repeatability and reproducibility. A case study of a Gage R&R will be performed in the next chapter. A good Gage R&R meets the standards presented in Figure 8.26.

Control plan—A control plan details the who, what, where, when, why, and how. The steps to a control plan are:

Collect existing documentation for the process.
Determine the scope of the process for the current control plan.
Form teams to update the control plan regularly.
Replace short-term capability studies with long-term capability results.
Complete control plan summaries.
Identify missing or inadequate components or gaps.
Review training, maintenance, and operational action plans.
Assign tasks to team members.
Verify compliance of actual procedures with documented procedures.
Retrain operators.
Collect sign-offs from all departments.
Verify effectiveness with long-term capabilities.

A control plan ensures that processes are consistent. It is intended to minimize variation, prevent operators from changing processes, assure the success of improvements, provide effective training, and manage maintenance calibration, and repair scheduling. Control plans should be reviewed quarterly. Figure 8.27 is a Six Sigma control plan. Planning is an essential step in using the tools and techniques described throughout the chapter. A proper control plan will produce results through data-driven activities.

Six Sigma Control Plan											
Product:				Core Team:					Date (orig):		
Key Contact:									Date (revised):		
Phone:											
Process	Process Step	Input	Output	Process Specs (LSL, USL, Target)	Pkg./Date	Measurement Technique	%P/T	Sample Size	Sample Frequency	Control Method	Reaction Plan

FIGURE 8.27
Template for Six Sigma control plan.

9

Project Control Case Studies

"Tell me, and I forget; show me, and I remember; involve me, and I understand."

Chinese proverb

Drawing an analogy from the quote above, case studies are effective for telling the story of a decision environment, showing the intricacies of the decision problem, and involving the audience in the outcome. This chapter presents a collection of case studies to illustrate the pursuit of project control.

Gage R&R

A Gage R&R analysis was performed to study measurements in a factory. The set-up was made as functional as it could be (Figure 9.1). It randomized the first six parts for operator 1. When operator 1 finished his first trial on each of the six parts, operator 2 performed his first trial on the exact same parts. After operator 2 finished his first trial, operator 1 started his second trial on the six parts in a different randomized pattern. This process continued for two trials. Both the A and B side measurements were taken, but only side B was analyzed because side A equaled side B. Six parts were used for the sample size due to availability of time and throughput of the parts. The first step was a cross tabulation of the numbers of columns and rows as shown in Figure 9.2.

Since this check was correct, a crossed Gage R&R was completed. Only one side of the study (B) was considered since a paired t-test proved that A = B for pressure measurements. The paired t-test was used since the same parts were compared (two different samples of the same parts). The hypothesis for all the samples in the paired t-tests is based on:

H_0 (null hypothesis) indicates the mean of the differences in the population is equal to the chosen reference value (0). H_0 = accept null hypothesis (A equals B).

Sample	Operator	Trial	Measurement A	Measurement B	Number
3	1	1	164.64	163.27	1
6	1	1	140.78	139.79	2
5	1	1	153.46	149.95	3
1	1	1	146.46	154.61	4
2	1	1	158.91	160.43	5
4	1	1	139.13	141.41	6
3	2	1	165.79	161.34	7
6	2	1	133.86	138.4	8
5	2	1	155.34	148.82	9
1	2	1	144.34	154.4	10
2	2	1	157.83	164.82	11
4	2	1	137.07	139.52	12
5	1	2	154.91	151.01	13
2	1	2	158.48	166.24	14
3	1	2	162.83	164.95	15
6	1	2	142.15	139.7	16
4	1	2	137.3	140.66	17
1	1	2	145.5	155.05	18
5	2	2	153.88	150.1	19
2	2	2	156.81	160.05	20
3	2	2	160.51	160.84	21
6	2	2	133.94	138.99	22
4	2	2	135.92	140.48	23
1	2	2	145.17	155.03	24

FIGURE 9.1
Gage R&R set-up.

Tabulated statistics: Trial, Operator

Rows: Trial   Columns: Operator

	Bob	Joe	All
1	6	6	12
2	6	6	12
All	12	12	24

Cell Contents: Count

FIGURE 9.2
Tabulated statistics.

H_a (alternative hypothesis) indicates the mean of the differences in the population is not equal to the chosen reference value. H_a = reject null hypothesis (A does not equal B).

The p value for side A was 0.089. The p value for side B was 0.988. The hypothesis was to see whether A equaled B. Side A was graphed for normality

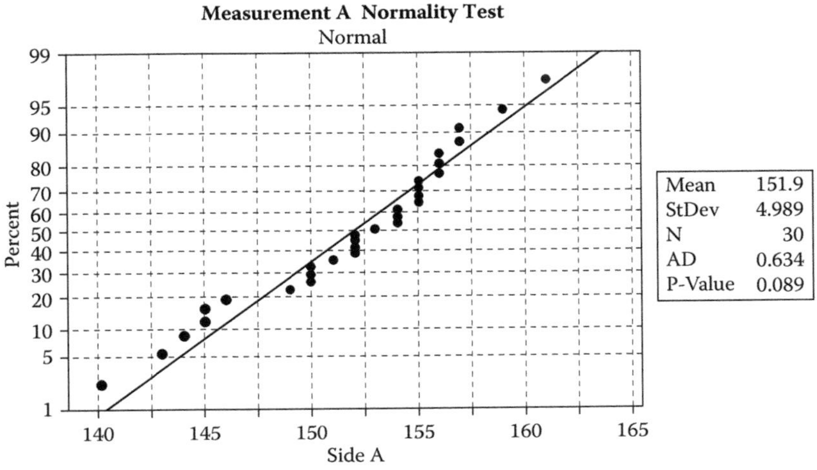

FIGURE 9.3
Side A normality test measurement.

(Figure 9.3). The p value for measurement of 0.089 provided enough evidence to suggest that the data follows a normal distribution. The normality check passed with a p value of 0.988—enough evidence to suggest that the data follows a normal distribution (Figure 9.4). The paired t-test and confidence interval (CI) figures from comparing side A and side B are:

Side	Mean	Standard Deviation	Standard Error	Mean
A	30	151.933	4.989	0.911
B	30	152.300	4.112	0.751
Difference	–0.366667	6.110947	1.115701	

The 95% CI for mean difference is –2.648532, 1.915199. The result of the t-test of mean difference = 0 (versus not 0): t value = –0.33; p value = 0.745. Ideally the p value should be above 0.05 and t should be above 2 to show significance. The p value of 0.745 indicates a 74.5% chance that the sample difference would be obtained. The 95% CI for mean differences also showed a crossing over 0. The conclusion from this test was that side A does equal side B. Potential variation may arise from other factors (parts, operators, or sides). The conclusion is accept the null hypothesis because side A equals side B for pressure measurements.

Figure 9.5 is the Gage run chart for side B. It shows consistency of measurement values within the operator results and between the operators. The next step is evaluating side B.

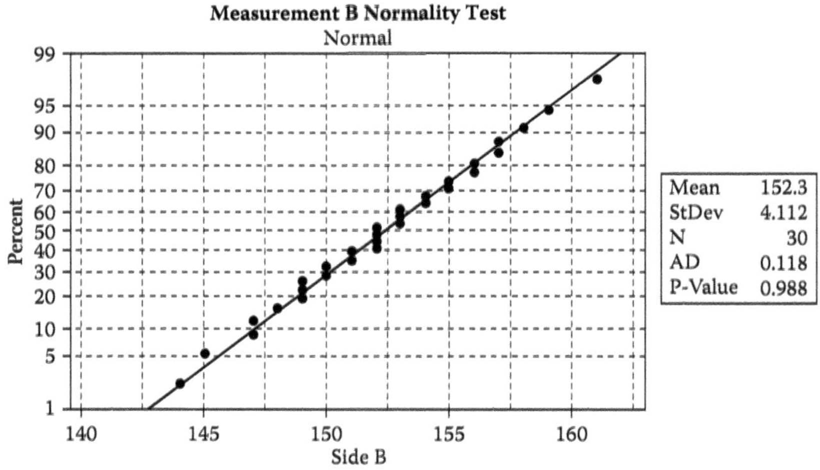

FIGURE 9.4
Side B normality test measurement.

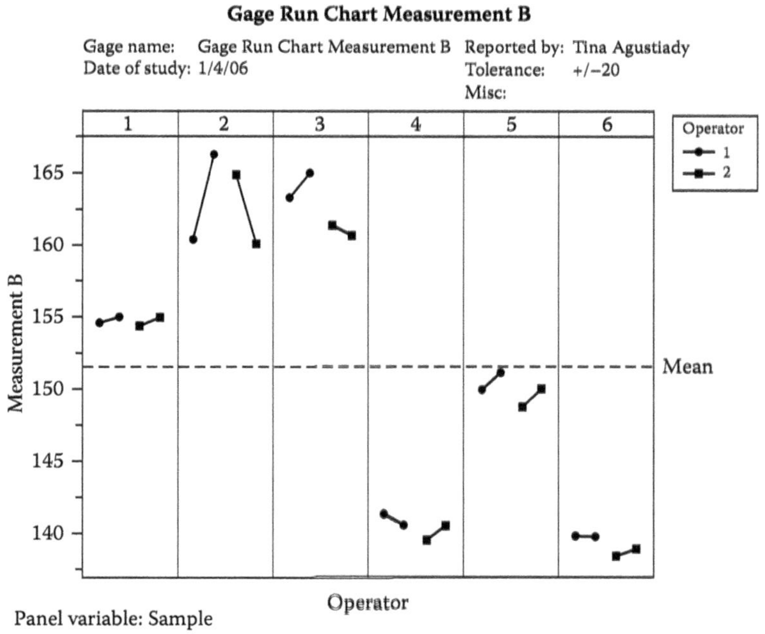

FIGURE 9.5
Side B Gage run chart.

The gage study is performed in MINITAB and the resulting statistics (table outputs from MINITAB) are shown below:

Gage R&R Study: ANOVA Method

Study name: Gage R&R, side B
Date of study: 1/4/06
Reported by: Tina Agustiady
Tolerance: +/–20
Misc:

Gage R&R Two-way ANOVA Method with Interaction

Source	DF	SS	MS	F	P
Sample	5	2148.55	429.710	460.948	0.000
Operator	1	8.50	8.497	9.114	0.029
Sample × operator	5	4.66	0.932	0.345	0.876
Repeatability	12	32.39	2.699		
Total	23	2194.09			

Gage R&R Two-way ANOVA Method without Interaction

Source	DF	SS	MS	F	P
Sample	5	2148.55	429.710	197.177	0.000
Operator	1	8.50	8.497	3.899	0.065
Repeatability	17	37.05	2.179		
Total	23	2194.09			

Gage R&R Variation Percentage Contributions

	%Contribution	
Source	**VarComp**	**%**
Total Gage R&R	2.706	2.47
Repeatability	2.179	1.99
Reproducibility	0.526	0.48
Operator	0.526	0.48
Part-to-part	106.883	97.53
Total variation	109.588	100.00

Standard Deviations and Variance Percentage for Gage R&R: Side B					
	Study	**Var %**	**Study**	**Var %**	**Tolerance**
Source	StdDev	(SD)	(6 × SD)	(%SV)	(SV/Toler)
Total Gage R&R	1.6449	9.8695	15.71	24.67	
Repeatability	1.4762	8.8575	14.10	22.14	
Reproducibility	0.7256	4.3534	6.93	10.88	
Operator	0.7256	4.3534	6.93	10.88	
Part-to-part	10.3384	62.0304	98.76	155.08	
Total variation	10.4684	62.8107	100.00	157.03	

Note: Distinct categories = 8.

Figure 9.6 illustrates the Gage R&R charts for side B. Side B was checked for normality (see Figure 9.4).

The interaction is checked for significance. An interaction is not significant unless the p value is below 0.05. In this case, the p value is 0.876.

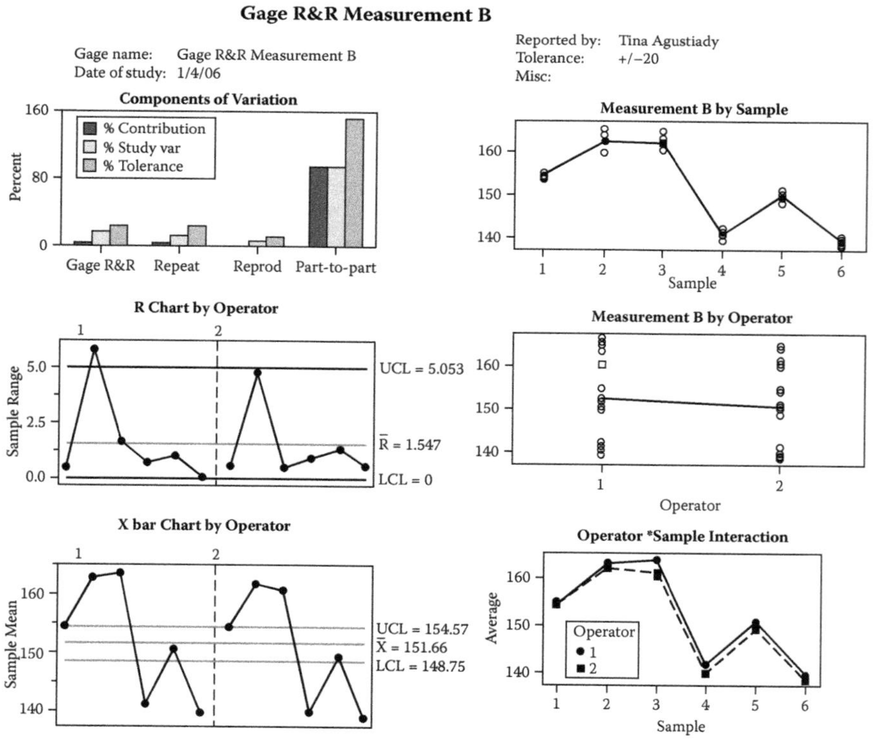

FIGURE 9.6
Side B Gage R&R chart.

Therefore the interaction is not significant—the result desired from this study. The two-way ANOVA table with interaction can be ignored due to the p value. The next step is to review the two-way ANOVA table without interaction. The p value for the sample is 0.0 instead of 0.876, so the sample is significant (it varied). The operator p value is 0.065 instead of 0.029 for the table without interaction (the sample changed). In the actual Gage R&R, the reproducibility of 0.48 seems to be sufficient. The repeatability is 1.99. The variation should be reduced from the repeatability. If the repeatability is too high, the data yields only one distinct category. The standard deviation of the Gage R&R for part-to-part is 10.33. Again, the minimization of noise is desired.

The study variation is 15.71—less than the allowable 30. The P/T ratio is 24.67 which is higher than 15 or less than ideal. The Gage consumes 24% of the total variation in the measurement system. The process tolerance was 40. The components of variation chart shows low noise to high signal. The Gage R&R is the noise variable. Repeatability and reproducibility constitute the Gage R&R; the part-to-part is the signal.

The X bar chart by operator (shown as the fifth plot in Figure 9.6) has points outside the control limits that indicate part-to-part variability. For the parts data, many points appear beyond the control limits, indicating that the measurement system is adequate. The measurement of side A by operator shows an approximate flat line which is ideal. The operator × sample interaction shows the operators reasonably parallel but far enough apart to show variability among parts.

The Gage R&R analysis indicates that pressure is a significant factor of the process. All the studies revealed variations. The P/T ratio is higher than it should be. Because many factors can affect pressure, the next steps involved investigating and resolving syringe and procedural issues. The study of side B passed the percent contribution, percent study variation, percent tolerance, and number of distinct categories tests. The percent tolerance of about 24 was close but still passed. Therefore this was a good Gage R&R (Figure 9.7).

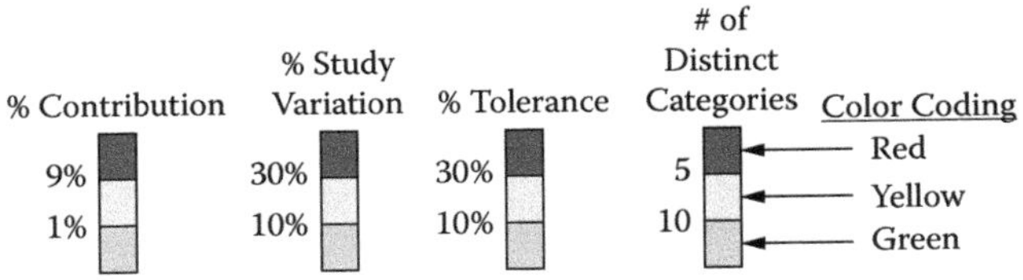

FIGURE 9.7
Good Gage study results.

Capability Studies

Thirty points of historical data were graphed for capability studies (Figure 9.8). The p value for side A was 0.089, suggesting that the data follows a normal distribution. The normality check passed with a p value of 0.988 for side B, again suggesting the data follows a normal distribution. A capability analysis was done for side A (Figure 9.9). The short-term process capability or C_{pk} is 1.41, indicating a capable process. The long-term process capability or P_{pk} is 1.20, again a capable result. A capability analysis was done for side B (Figure 9.10). The C_{pk} is 1.26. The P_{pk} is 1.42. Both figures indicate a capable process.

	Part 1		Part 2
A	157	B	157
A	156	B	153
A	155	B	151
A	152	B	147
A	155	B	152
A	156	B	144
A	150	B	150
A	156	B	161
A	145	B	149
A	154	B	151
A	154	B	152
A	157	B	157
A	153	B	149
A	143	B	154
A	144	B	153
A	154	B	156
A	159	B	155
A	155	B	152
A	161	B	149
A	149	B	152
A	152	B	158
A	150	B	145
A	140	B	154
A	150	B	156
A	151	B	153
A	152	B	155
A	146	B	147
A	145	B	150
A	152	B	159
A	155	B	148

FIGURE 9.8
Capability study set-up.

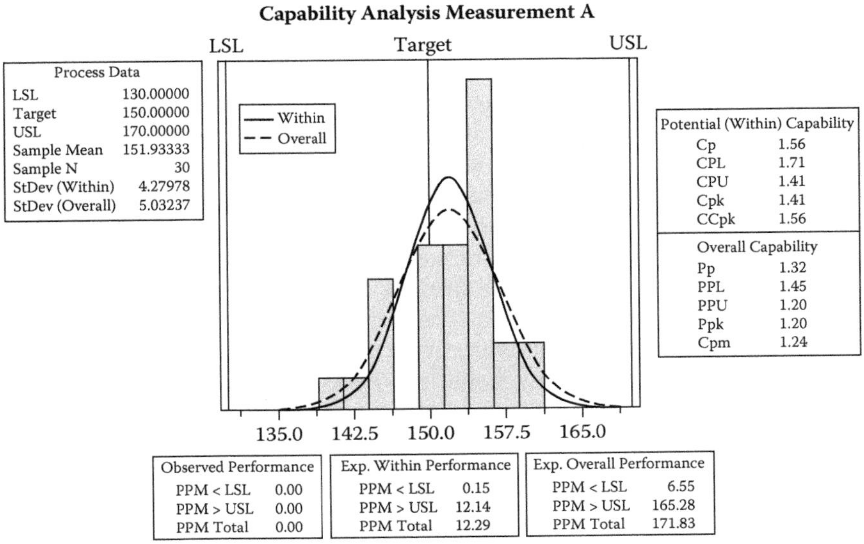

FIGURE 9.9
Side A capability study.

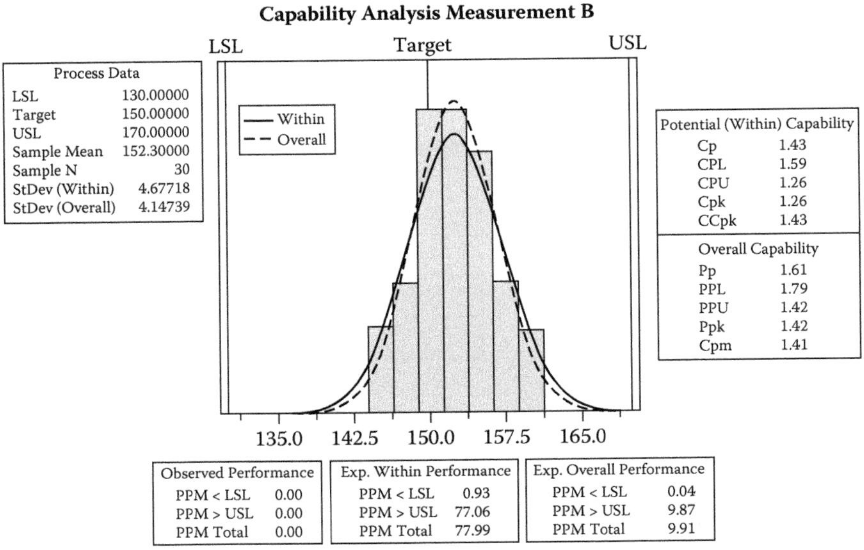

FIGURE 9.10
Side B capability study.

Analytical Studies

This section covers hypothesis testing, basic statistics, and graphical analysis. The data from the capability studies were used in the *measure* phase to run normality tests. Descriptive statistics of the measurements were compiled before the paired t-tests. Figure 9.11 shows the descriptive statistics. The median (center) is 152.50 and is in an acceptable range. The target or desired measurement is 150; the range of measurements is 130 to 170. The shape of a box plot tells a lot about data distribution. The box represents the middle 50% of the differences. The line through the box represents the median difference. The lines extending from the box represent the upper and lower 25% of differences (excluding outliers). Outliers are represented by asterisks (*) as shown in Figure 9.12. Figure 9.13 is a box plot for side B.

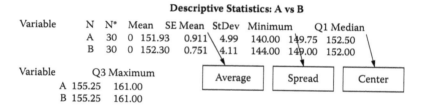

Descriptive Statistics: A vs B

Variable	N	N*	Mean	SE Mean	StDev	Minimum	Q1	Median
A	30	0	151.93	0.911	4.99	140.00	149.75	152.50
B	30	0	152.30	0.751	4.11	144.00	149.00	152.00

Variable	Q3	Maximum
A	155.25	161.00
B	155.25	161.00

Average → StDev; Spread → Q1; Center → Median

FIGURE 9.11
Descriptive statistics.

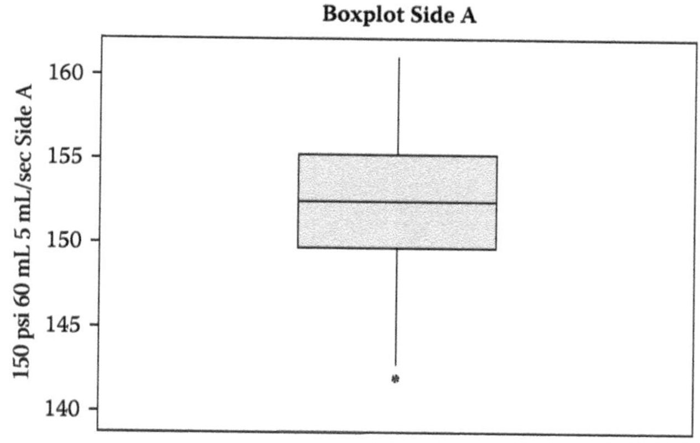

FIGURE 9.12
Side A box plot.

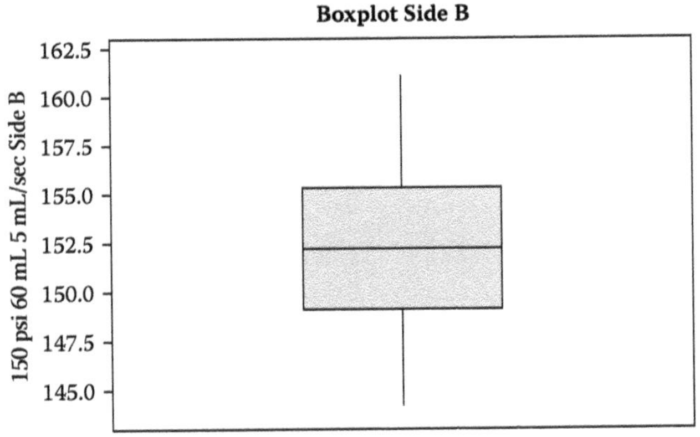

FIGURE 9.13
Side B box plot.

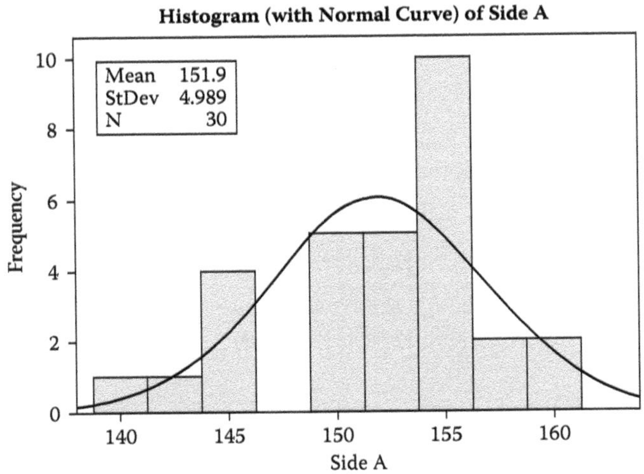

FIGURE 9.14
Side A histogram.

Figure 9.14 is a histogram for side A showing a normal curve. Figure 9.15 is the same type of chart for side B. The normality graphs were discussed earlier in the chapter. We have enough evidence to suggest that the data for sides A and B follow a normal distribution. An F test was done to see whether the samples revealed equal variances.

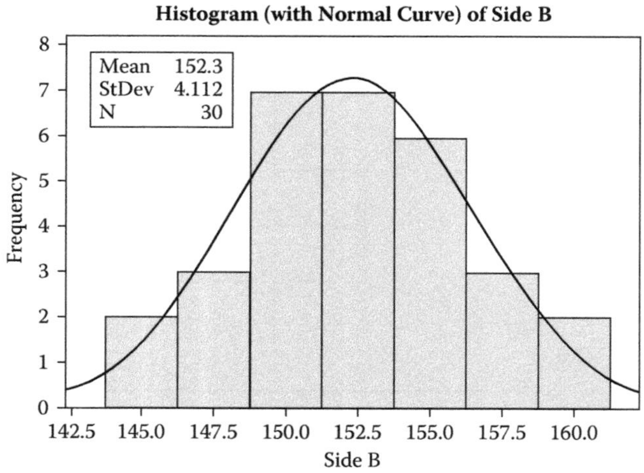

FIGURE 9.15
Side B histogram.

Test for Equal Variances

95% Bonferroni Confidence Intervals for Standard Deviations

CI	N	Lower	Standard Deviation	Upper
A	30	3.85107	4.98918	7.01755
B	30	3.17383	4.11180	5.78346

F test (normal distribution): Test statistic = 1.47, p value = 0.303

Levene's test (any continuous distribution): Test statistic = 0.75, p value = 0.389.

Since the p value was over 0.05, the test confirms that the two samples had equal variances (Figure 9.16). Three graphs were chosen to graph the data. The first was a box plot of differences (Figure 9.17) that showed the following data properties:

Mean: The point below the plot represents the mean difference between paired observations.

Confidence interval: The line and brackets on the sides of the mean represent the confidence interval.

The point labeled H_0 represents the value specified in the null hypothesis.

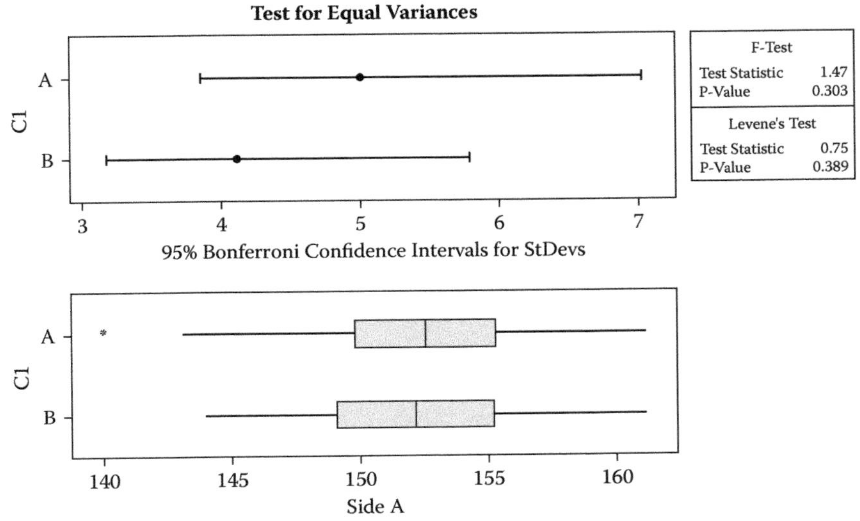

FIGURE 9.16
Test for equal variances graph.

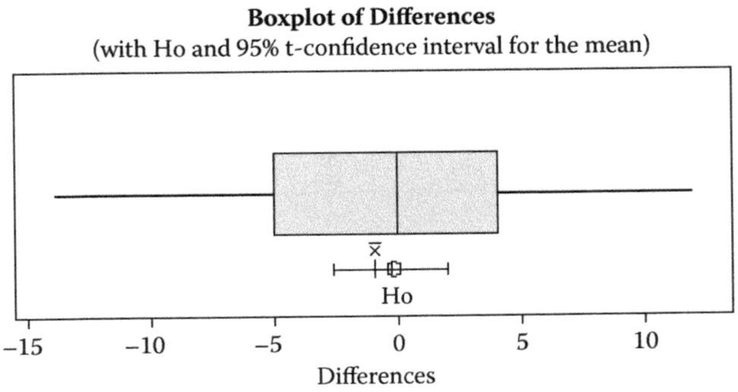

FIGURE 9.17
Box plot of differences.

The next graph was the individual value scatter plot of differences (Figure 9.18) that shows the same conclusion. The last graph of the paired t-test was a histogram of differences (Figure 9.19). The histogram indicates the differences are roughly normally distributed; the range does cross 0. Since H_o lies inside the confidence interval, we can say that the 95% CI for mean differences is between –2.65 and 1.92.

An important assembly issue involves measurements out of specification. A parts life study was done to see whether parts degraded after a specified

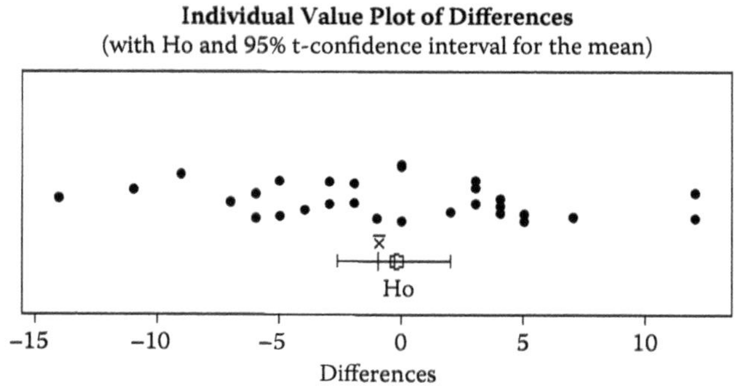

FIGURE 9.18
Individual value plot of differences.

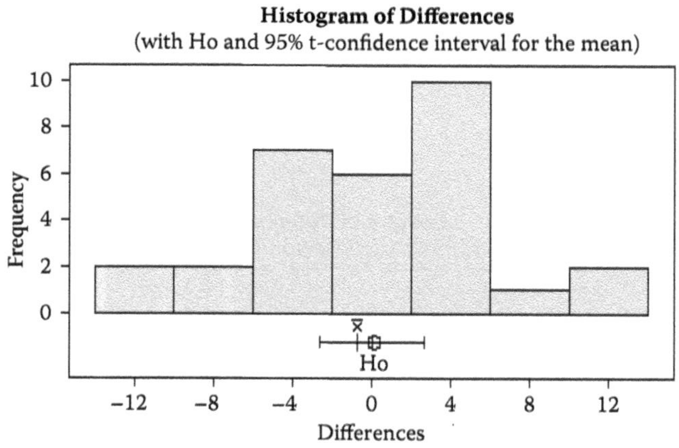

FIGURE 9.19
Histogram of differences.

duration of use before they produced a measurement out of specification. This first part was run repeatedly until a problem occurred with the part. The operator recorded each measurement for the part.

Trend Analysis

A trend analysis for the time series of the parts was graphed to determine a significant point where the measurements started to drop (Figure 9.20). The quadratic method was used to find a significant drop-off point of the measurements. The graph indicates that the measurements begin to drop around the 64th measurement. The 64th measure was considered significant because it shows about a five-point decrease. The decrease became the

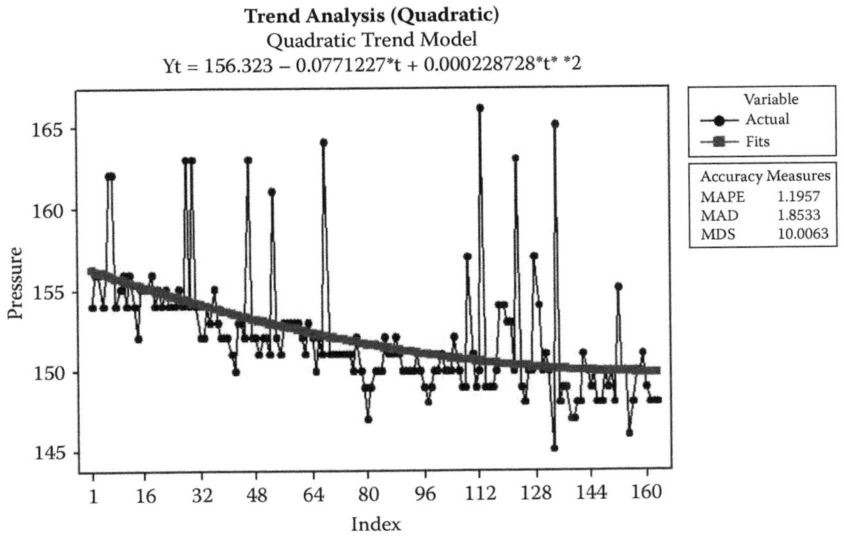

FIGURE 9.20
Quadratic trend model.

baseline because it represents about 10% of the specification range. The study determined that the life cycle of the parts should be around the 64th measurement. The parts were also subjected to basic statistics, graphical analysis, hypothesis testing, and multivariate studies.

To investigate the degradation of multiple parts over time, a degradation study was done. To do the study, an operator tested a part in a factory that historically resulted in the most problems. The part was run repeatedly until a problem occurred with the syringe. Five trials of this were done. The operator recorded each measurement. Basic descriptive statistics were generated for each of the points as summarized below.

Variable Measurement	N	N*	Mean Standard Error	Mean	Standard Deviation	Minimum Q1	Median	Q3
127	0	292.44	0.313	3.52	282.00	290.00	293.00	298.00

Note: Variable maximum measurement = 295.00.

The mean, median, minimum, and maximum of the measurements were all in specification (275 to 325). Line graphs of the measurements are shown in Figure 9.21. Some of the lower measurements were investigated and showed unidentifiable causes. No operator changes or major problems were found. Since the measurements were still in specification, the team decided

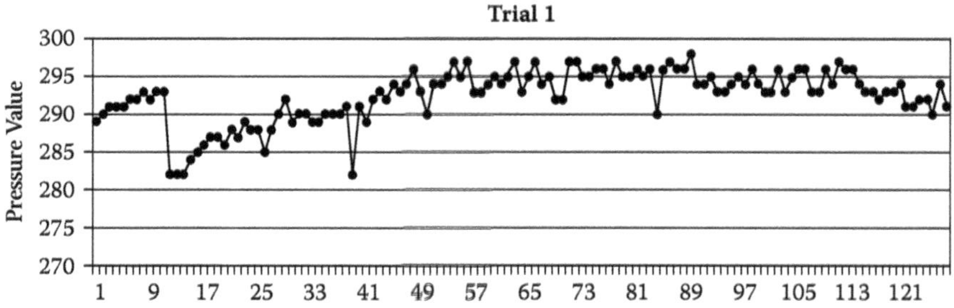

FIGURE 9.21
Line graph for trial 1.

to continue. An error occurred after the 122nd trial. A second analysis was performed with the descriptive statistics summarized below.

Variable Measurement	N	N*	Mean Standard Error	Mean Standard Deviation	Minimum	Q1	Median	Q3
53	0	294.25	0.170	1.24	290.00	294.00	294.00	295.00

Note: Variable maximum measurement = 297.00.

Again, the measurements were in specification for all the trials. Figure 9.22 is the line graph. This time, the part was unable to perform after the 53rd trial. A study was performed as summarized below.

Variable Measurement	N	N*	Mean Standard Error	Mean Standard Deviation	Minimum	Q1	Median	Q3
52	0	295.08	0.307	2.21	291.00	294.00	295.00	297.00

Note: Variable maximum measurement = 300.00.

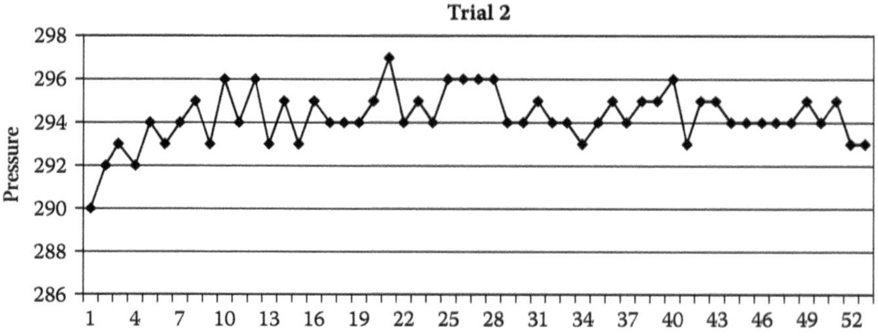

FIGURE 9.22
Line graph for trial 2.

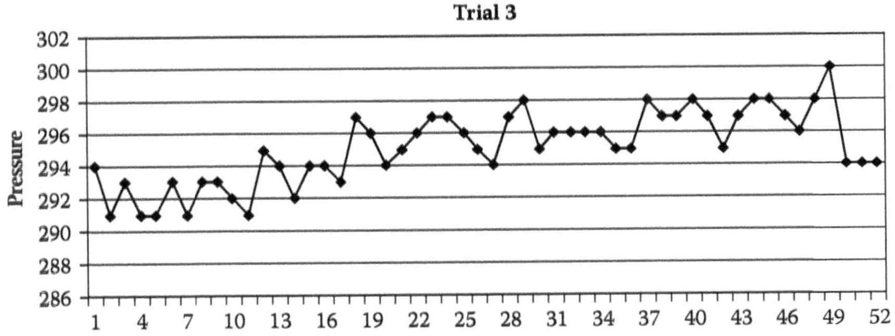

FIGURE 9.23
Line graph for trial 3.

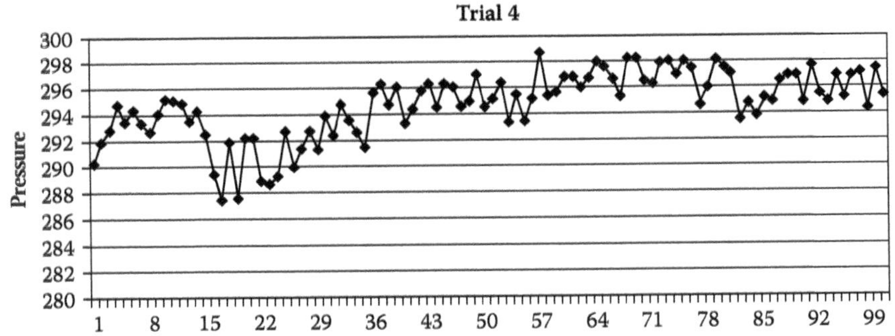

FIGURE 9.24
Line graph for trial 4.

The basic descriptive statistics proved that the measurements were all in specification as shown in Figure 9.23. This time, the part was unable to perform after the 52nd trial. A fourth trial was done. The descriptive statistics are summarized below. Figure 9.24 is the graph for trial 4.

Variable Measurement	N	N*	Mean Standard Error	Mean Standard Deviation	Minimum	Q1	Median	Q3
100	0	294.72	0.247	2.47	287.46	293.44	295.06	296.61

Note: Variable maximum measurement = 298.72.

An error occurred after the 100th trial. A fifth and final study was done as summarized below. Figure 9.25 shows the line graph.

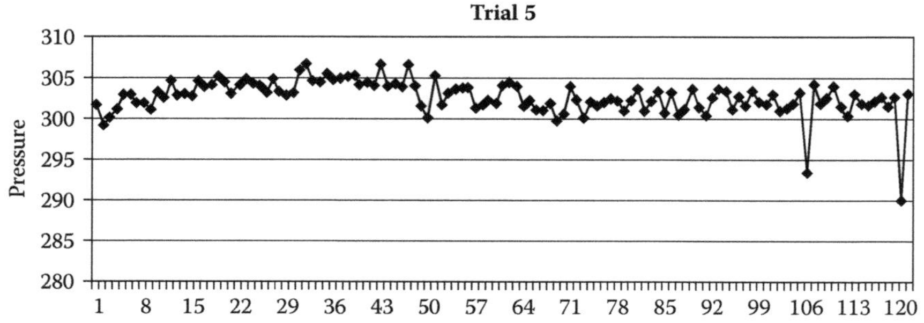

FIGURE 9.25
Line graph for trial 5.

Variable Measurement	N	N*	Mean Standard Error	Mean Standard Deviation	Minimum	Q1	Median	Q3
121	0	302.76	0.193	2.12	290.06	301.73	302.89	304.11

Note: Variable maximum measurement = 306.88.

The error occurred after the 121st trial. Figure 9.26 graphs all five trials. The trials overlap each other in most situations, but trial 5 scored higher than the rest, possibly due to the design of the part. This was not investigated further to keep the scope of the project within reach.

Each of these tests involved extreme circumstances and did not correlate well to real-life situations. For that reason, reliability analyses were performed to determine a baseline for replacing the parts. The data collection was simple: recording each failure. An analysis was performed to determine at what point it was cheaper to replace the part than it would be to keep using it in production. The analysis was done with Weibull++ software.

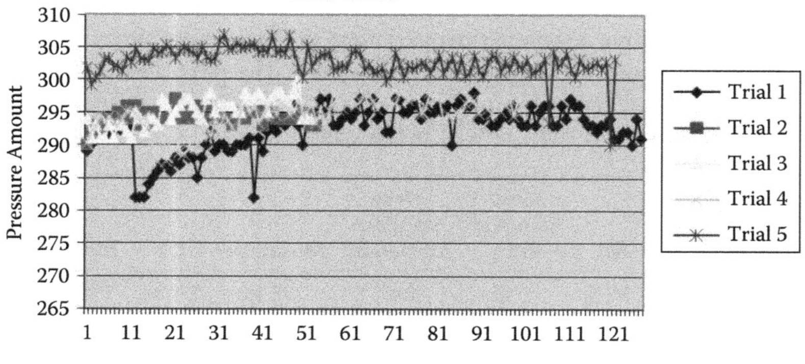

FIGURE 9.26
Graph of five trials.

Time Units	Cost/Unit Time
1.00	3.05
2.00	1.78
3.00	1.44
4.00	1.33
5.00	1.31
6.00	1.32
7.00	1.35
8.00	1.39
9.00	1.44
10.00	1.48
11.00	1.53
12.00	1.57
13.00	1.60
14.00	1.63
15.00	1.66
16.00	1.68
17.00	1.70
18.00	1.72
19.00	1.73
20.00	1.74
21.00	1.75
22.00	1.76
23.00	1.76

FIGURE 9.27
Weibull software analysis.

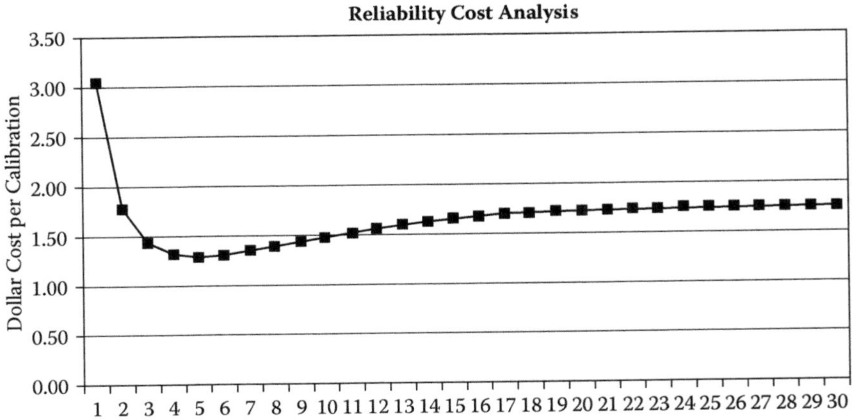

FIGURE 9.28
Reliability cost analysis.

Time in units was found based on the cheapest cost per unit time as can be seen in Figure 9.27. The parts should be replaced after five uses as shown in Figure 9.28 to prevent errors and degradation of measurements. Figure 9.29 is a probability plot from the Weibull distribution. The correlation of 0.970 indicates a good representation.

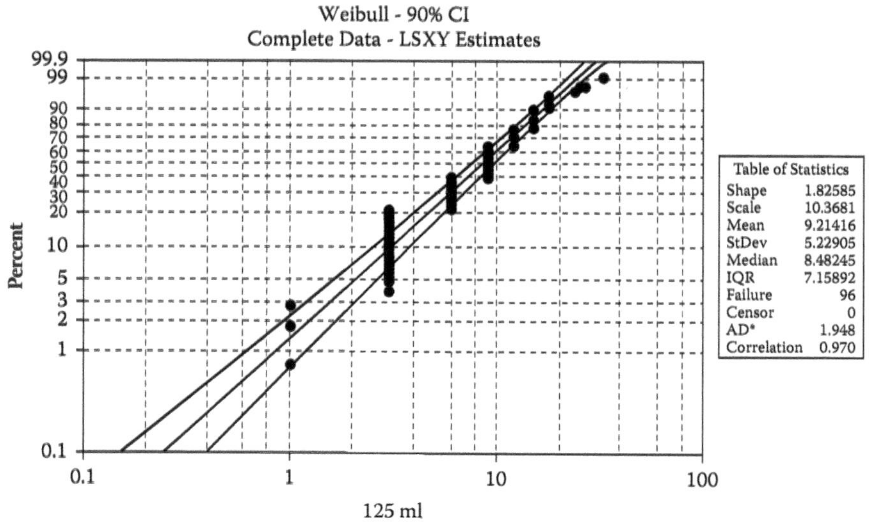

FIGURE 9.29
Probability plot.

Simple Linear Regression

A simple linear regression was done on trial 1 because a slope was noted and a linear relationship was sought. The team wanted to find a correlation after noting linear relationship between the variables. The y represented the measurement and x was the time.

Correlations: Trial 1 and time
Pearson correlation of Trial 1 (150 ml) and time = −0.994
p = 0.000

The correlation coefficient of r = −0.994 shows a high positive dependence. The p value below 0.05 also shows that the correlation coefficient is significant. A fitted line plot was prepared to see the inverse relationship as shown in Figure 9.30, then regression analysis was done.

Regression analysis: Trial 1 versus time
Regression equation: Trial 1 = 1175 – 1.56 time

Predictor	Coef	SE Coef	T	P
Constant	1174.89	0.74	1591.47	0.000
Time	−1.56360	0.02623	−59.61	0.000

Note: S = 2.51748. R^2 = 98.7%. R^2 (adjusted) = 98.7%.

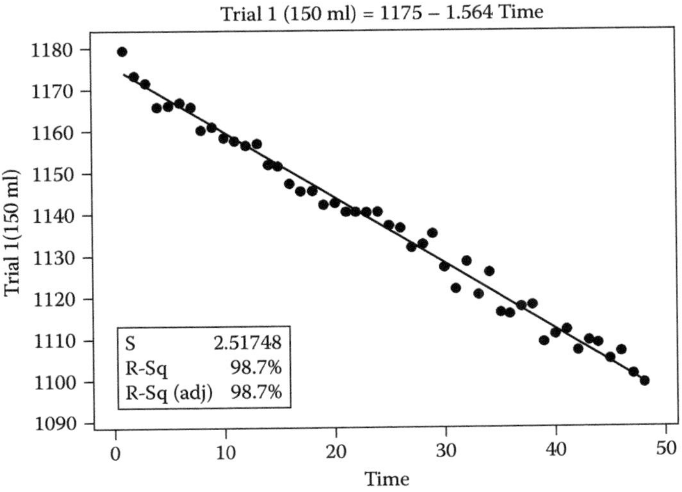

FIGURE 9.30
Fitted line plot.

The ANOVA results are shown in Figure 9.31. R denotes an observation with a large standardized residual. Each time showed a 1.5 measurement of degradation according to the equation $Y_1 = \beta o + \beta_1 X_1$. The slope equals 1.564 and is negative. In this situation, slope becomes critical after a loss of more than 10% of the measurement in the specification range. According to the graph, a degradation of about 15 occurs about every 10 trials.

Analysis of Variance

Source	DF	SS	MS	F	P
Regression	1	22522	22522	3553.63	0.000
Residual Error	46	292	6		
Total	47	22813			

Unusual Observations

Obs	Time	Trial 1 (150ml)	Fit	SE Fit	Residual	St Resid
1	1.0	1179.00	1173.33	0.72	5.67	2.35R
29	29.0	1135.10	1129.55	0.38	5.55	2.23R

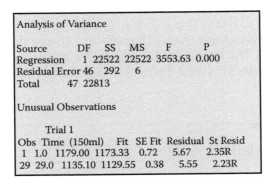

FIGURE 9.31
Analysis of variances and constants over time.

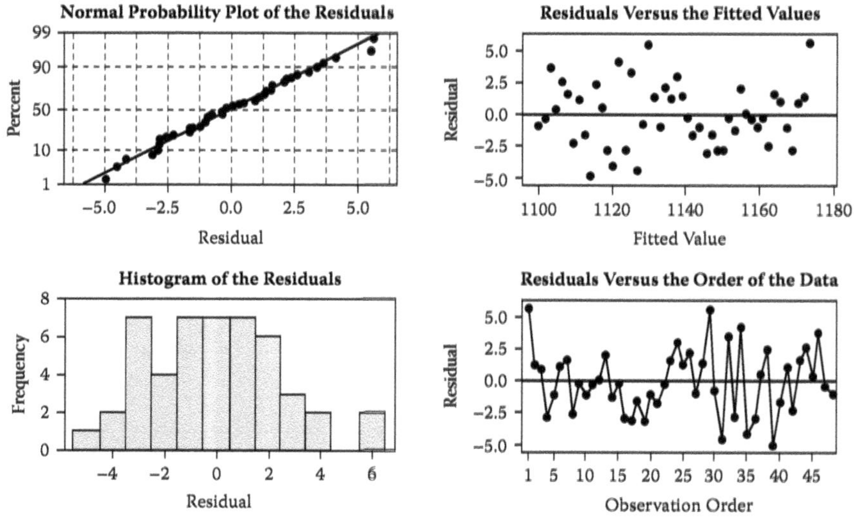

FIGURE 9.32
Residual plot.

Hypothesis Testing

The p value shows that the loss rate is not normal. H_o = accept the null hypothesis ($\beta = 0$, no correlation). H_a = reject the null hypothesis ($\beta \neq 0$, correlation is present). In this case, we reject the null hypothesis based on correlation. The R^2 value above is 98.7% which means 98.7% of the Y variables (pressures) can be explained by the model (regression equation). The residuals are then evaluated as shown in Figure 9.32. The normality of the residuals is plotted in Figure 9.33. The normality test passed with a value of 0.850. The residuals are in control as shown in Figure 9.34. The residuals are contained in a straight band, with no obvious pattern in the graph showing that this model is adequate.

The conclusion is to reject the null hypothesis because the slope of the line does not equal 0. The linear relationship in the measurement versus time indicates correlation. This model proves to be adequate based on testing.

Multivariate Studies

A multivariate analysis was based on principal components (used to reduce data into a smaller number of components) and factor analysis (to describe

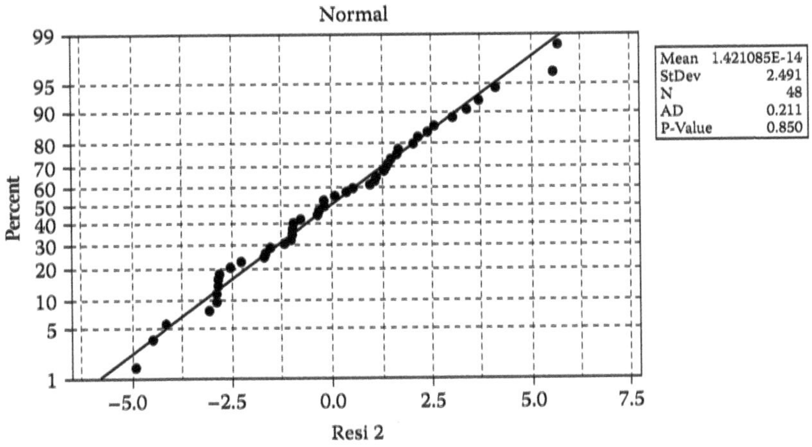

FIGURE 9.33
Normality of residuals.

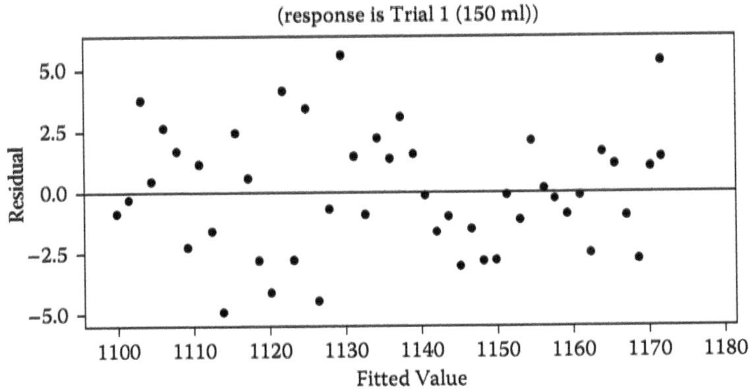

FIGURE 9.34
Residual versus fitted values.

the covariance among variables in terms of a few underlying factors). Two technicians, two quality control technicians, two software versions, and a part measurement were studied. The first step was to identify the key factors of the study: the personnel and the software version may have affected the measurement during quality control. First, descriptive statistics were generated. The results are summarized below and Figure 9.35 is a box plot of the analysis.

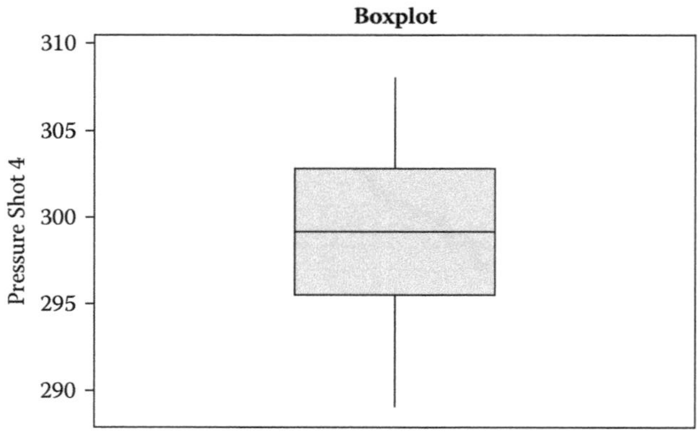

FIGURE 9.35
Descriptive statistics.

Variable Measurement	N	N*	Mean Standard Error	Mean Standard Deviation	Minimum	Q1	Median	Q3
30	0	299.37	0.812	4.44	289.00	295.75	299.50	308.00

Note: Variable maximum measurement = 303.00.

The measurement is within specification. A normality check was done next and is plotted in Figure 9.36. The normality test passed with a value of 0.924. A main effects plot was generated to make comparisons as shown in Figure 9.37. Additional comparisons were done for a multivariate study (Figure 9.38). The figure shows the means of each stratified group. Testing was done to check for equal variances between the two samples (Figure 9.39).

Test for Equal Variances: Measurement versus Software

The 95% Bonferroni confidence intervals for standard deviations are summarized below:

Software Version	N	Lower	Standard Deviation	Upper
1	15	2.89132	4.11964	6.98257
2	15	2.91636	4.15532	7.04304

Note: F test (normal distribution).
Test statistic = 0.98, p = 0.975.
Levene's test (any continuous distribution).
Test statistic = 0.00; p = 1.000.

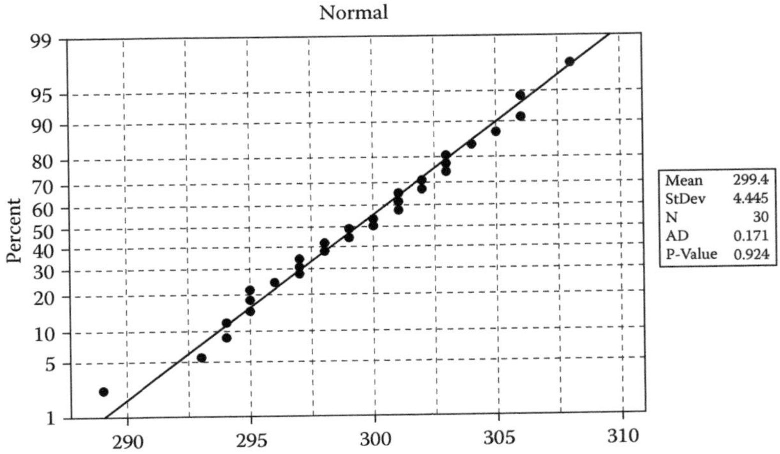

FIGURE 9.36
Normality check.

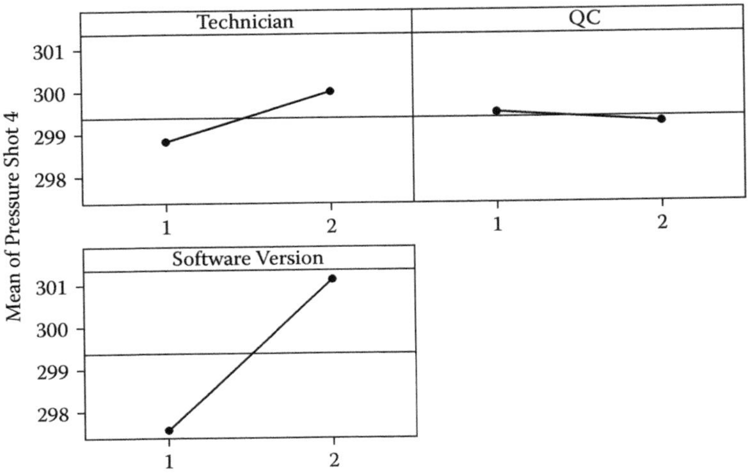

FIGURE 9.37
Main effects plot.

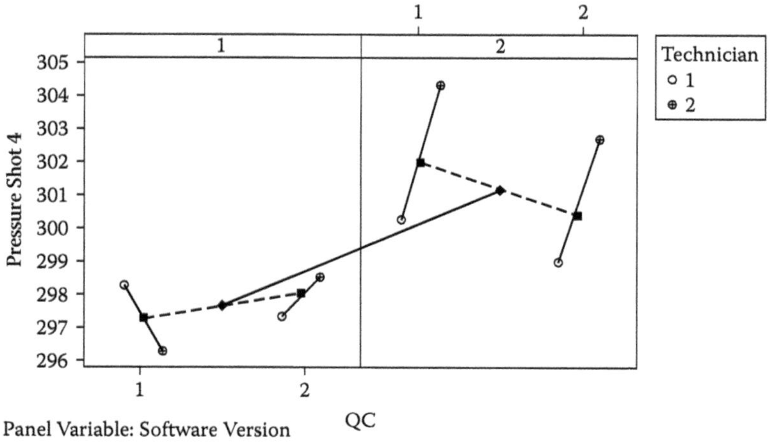

FIGURE 9.38
Multivariate chart.

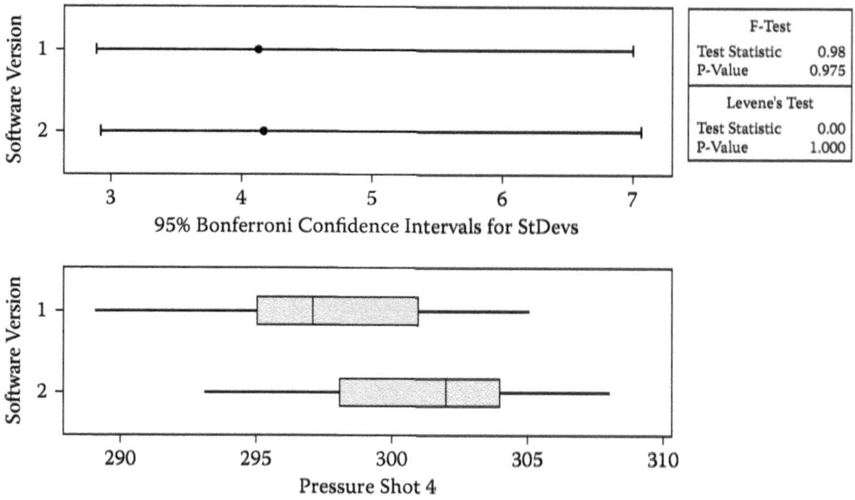

FIGURE 9.39
Test for equal variances.

Since the p value was over 0.05, the test supports the equal variances of the two samples. A general linear model was devised to analyze measurement versus technicians, quality control technicians, and software versions.

General Linear Model

General Linear Model: Measurement, Technician, QC, and Software

Factor	Type	Level	Value
Technician	Fixed	2	1, 2
QC	Fixed	2	1, 2
Software	Fixed	2	1, 2

ANOVA Table Based on Using SS

Analysis of Variance for Pressure Shot 4 (Using Adjusted SS)

Source	DF Seq	SS	Adj SS	Adj MS	F	P
Technician	1	10.53	20.48	20.48	1.15	0.294
QC	1	0.30	0.56	0.56	0.03	0.861
Software	1	105.30	114.40	114.40	6.43	0.018
Technician × QC	1	6.91	3.58	3.580	20	0.658
Technician × Software	1	33.42	33.92	33.92	1.91	0.181
QC × Software	1	7.42	7.42	7.42	0.42	0.525
Error	23	409.09	409.09	17.79		

Note: Total = 29 572.97. S = 4.21739. R^2 = 28.60%. R^2 (adjusted) = 9.98%.

Residuals

Unusual Observations for Pressure Shot 4

Observation	Shot 4	Fit Standard Error	Fit	Residual	Standard Residual
2	289.000	296.657	1.985	−7.657	−2.06 R

R denotes an observation with a large standardized residual. After the model was run, the variables were analyzed and then simplified. The software result was significant with a p value of 0.018. Another general linear model was run with only the software as the variable to be analyzed; the measurement was the y variable.

General Linear Model: Measurement versus Software

Factor	Type	Levels	Values
Software	Fixed	2	1, 2

ANOVA Table Based on Using SS

Analysis of Variance for Pressure Shot 4 Using Adjusted SS)

Source	DF Seq	SS	Adjusted SS	Adjusted MS	F	P
Software	1	93.63	93.63	93.63	5.47	0.027
Error	28	479.33	479.33	17.12		

Note: Total = 29 572.97. S = 4.13752. R^2 = 16.34%. R^2 (adjusted) = 13.35%.

Residuals

Unusual Observations for Pressure Shot 4

Observation	Shot 4	Fit Standard Error	Fit	Residual	Standard Residual
2	289.000	297.600	1.068	−8.600	−2.15 R
27	293.000	301.133	1.068	−8.133	−2.03 R

R denotes an observation with a large standardized residual. The results indicate that the software is significant based on a p value of 0.027. The R^2-squared value of 16.34% indicates that only about 16% of the proportion of the variation is explained by the model and other factors may be causing variations of the data. Points 2 and 27 were reviewed but no identifiable causes were found. Box plots were graphed for the two measurement values versus the software as shown in Figure 9.40. The residuals were stored for further analysis as shown in Figure 9.41 and were checked for normality as shown in Figure 9.42. The normality check passed with a p of 0.833.

Hypothesis testing [H_o: $\mu 1 = \mu 2$; H_a: $\mu 1 \neq \mu 2$ (at least one is different)] led to two conclusions: (1) accept the alternate hypothesis and (2) the software impacted the measurements.

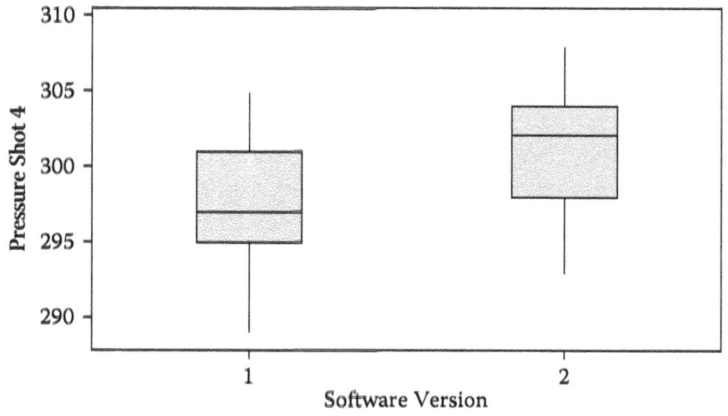

FIGURE 9.40
Box plot of measurement versus software version.

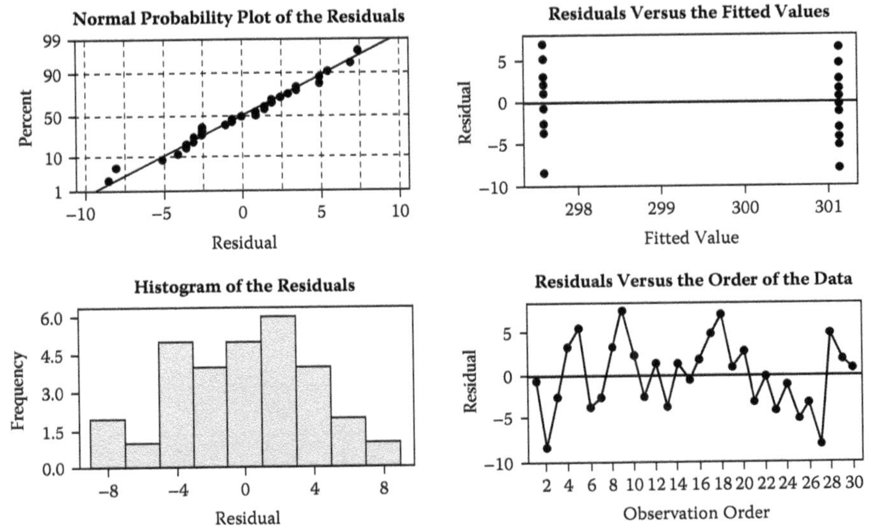

FIGURE 9.41
Residual plots.

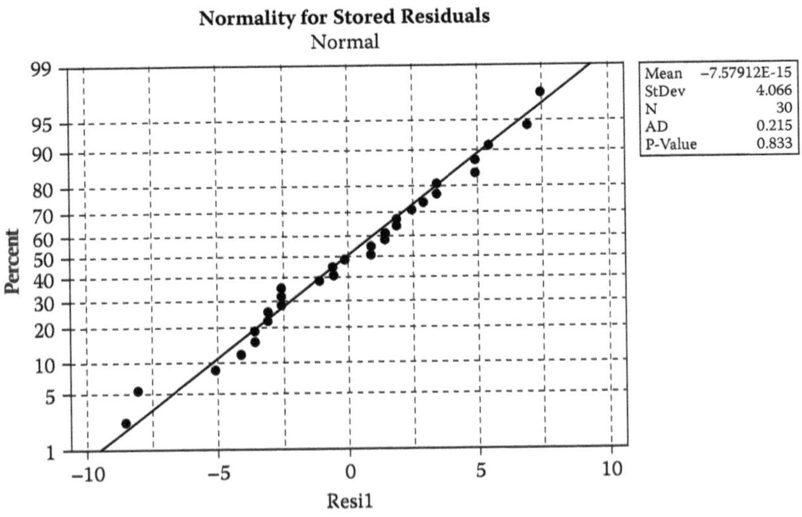

FIGURE 9.42
Normality of residual plot.

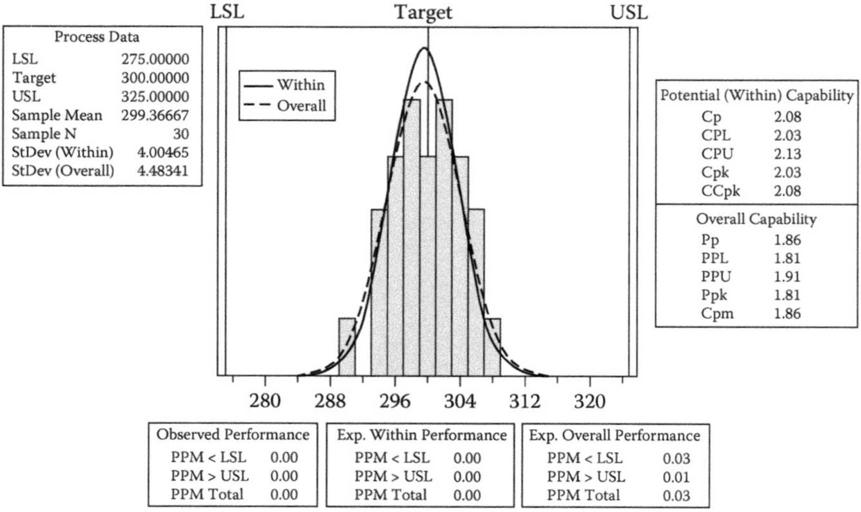

FIGURE 9.43
Capability analysis.

Finally a capability study of the measurements was done (Figure 9.43). The C_{pk} of 2.03 indicated the process was capable. Based on multivariate study, the software impacted the results. The software centered the process somewhat better (was a slight shift) and may have impacted the measurements. The variation in the process seemed to arise from the part. The technician variation decreased after the software change.

10

Management Support for Project Control

"If gold should rust, what shall iron do?"

Anecdote about leadership

Management must set the tone, example, and leadership for project execution. As the quote above suggests, management practices must be above reproach. For effective and sustainable project control, management must create an environment for continual monitoring and mentoring of tasks. Mentoring in this case relates to providing support and guidance for project execution. Management must provide leadership to take actions when needed for project control purposes. Management support is crucial for any type of project to achieve successful results. Top-down approaches are the most results oriented because people believe in changes if management believes in the changes. A deep commitment from the highest levels of management is necessary for the trust to occur. Senior management not only has to be committed, but has to drive the process throughout the entire organization, and select projects based on data-driven exercises. The traditional methods will change. The implementation strategy consists of top management support and participation, project identification, resource allocation, data-based decision making techniques, and finally measurement and feedback.

Ensuring people are comfortable with change is a key component. The changes should also be sustained so that "the old way of doing things" is left behind. Successful projects managers make projects their top priorities to which they devote time, energy, and resources. Challenging employees in a productive way also makes a successful leader. Encouragement during a project from beginning to completion is critical.

The control phase is where ongoing support of a process occurs. Usually SOPs are put in place along with a monitoring plan to ensure sustainability. Justifying improvement as a philosophy of management is a key component of project control. The process should be treated as a project strategy rather than a set of rules.

The knowledge of systems is key for stability in project controls. A system is a network of interdependent components that work together to accomplish a goal. These systems must be management's responsibilities. Management must own the systems, manage them over time, and implement improvements and optimizations. The system variations from common causes should motivate management to take action.

Many types of variation occur over time. Special cause variation and common cause variation are the two main types. Special cause variations during processes are "out of the ordinary." Common cause variations are ongoing. Special cause variation should be found and addressed to prevent recurrence. Common cause variations require management action. A common cause variation is within three standard deviations of the mean where we expect to find about 99.73% of the values. These values must be normally stable and predictable to produce common cause variations.

Active involvement and regular reviews of a project during key milestones are additional management responsibilities. Key decisions normally are made when milestones are reached. Management must understand the entire process or project so that decisions are not based solely on monetary impacts. Utilizing data for these decisions enables management to make commitments. However, sometimes judgment calls must be made and wrong decisions are made. Having adequate data to support decisions will help management make proper assessments.

Proper management support for project control helps employees feel comfortable and capable of making required decisions. The effects of proper management are ongoing if management uses a top-down approach to continue the momentum. The impacts of management on the team affect a business as a whole. Specific examples must be used to convey the impacts to the team. Employees must never see managers as "mean" or only as decision makers. Employees must be engaged and management should ask frequently what can be done to enhance that engagement. Managers should see themselves not as power players but as top-down support using leadership to influence others.

Communication is a main aspect of management. Two-way communication is required. Employees must be able to confide in management about difficult situations; management must be able to explain expectations and timelines. Communication can boost positive morale throughout an organization. Positive reinforcement helps employees maintain structure and commitment and builds self-esteem. Positive attitudes are contagious.

Cost cutting is a tricky situation for management. If the need is demonstrated properly, employees will understand. If employees understand the company's direction they will understand the reasons for the cutbacks. Explaining that the culture will remain positive through the initiatives will keep the employees secure. Most importantly, the use by senior management of a top-down technique will build a successful and operational organizations and project controls to be sustained.

Management improvement is the factor that distinguishes quality management techniques in a Six Sigma effort. Project management is a key improvement tool for management to support and may involve managing other project managers to ensure ongoing support of projects. This control activity is critical to success. Project management at this level does not necessarily mean attending every team meeting; it serves as support structure to motivate employees to remove obstacles and take responsibility. Management support includes prioritizing projects by participation and by dedicating key employees to projects intended to grow the business. Management support also includes basic common sense techniques such as encouragement, understanding of processes, and knowledge of methods and analyses.

Leadership

Leadership is a key aspect of management. Leadership requires self direction to inspire, set goals, supervise, and empower relationships. An effective leader must have extraordinary communication skills. Communicating business plans, strategies, and objectives in a detailed manner conveys the top-down vision of a business. A leader must be responsible and never pass blame onto others. He or she should find the root cause of a problem and analyze the reasons for failures. Delegating responsibility for tasks also shows leadership capabilities. After a task is communicated, it is important for a leader to ensure the employees understood the task. The "five why" tool is useful at this point. Asking *why* five times drills down to the true understanding of a task. The answers reveal knowledge of the process, systems, and business.

Dr. W. E. Deming proposed the "Theory of Profound Knowledge" and said it was key for any business that desired to be competitive. According to Deming, for employees or a business to be successful, the system depends on management's ability to balance each component to optimize every system. The four categories of profound knowledge are:

- Systems (processes)
- Statistics (variation)
- Psychology (motivation)
- Knowledge

The knowledge of systems involves having outside management improve a system because a system has no knowledge. Quality management is the key for a system to ensure the end goal is met. If customers are not happy with the existing system, they can easily go to a competitor. Constantly improving a system by understanding its components will demonstrate continuous improvement.

Statistics knowledge reveals details of variations. Variation happens over time and should be predictable. Statistics shows different patterns and types of variations. The end goal is to reduce variation as much as possible by identifying common cause and special cause variations. Management should be aware of common cause variations and when they may occur.

The knowledge of psychology is important for leaders because they want to motivate employees, and all employees want to be motivated. If employees do good work, they want to be recognized. It is important for a leader to prevent employees from fearing that they will lose their jobs, their value, and their recognition. Leaders must demonstrate trust of their employees and other leaders and must ensure that employees trust one another.

The knowledge of knowledge consists of the facts and information that employees gain over time after they learn a job or a system. The knowledge can also be taught by others or through formal training. Knowledge should be transferred; no single person should hold all the knowledge. Many people should understand a system so that if a key stakeholder is absent, an entire process does not stop, fail, or incur more variation.

Deming's philosophy utilized (a) versus (b) comparisons: (a) when people and organizations focus primarily on quality, defined by:

$$\text{Quality} = \frac{\text{Results of work efforts}}{\text{Total costs}},$$

quality tends to increase and costs fall over time; (b) when people and organizations focus primarily on *costs*, costs tend to rise and quality declines over time. Deming developed a cycle for this principle from the work of Walter A. Shewhart, an American physicist, engineer, and statistician considered the "Father of Statistics." The cycle is called the Deming cycle or the Shewhart cycle and involves four steps: plan, do, check, act (PDCA). The PDCA steps for a project are planning the study or designing the experiment, performing the required steps, checking the results and testing the information, and acting on the data-driven results.

Shewhart developed a philosophy to increase productivity through statistical process control. Assume a process has upper and lower specifications of 615 and 585, respectively. The process has a mean of 600 and a standard deviation of 6. Management wants to know the percentage of nonconforming parts, the number of defects per million opportunities, and the sigma level. The formula is:

$$Z_u = \frac{\text{USL} - \text{x-bar}}{S} = \frac{615 - 600}{6} = 2.5$$

$$Z_l = \frac{\text{LSL} - \text{x-bar}}{S} = \frac{|585 - 600|}{6} = 2.5.$$

TABLE 10.1

Normal Table with Z Values and Tail Proportions

Z Value	Tail Proportion	Z Value	Tail Proportion	Z Value	Tail Proportion
2.41	0.0080	2.71	0.0034	3.10	0.00096760
2.42	0.0078	2.72	0.0033	3.20	0.00068714
2.43	0.0075	2.73	0.0032	3.30	0.00048342
2.44	0.0073	2.74	0.0031	3.40	0.00033693
2.45	0.0071	2.75	0.0030	3.50	0.00023263
2.46	0.0069	2.76	0.0029	3.60	0.00015911
2.47	0.0068	2.77	0.0028	3.70	0.00010780
2.48	0.0066	2.78	0.0027	3.80	0.00007235
2.49	0.0064	2.79	0.0026	3.90	0.00004810
2.50	0.0062	2.80	0.0026	4.00	0.00003167
2.51	0.0060	2.81	0.0025	4.10	0.00002066
2.52	0.0059	2.82	0.0024	4.20	0.00001335
2.53	0.0057	2.83	0.0023	4.30	0.00000854
2.54	0.0055	2.84	0.0023	4.40	0.00000541
2.55	0.0054	2.85	0.0022	4.50	0.00000340
2.56	0.0052	2.86	0.0021	4.60	0.00000211
2.57	0.0051	2.87	0.0021	4.70	0.00000130
2.58	0.0049	2.88	0.0020	4.80	0.00000079
2.59	0.0048	2.89	0.0019	4.90	0.00000048
2.60	0.0047	2.90	0.0019	5.00	0.00000029
2.61	0.0045	2.91	0.0018	5.10	0.00000017
2.62	0.0044	2.92	0.0018	5.20	0.00000010
2.63	0.0043	2.93	0.0017	5.30	0.00000006
2.64	0.0041	2.94	0.0016	5.40	0.00000003
2.65	0.0040	2.95	0.0016	5.50	0.00000020
2.66	0.0039	2.96	0.0015	6.00	0.00000001
2.67	0.0038	2.97	0.0015	6.50	0.00000000
2.68	0.0037	2.98	0.0014	7.00	0.00000000
2.69	0.0036	2.99	0.0014	7.50	0.00000000
2.70	0.0035	3.00	0.0013	8.00	0.00000000

The Z values in Table 10.1 are used to find tail proportions. Total nonconforming = 0.0062 + 0.0062 = 0.0124 or 1.24%. Defects per million opportunities = proportion nonconforming × 1,000,000 = 0.0124 × 1,000,000 = 12,400. Sigma level represents the smaller of two Z values which is a measure of the number of standard deviation measurements unit that will fit between the process mean and the nearer of the two specification limits. The 2.5 sigma level shown above indicates that the process does not conform to a satisfactory level. A conforming sigma level is 3 or more.

SigmSResource Allocation by Management

Resource allocation is one of the most difficult management decisions due to conflicting priorities. Management must ensure that proper resources are allocated to significant projects when they are needed. Projects should be prioritized by need, cost savings, and significance to the business to determine whether resources should be sourced out. A project must have key stakeholders, not entire teams of management. An essential and effective team consists of fewer than 10 team members with varied backgrounds. Normally a key stakeholder from each major department (operations, quality, engineering, etc.) should be utilized. A project champion should be assigned to remove barriers, motivate employees, ensure the project is on track, and provide resources. Champions should be fully trained and have an in-depth knowledge of the processes and measurement systems. Champions understand key quality measures and customer requirements. Champions support the business by finding projects with financial paybacks and high returns on investments. Quality should also be a key consideration to reduce the costs of non-quality results.

Key projects can be determined by performing Pareto analysis. This statistical tool implies that by doing 20% of the work, 80% of the advantage can be generated. When applied to quality, this philosophy states that 80% of problems stem from 20% of key causes. Pareto analyses are guides to prioritizing and determining key opportunities. Figure 10.1 illustrates a Pareto analysis.

The chart shows priorities and eliminates areas that exhibit low values. Trendlines can also be added to show the linear trending of a process as shown in Figure 10.2.

Design for Six Sigma (DFSS) techniques are also used for determining priorities. The phases for this are similar to DMAIC (define, measure, analyze, improve, control). The DFSS phases are define, measure, analyze, design,

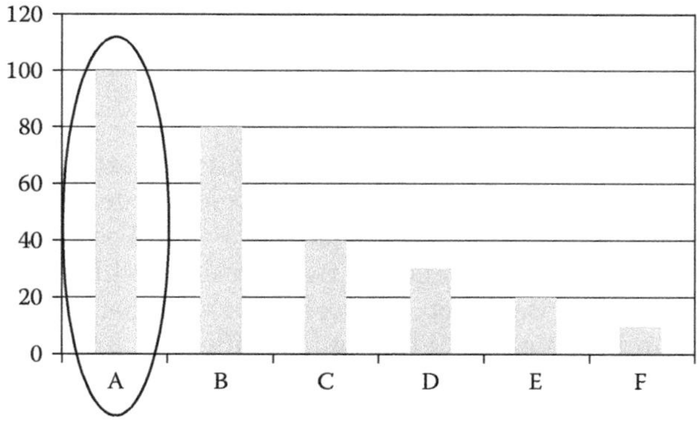

FIGURE 10.1
Pareto analysis for project prioritization.

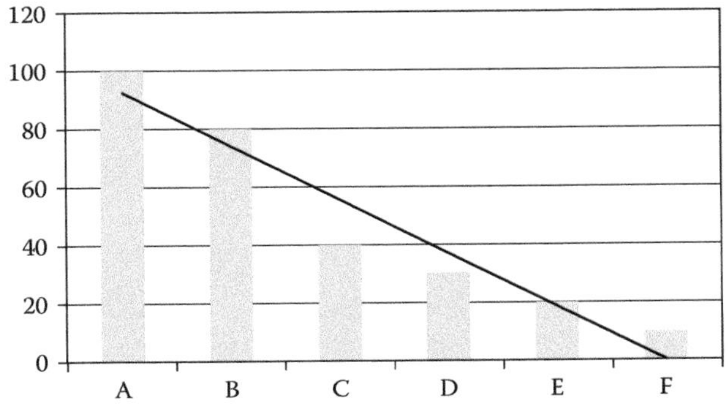

FIGURE 10.2
Trend-line for project analysis.

verify. DFSS techniques are used to establish proper customer requirements for successful Six Sigma performance.

Resource allocation can be performed using tools such as Microsoft Project. The main principle is to assign resources to particular projects based on tasks and subtasks with due dates. Certain tasks must be completed before others and should be listed on a complete timeline for a project. Resource allocation utilizes strategic methods to sustain a results-oriented culture. The steering committee of an organization normally allocates resources and selects projects, identifies Six Sigma green belts or black belts to utilize, monitors progress, and establishes implementation strategies and policies. Each project must be reviewed to ensure the steering committee is on the right track.

Over-allocation of resources must not occur. If one employee is involved in every project deemed a priority by the steering committee, the employee is over-allocated. A steering committee should spread resources evenly. The knowledge of knowledge principle dictates that no single resource should be a single stakeholder. Many individuals should participate to ensure that projects are ongoing and activities are completed normally.

Appendix: Useful Statistical Distributions

Industrial engineering uses statistical distributions and methods extensively in design and process improvement applications. The most common distributions are summarized in Table A.1.

Discrete Distributions

Probability mass function, $p(x)$

Mean, μ

Variance, σ^2

Coefficient of skewness, β_1

Coefficient of kurtosis, β_2

Moment-generating function, $M(t)$

Characteristic function, $\phi(t)$

Probability-generating function, $P(t)$

Bernoulli Distribution

$$p(x) = p^x q^{x-1} \qquad x = 0,1 \qquad 0 \le p \le 1 \qquad q = 1 - p$$

$$\mu = p \qquad \sigma^2 = pq \qquad \beta_1 = \frac{1-2p}{\sqrt{pq}} \qquad \beta_2 = 3 + \frac{1-6pq}{pq}$$

$$M(t) = q + pe^t \qquad \phi(t) = q + pe^{it} \qquad P(t) = q + pt$$

TABLE A.1

Summary of Common Statistical Distributions

Distribution of Random Variable x	Functional Form	Parameters	Mean	Variance	Range
Binomial	$P_x(k) = \frac{n!}{k!(n-k)!} p^k (1-p)^{n-k}$	n, p	np	$np(1-p)$	$0, 1, 2, \ldots n$
Poisson	$P_x(k) = \frac{\lambda^k e^{-\lambda}}{k!}$	λ	λ	λ	$0, 1, 2, \ldots, n$
Geometric	$P_x(k) = p(1-p)^{k-1}$	p	$1/p$	$\frac{1-p}{p^2}$	$1, 2, \ldots, n$
Exponential	$f_x(y) = \frac{1}{\theta} e^{-y/\theta}$	θ	θ	θ^2	$(0, \infty)$
Gamma	$f_x(y) = \frac{1}{\Gamma(\alpha)\beta^\alpha} y^{(\alpha-1)} e^{-y/\beta}$	α, β	$\alpha\beta$	$\alpha\beta^2$	$(0, \infty)$

Beta	$f_x(y) = \frac{\Gamma(\alpha+\beta)}{\Gamma(\alpha)\Gamma(\beta)} y^{(\alpha-1)}(1-y)^{(\beta-1)}$	α, β	$\frac{\alpha}{\alpha+\beta}$	$\frac{\alpha\beta}{(\alpha+\beta)^2(\alpha+\beta+1)}$	$(0, 1)$
Normal	$f_x(y) = \frac{1}{\sqrt{2\pi}\sigma} e^{-(y-\mu)^2/2\sigma^2}$	μ, σ	μ	σ^2	$(-\infty, \infty)$
Student t	$f_x(y) = \frac{1}{\sqrt{\pi v}} \frac{\Gamma\left(\frac{v+1}{2}\right)}{\Gamma(v/2)} (1+y^2/v)^{-(v+1)/2}$	v	0 for $v > 1$	$\frac{v}{v-2}$ for $v > 2$	$(-\infty, \infty)$
Chi square	$f_x(y) = \frac{1}{2^{v/2}\Gamma(v/2)} y^{(v-2)/2} e^{-y/2}$	v	v	$2v$	$(0, \infty)$
F	$f_x(y) = \frac{\Gamma\left(\frac{v_1+v_2}{2}\right) v_1^{v_1/2} v_2^{v_2/2}}{\Gamma\left(\frac{v_1}{2}\right)\Gamma\left(\frac{v_2}{2}\right)} \frac{(y)^{(v_1/2)-1}}{(v_2+v_1 y)^{\frac{v_1+v_2}{2}}}$	v_1, v_2	$\frac{v_2}{v_2-2}$ for $v_2>2$	$\frac{v_2^2(2v_2+2v_1-4)}{v_1(v_2-2)^2(v_2-4)}$ for $v_2>4$	$(0, \infty)$

Beta Binomial Distribution

$$p(x)=\frac{1}{n+1}\frac{B(a+x,b+n-x)}{B(x+1,n-x+1)B(a,b)} \qquad x=0,1,2,\ldots,n \qquad a>0 \qquad b>0$$

$$\mu=\frac{na}{a+b} \qquad \sigma^2=\frac{nab(a+b+n)}{(a+b)^2(a+b+1)}$$ $B(a,b)$ is the beta function.

Beta Pascal Distribution

$$p(x)=\frac{\Gamma(x)\Gamma(\nu)\Gamma(\rho+\nu)\Gamma(\nu+x-(\rho+r))}{\Gamma(r)\Gamma(x-r+1)\Gamma(\rho)\Gamma(\nu-\rho)\Gamma(\nu+x)} \qquad x=r,r+1,\ldots \quad \nu>p>0$$

$$\mu=r\frac{\nu-1}{\rho-1},\ \rho>1 \qquad \sigma^2=r(r+\rho-1)\frac{(\nu-1)(\nu-\rho)}{(\rho-1)^2(\rho-2)},\ \rho>2$$

Binomial Distribution

$$p(x)=\binom{n}{x}p^xq^{n-x} \quad x=0,1,2,\ldots,n \quad 0\le p\le 1 \quad q=1-p$$

$$\mu=np \qquad \sigma^2=npq \qquad \beta_1=\frac{1-2p}{\sqrt{npq}} \qquad \beta_2=3+\frac{1-6pq}{npq}$$

$$M(t)=\left(q+pe^t\right)^n \qquad \phi(t)=\left(q+pe^{it}\right)^n \qquad P(t)=(q+pt)^n$$

Discrete Weibull Distribution

$$p(x)=(1-p)^{x^\beta}-(1-p)^{(x+1)^\beta} \quad x=0,1,\ldots \quad 0\le p\le 1 \quad \beta>0$$

Geometric Distribution

$$p(x)=pq^{1-x} \quad x=0,1,2,\ldots \quad 0\le p\le 1 \quad q=1-p$$

$$\mu = \frac{1}{p} \qquad \sigma^2 = \frac{q}{p^2} \qquad \beta_1 = \frac{2-p}{\sqrt{q}} \qquad \beta_2 = \frac{p^2+6q}{q}$$

$$M(t) = \frac{p}{1-qe^t} \qquad \phi(t) = \frac{p}{1-qe^{it}} \qquad P(t) = \frac{p}{1-qt}$$

Hypergeometric Distribution

$$p(x) = \frac{\binom{M}{x}\binom{N-M}{n-x}}{\binom{N}{n}} \qquad x = 0,1,2,\ldots,n \quad x \le M \quad n-x \le N-M$$

$$n, M, N, \in N \qquad 1 \le n \le N \qquad 1 \le M \le N \qquad N = 1,2,\ldots$$

$$\mu = n\frac{M}{N} \qquad \sigma^2 = \left(\frac{N-n}{N-1}\right) n \frac{M}{N}\left(1-\frac{M}{N}\right) \qquad \beta_1 = \frac{(N-2M)(N-2n)\sqrt{N-1}}{(N-2)\sqrt{nM(N-M)(N-n)}}$$

$$\beta_2 = \frac{N^2(N-1)}{(N-2)(N-3)nM(N-M)(N-n)}$$

$$\left\{N(N+1) - 6n(N-n) + 3\frac{M}{N^2}(N-M)\left[N^2(n-2) - Nn^2 + 6n(N-n)\right]\right\}$$

$$M(t) = \frac{(N-M)!(N-n)!}{N!} F(.,e^t)$$

$$\phi(t) = \frac{(N-M)!(N-n)!}{N!} F(.,e^{it})$$

$$P(t) = \left(\frac{N-M}{N}\right)^n F(.,t)$$

$F(\alpha, \beta, \gamma, x)$ is the hypergeometric function. $\alpha = -n$; $\beta = -M$; $\gamma = N - M - n + 1$.

Negative Binomial Distribution

$$p(x) = \binom{x+r-1}{r-1} p^r q^x \qquad x = 0,1,2,\ldots \quad r = 1,2,\ldots \quad 0 \le p \le 1 \quad q = 1-p$$

$$\mu = \frac{rq}{p} \quad \sigma^2 = \frac{rq}{p^2} \quad \beta_1 = \frac{2-p}{\sqrt{rq}} \quad \beta_2 = 3 + \frac{p^2+6q}{rq}$$

$$M(t) = \left(\frac{p}{1-qe^t}\right)^r \quad \phi(t) = \left(\frac{p}{1-qe^{it}}\right)^r \quad P(t) = \left(\frac{p}{1-qt}\right)^r$$

Poisson Distribution

$$p(x) = \frac{e^{-\mu}\mu^x}{x!} \qquad x = 0,1,2,\ldots \qquad \mu > 0$$

$$\mu = \mu \quad \sigma^2 = \mu \quad \beta_1 = \frac{1}{\sqrt{\mu}} \quad \beta_2 = 3 + \frac{1}{\mu}$$

$$M(t) = \exp\left[\mu\left(e^t - 1\right)\right] \quad \sigma(t) = \exp\left[\mu\left(e^{it} - 1\right)\right] \quad P(t) = \exp\left[\mu(t-1)\right]$$

Rectangular (Discrete Uniform) Distribution

$$p(x) = 1/n \qquad x = 1,2,\ldots,n \qquad n \in N$$

$$\mu = \frac{n+1}{2} \quad \sigma^2 = \frac{n^2-1}{12} \quad \beta_1 = 0 \quad \beta_2 = \frac{3}{5}\left(3 - \frac{4}{n^2-1}\right)$$

$$M(t) = \frac{e^t\left(1-e^{nt}\right)}{n\left(1-e^t\right)} \quad \phi(t) = \frac{e^{it}\left(1-e^{nit}\right)}{n\left(1-e^{it}\right)} \quad P(t) = \frac{t\left(1-t^n\right)}{n(1-t)}$$

Continuous Distributions

Probability density function, $f(x)$
Mean, μ
Variance, σ^2
Coefficient of skewness, β_1

Coefficient of kurtosis, β_2
Moment-generating function, $M(t)$
Characteristic function, $\phi(t)$

Arcsin Distribution

$$f(x)=\frac{1}{\pi\sqrt{x(1-x)}} \qquad 0<x<1$$

$$\mu=\frac{1}{2} \qquad \sigma^2=\frac{1}{8} \qquad \beta_1=0 \qquad \beta_2\frac{3}{2}$$

Beta Distribution

$$f(x)=\frac{\Gamma(\alpha+\beta)}{\Gamma(\alpha)\Gamma(\beta)}x^{\alpha-1}(1-x)^{\beta-1} \qquad 0<x<1 \qquad \alpha,\beta>0$$

$$\mu=\frac{\alpha}{\alpha+\beta} \qquad \sigma^2=\frac{\alpha\beta}{(\alpha+\beta)^2(\alpha+\beta+1)} \qquad \beta_1=\frac{2(\beta-\alpha)\sqrt{\alpha+\beta+1}}{\sqrt{\alpha\beta}(\alpha+\beta+2)}$$

$$\beta_2=\frac{3(\alpha+\beta+1)\left[2(\alpha+\beta)^2+\alpha\beta(\alpha+\beta-6)\right]}{\alpha\beta(\alpha+\beta+2)(\alpha+\beta+3)}$$

Cauchy Distribution

$$f(x)=\frac{1}{b\pi\left(1+\left(\frac{x-a}{b}\right)^2\right)} \qquad -\infty<x<\infty \qquad -\infty<a<\infty \qquad b>0$$

μ, σ^2, β_1, β_2, $M(t)$ do not exist; $f(t)=\exp[ait-b|t|]$.

Chi Distribution

$$f(x)=\frac{x^{n-1}e^{-x^2/2}}{2^{(n/2)-1}\Gamma(n/2)} \qquad x\geq 0 \qquad n\in N$$

$$\mu = \frac{\Gamma\left(\frac{n+1}{2}\right)}{\Gamma\left(\frac{n}{2}\right)} \qquad \sigma^2 = \frac{\Gamma\left(\frac{n+2}{2}\right)}{\Gamma\left(\frac{n}{2}\right)} - \left[\frac{\Gamma\left(\frac{n+1}{2}\right)}{\Gamma\left(\frac{n}{2}\right)}\right]^2$$

Chi-Square Distribution

$$f(x) = \frac{e^{-x/2}x^{(\nu/2)-1}}{2^{\nu/2}\Gamma(\nu/2)} \qquad x \geq 0 \qquad \nu \in N$$

$$\mu = \nu \quad \sigma^2 = 2\nu \quad \beta_1 = 2\sqrt{2/\nu} \quad \beta_2 = 3 + \frac{12}{\nu} \quad M(t) = (1-2t)^{-\nu/2},\ t < \frac{1}{2}$$

$$\phi(t) = (1-2it)^{-\nu/2}$$

Erlang Distribution

$$f(x) = \frac{1}{\beta^{n(n-1)!}} x^{n-1}e^{-x/\beta} \qquad x \geq 0 \qquad \beta > 0 \qquad n \in N$$

$$\mu = n\beta \qquad \sigma^2 = n\beta^2 \qquad \beta_1 = \frac{2}{\sqrt{n}} \qquad \beta_2 = 3 + \frac{6}{n}$$

$$M(t) = (1-\beta t)^{-n} \qquad \phi(t) = (1-\beta it)^{-n}$$

Exponential Distribution

$$f(x) = \lambda e^{-\lambda x} \qquad x \geq 0 \qquad \lambda > 0$$

$$\mu = \frac{1}{\lambda} \qquad \sigma^2 = \frac{1}{\lambda^2} \qquad \beta_1 = 2 \qquad \beta_2 = 9 \qquad M(t) = \frac{\lambda}{\lambda - t}$$

$$\phi(t) = \frac{\lambda}{\lambda - it}$$

Extreme-Value Distribution

$$f(x)=\exp\left[-e^{-(x-\alpha)/\beta}\right] \qquad -\infty<x<\infty \qquad -\infty<\alpha<\infty \qquad \beta>0$$

$$\mu=\alpha+\gamma\beta,\quad \gamma \doteq .5772\ldots \text{ is Euler's constant } \sigma^2=\frac{\pi^2\beta^2}{6}$$

$$\beta_1=1.29857 \qquad \beta_2=5.4$$

$$M(t)=e^{\alpha t}\Gamma(1-\beta t),\quad t<\frac{1}{\beta} \qquad \phi(t)=e^{\alpha it}\Gamma(1-\beta it)$$

F Distribution

$$f(x)\frac{\Gamma\left[(\nu_1+\nu_2)/2\right]\nu_1^{\nu_1/2}\nu_2^{\nu_2/2}}{\Gamma(\nu_1/2)\Gamma(\nu_2/2)}x^{(\nu_1/2)-1}(\nu_2+\nu_1x)^{-(\nu_1+\nu_2)/2}$$

$$x>0 \qquad \nu_{1,}\ \nu_2\in N$$

$$\mu=\frac{\nu_2}{\nu_2-2},\ \nu_2\ge 3 \qquad \sigma^2=\frac{2\nu_2^{\ 2}(\nu_1+\nu_2-2)}{\nu_1(\nu_2-2)^2(\nu_2-4)},\quad \nu_2\ge 5$$

$$\beta_1=\frac{(2\nu_1+\nu_2-2)\sqrt{8(\nu_2-4)}}{\sqrt{\nu_1}(\nu_2-6)\sqrt{\nu_1+\nu_2-2}},\quad \nu_2\ge 7$$

$$\beta_2=3+\frac{12\left[(\nu_2-2)^2(\nu_2-4)+\nu_1(\nu_1+\nu_2-2)(5\nu_2-22)\right]}{\nu_1(\nu_2-6)(\nu_2-8)(\nu_1+\nu_2-2)},\quad \nu_2\ge 9$$

$M(t)$ does not exist. $$\phi\left(\frac{\nu_1}{\nu_2}t\right)=\frac{G(\nu_1,\nu_2,t)}{B(\nu_1/2,\nu_2/2)}$$

$B(a,b)$ is the beta function. G is defined by

$$(m+n-2)G(m,n,t)=(m-2)G(m-2,n,t)+2itG(m,n-2,t),\quad m,n>2$$

$$mG(m,n,t)=(n-2)G(m+2,n-2,t)-2itG(m+2,n-4,t),\quad n>4$$

$$nG(2,n,t)=2+2itG(2,n-2,t),\quad n>2$$

Gamma Distribution

$$f(x)=\frac{1}{\beta^{\alpha}\Gamma(\alpha)}x^{\alpha-1}e^{-x/\beta} \qquad x\ge 0 \qquad \alpha,\beta>0$$

$$\mu=\alpha\beta \qquad \sigma^2=\alpha\beta^2 \qquad \beta_1=\frac{2}{\sqrt{\alpha}} \qquad \beta_2=3\left(1+\frac{2}{\alpha}\right)$$

$$M(t)=(1-\beta t)^{-\alpha} \qquad \phi(t)=(1-\beta it)^{-\alpha}$$

Half-Normal Distribution

$$f(x)=\frac{2\theta}{\pi}\exp\left[-\left(\theta^2x^2/\pi\right)\right] \qquad x\ge 0 \qquad \theta>0$$

$$\mu=\frac{1}{\theta} \qquad \sigma^2=\left(\frac{\pi-2}{2}\right)\frac{1}{\theta^2} \qquad \beta_1=\frac{4-\pi}{\theta^3} \qquad \beta_2=\frac{3\pi^2-4\pi-12}{4\theta^4}$$

Laplace (Double Exponential) Distribution

$$f(x)=\frac{1}{2\beta}\exp\left[-\frac{|x-\alpha|}{\beta}\right] \qquad -\infty<x<\infty \qquad -\infty<\alpha<\infty \qquad \beta>0$$

$$\mu=\alpha \qquad \sigma^2=2\beta^2 \qquad \beta_1=0 \qquad \beta_2=6$$

$$M(t)=\frac{e^{\alpha t}}{1-\beta^2t^2} \qquad \phi(t)=\frac{e^{\alpha it}}{1+\beta^2t^2}$$

Logistic Distribution

$$f(x)=\frac{\exp[(x-\alpha)/\beta]}{\beta\left(1+\exp[(x-\alpha)/\beta]\right)^2}$$

$$-\infty<x<\infty \qquad -\infty<\alpha<\infty \qquad -\infty<\beta<\infty$$

$$\mu=\alpha \qquad \sigma^2=\frac{\beta^2\pi^2}{3} \qquad \beta_1=0 \qquad \beta_2=4.2$$

$$M(t)=e^{\alpha t}\pi\beta t\csc(\pi\beta t) \qquad \phi(t)=e^{\alpha it}\pi\beta it\csc(\pi\beta it)$$

Lognormal Distribution

$$f(x)=\frac{1}{\sqrt{2\pi}\sigma x}\exp\left[-\frac{1}{2\sigma^2}(1nx-\mu)^2\right]$$

$$x>0 \qquad -\infty<\mu<\infty \qquad \sigma>0$$

$$\mu=e^{\mu+\sigma^2/2} \qquad \sigma^2=e^{2\mu+\sigma^2}\left(e^{\sigma^2}-1\right)$$

$$\beta_1=\left(e^{\sigma^2}+2\right)\left(e^{\sigma^2}-1\right)^{1/2} \qquad \beta_2=\left(e^{\sigma^2}\right)^4+2\left(e^{\sigma^2}\right)^3+3\left(e^{\sigma^2}\right)^2-3$$

Noncentral Chi-Square Distribution

$$f(x)=\frac{\exp\left[-\frac{1}{2}(x+\lambda)\right]}{2^{\nu/2}}\sum_{j=0}^{\infty}\frac{x^{(\nu/2)+j-1\lambda j}}{\Gamma\left(\frac{\nu}{2}+j\right)2^{2j}j!}$$

$$x>0 \qquad \lambda>0 \qquad \nu\in N$$

$$\mu=\nu+\lambda \qquad \sigma^2=2(\nu+2\lambda) \qquad \beta_1=\frac{\sqrt{8}(\nu+3\lambda)}{(\nu+2\lambda)^{3/2}} \qquad \beta_2=3+\frac{12(\nu+4\lambda)}{(\nu+2\lambda)^2}$$

$$M(t)=(1-2t)^{-\nu/2}\exp\left[\frac{\lambda t}{1-2t}\right] \qquad \phi(t)=(1-2it)^{-\nu/2}\exp\left[\frac{\lambda it}{1-2it}\right]$$

Noncentral F Distribution

$$f(x)=\sum_{i=0}^{\infty}\frac{\Gamma\left(\frac{2i+\nu_1+\nu_2}{2}\right)\left(\frac{\nu_1}{\nu_2}\right)^{(2i+\nu_1)/2}x^{(2i+\nu_1-2)/2}e^{-\lambda/2}\left(\frac{\lambda}{2}\right)}{\Gamma\left(\frac{\nu_2}{2}\right)\Gamma\left(\frac{2i+\nu_1}{2}\right)\nu_1!\left(1+\frac{\nu_1}{\nu_2}x\right)^{(2i+\nu_1+\nu_2)/2}}$$

$$x>0 \qquad \nu_1,\nu_2\in N \qquad \lambda>0$$

$$\mu = \frac{(\nu_1 + \lambda)\nu_2}{(\nu_2 - 2)\nu_1}, \quad \nu_2 > 2$$

$$\sigma^2 = \frac{(\nu_1 + \lambda)^2 + 2(\nu_1 + \lambda)\nu_2^2}{(\nu_2 - 2)(\nu_2 - 4)v_1^2} - \frac{(\nu_1 + \lambda)^2 \nu_2^2}{(\nu_2 - 2)^2 \nu_1^2}, \quad \nu_2 > 4$$

Noncentral t Distribution

$$f(x) = \frac{\nu^{\nu/2}}{\Gamma\left(\frac{\nu}{2}\right)} \frac{e^{-\delta^2/2}}{\sqrt{\pi}\left(\nu + x^2\right)^{(\nu+1)/2}} \sum_{i=0}^{\infty} \Gamma\left(\frac{\nu + i + 1}{2}\right)\left(\frac{\delta^i}{i!}\right)\left(\frac{2x^2}{\nu + x^2}\right)^{i/2}$$

$$-\infty < x < \infty \qquad -\infty < \delta < \infty \qquad \nu \in N$$

$$\mu_r' = c_r \frac{\Gamma\left(\frac{\nu - r}{2}\right)\nu^{r/2}}{2^{r/2}\Gamma\left(\frac{\nu}{2}\right)}, \quad \nu > r, \qquad c_{2r-1} = \sum_{i=1}^{r} \frac{(2r-1)!\delta^{2r-1}}{(2i-1)!(r-i)!2^{r-i}},$$

$$c_{2r} = \sum_{i=0}^{r} \frac{(2r)!\delta^{2i}}{(2i)!(r-i)!2^{r-i}}, \qquad r = 1, 2, 3, \ldots$$

Normal Distribution

$$f(x) = \frac{1}{\sigma\sqrt{2\pi}} \exp\left[-\frac{(x-\mu)^2}{2\sigma^2}\right]$$

$$-\infty < x < \infty \qquad -\infty < \mu < \infty \qquad \sigma > 0$$

$$\mu = \mu \qquad \sigma^2 = \sigma^2 \qquad \beta_1 = 0 \qquad \beta_2 = 3 \qquad M(t) = \exp\left[\mu t + \frac{t^2\sigma^2}{2}\right]$$

$$\phi(t) = \exp\left[\mu i t - \frac{t^2\sigma^2}{2}\right]$$

Pareto Distribution

$$f(x) = \theta a^{\theta} / x^{\theta+1} \qquad x \geq a \qquad \theta > 0 \qquad a > 0$$

$$\mu = \frac{\theta a}{\theta - 1}, \quad \theta > 1 \qquad \sigma^2 = \frac{\theta a^2}{(\theta-1)^2(\theta-2)}, \quad \theta > 2$$

M(*t*) does not exist.

Rayleigh Distribution

$$f(x) = \frac{x}{\sigma^2}\exp\left[-\frac{x^2}{2\sigma^2}\right] \qquad x \geq 0 \qquad \sigma = 0$$

$$\mu = \sigma\sqrt{\pi/2} \qquad \sigma^2 = 2\sigma^2\left(1 - \frac{\pi}{4}\right) \qquad \beta_1 = \frac{\sqrt{\pi}}{4}\frac{(\pi - 3)}{\left(1 - \frac{\pi}{4}\right)^{3/2}}$$

$$\beta_2 = \frac{2 - \frac{3}{16}\pi^2}{\left(1 - \frac{\pi}{4}\right)^2}$$

t Distribution

$$f(x) = \frac{1}{\sqrt{\pi\nu}}\frac{\Gamma\left(\frac{\nu+1}{2}\right)}{\Gamma\frac{\nu}{2}}\left(1 + \frac{x^2}{\nu}\right)^{-(\nu+1)/2} \qquad -\infty < x < \infty \qquad \nu \in N$$

$$\mu = 0, \quad \nu \geq 2 \qquad \sigma^2 = \frac{\nu}{\nu - 2}, \quad \nu \geq 3 \qquad \beta_1 = 0, \quad \nu \geq 4$$

$$\beta_2 = 3 + \frac{6}{\nu - 4}, \quad \nu \geq 5$$

M(*t*) does not exist.

$$\phi(t) = \frac{\sqrt{\pi}\Gamma\left(\frac{\nu}{2}\right)}{\Gamma\left(\frac{\nu+1}{2}\right)}\int_{-\infty}^{\infty}\frac{e^{itz\sqrt{\nu}}}{\left(1+z^2\right)^{(\nu+1)/2}}dz$$

Triangular Distribution

$$f(x)=\begin{cases}0 & x\le a\\ 4(x-a)/(b-a)^2 & a<x\le(a+b)/2\\ 4(b-x)/(b-a)^2 & (a+b)/2<x<b\\ 0 & x\ge b\end{cases}$$

$$-\infty<a<b<\infty$$

$$\mu=\frac{a+b}{2}\qquad \sigma^2=\frac{(b-a)^2}{24}\qquad \beta_1=0\qquad \beta_2=\frac{12}{5}$$

$$M(t)=-\frac{4\left(e^{at/2}-e^{bt/2}\right)^2}{t^2(b-a)^2}\qquad \phi(t)=\frac{4\left(e^{ait/2}-e^{bit/2}\right)^2}{t^2(b-a)^2}$$

Uniform Distribution

$$f(x)=\frac{1}{b-a}\qquad a\le x\le b\qquad -\infty<a<b<\infty$$

$$\mu=\frac{a+b}{2}\qquad \sigma^2=\frac{(b-a)^2}{12}\qquad \beta_1=0\qquad \beta_2=\frac{9}{5}$$

$$M(t)=\frac{e^{bt}-e^{at}}{(b-a)t}\qquad \phi(t)=\frac{e^{bit}-e^{ait}}{(b-a)it}$$

Weibull Distribution

$$f(x)=\frac{\alpha}{\beta^{\alpha}}x^{\alpha-1}e^{-(x/\beta)^{\alpha}}\qquad x\ge 0\qquad \alpha,\beta>0$$

$$\mu=\beta\Gamma\left(1+\frac{1}{\alpha}\right)\qquad \sigma^2=\beta^2\left[\Gamma\left(1+\frac{2}{\alpha}\right)-\Gamma^2\left(1+\frac{1}{\alpha}\right)\right]$$

$$\beta_1 = \frac{\Gamma\left(1+\frac{3}{\alpha}\right) - 3\Gamma\left(1+\frac{1}{\alpha}\right)\Gamma\left(1+\frac{2}{\alpha}\right) + 2\Gamma^3\left(1+\frac{1}{\alpha}\right)}{\left[\Gamma\left(1+\frac{2}{\alpha}\right) - \Gamma^2\left(1+\frac{1}{\alpha}\right)\right]^{3/2}}$$

$$\beta_2 = \frac{\Gamma\left(1+\frac{4}{\alpha}\right) - 4\Gamma\left(1+\frac{1}{\alpha}\right)\Gamma\left(1+\frac{3}{\alpha}\right) + 6\Gamma^2\left(1+\frac{1}{\alpha}\right)\Gamma\left(1+\frac{2}{\alpha}\right) - 3\Gamma^4\left(1+\frac{1}{\alpha}\right)}{\left[\Gamma\left(1+\frac{2}{\alpha}\right) - \Gamma^2\left(1+\frac{1}{\alpha}\right)\right]^{2}}$$

Distribution Parameters

Average

$$\bar{x} = \frac{1}{n}\sum_{i=1}^{n} x_i$$

Variance

$$s^2 = \frac{1}{n-1}\sum_{i=1}^{n}(x_i - \bar{x})^2$$

Standard deviation

$$s = \sqrt{s^2}$$

Standard error

$$\frac{s}{\sqrt{n}}$$

Skewness (missing if $s = 0$ or $n < 3$)

$$\frac{n\sum_{i=1}^{n}(x_i - \bar{x})^3}{(n-1)(n-2)s^3}$$

Standardized skewness

$$\frac{\text{skewness}}{\sqrt{\frac{6}{n}}}$$

Kurtosis (missing if $s = 0$ or $n < 4$)

$$\frac{n(n+1)\sum_{i=1}^{n}(x_i-\bar{x})^4}{(n-1)(n-2)(n-3)s^4}-\frac{3(n-1)^2}{(n-2)(n-3)}$$

Standardized kurtosis

$$\frac{\text{Kurtosis}}{\sqrt{\dfrac{24}{n}}}$$

Weighted average

$$\frac{\sum_{i=1}^{n}x_iw_i}{\sum_{i=1}^{n}w_i}$$

Estimation and Testing

100 (1 – α)% confidence interval for mean

$$\bar{x}\ \pm\ t_{n-1;\alpha/2}\frac{s}{\sqrt{n}}$$

100 (1 – α)% confidence interval for variance

$$\left[\frac{(n-1)s^2}{\chi^2_{n-1;\alpha/2}},\ \frac{(n-1)s^2}{\chi^2_{n-1;1-\alpha/2}}\right]$$

100 (1 – α)% confidence interval for difference in means:

Equal variance

$$(\bar{x}_1-\bar{x}_2)\pm t_{n_1+n_2-2;\alpha/2}\ s_p\sqrt{\frac{1}{n_1}+\frac{1}{n_2}}$$

where

$$s_p=\sqrt{\frac{(n_1-1)s_1^2+(n_2-1)s_2^2}{n_1+n_2-2}}$$

Unequal variance

$$\left[(\bar{x}_1 - \bar{x}_2) \pm t_{m;\alpha/2}\sqrt{\frac{s_1^2}{n_1} + \frac{s_2^2}{n_2}}\right]$$

where

$$\frac{1}{m} = \frac{c^2}{n_1 - 1} + \frac{(1-c)^2}{n_2 - 1}$$

and

$$c = \frac{\dfrac{s_1^2}{n_1}}{\dfrac{s_1^2}{n_1} + \dfrac{s_2^2}{n_2}}$$

100 (1 – α)% confidence interval for ratio of variances

$$\left(\left(\frac{s_1^2}{s_2^2}\right)\left(\frac{1}{F_{n_1-1,n_2-1;\alpha/2}}\right), \left(\frac{s_1^2}{s_2^2}\right)\left(\frac{1}{F_{n_1-1,n_2-1;\alpha/2}}\right)\right)$$

Normal Probability Plot

The data are sorted from the smallest to the largest value to compute order statistics. A scatter plot is generated where:

Horizontal position = $x_{(i)}$

$$\text{Vertical position} = \Phi\left(\frac{i - 3/8}{n + 1/4}\right)$$

The labels for the vertical axis are based upon the probability scale using:

$$100\left(\frac{i - 3/8}{n + 1/4}\right)$$

Comparison of Poisson Rates

$$n_j = \#\text{of events in sample } j$$

$$t_j = \text{length of sample } j$$

Rate estimates: $r_j = \frac{n_j}{t_j}$

Rate ratio: $\frac{r_1}{r_2}$

Test statistic: $$z = \max\left(0, \frac{\left|n_1 - \frac{(n_1+n_2)}{2}\right| - \frac{1}{2}}{\sqrt{\frac{(n_1+n_2)}{4}}}\right)$$

where z follows the standard normal distribution.

Distribution Functions: Parameter Estimation

Bernoulli

$$\hat{p} = \bar{x}$$

Binomial

$$\hat{p} = \frac{\bar{x}}{n}$$

where n is the number of trials.

Discrete uniform

$$\hat{a} = \min x_i$$

$$\hat{b} = \max x_i$$

Geometric

$$\hat{p} = \frac{1}{1+\bar{x}}$$

Negative binomial

$$\hat{p} = \frac{k}{\bar{x}}$$

where k is the number of successes.

Poisson

$$\hat{\beta} = \bar{x}$$

Beta

$$\hat{\alpha} = \bar{x}\left[\frac{\bar{x}(1-\bar{x})}{s^2} - 1\right]$$

$$\hat{\beta} = (1-\bar{x})\left(\frac{\bar{x}(1-\bar{x})}{s^2} - 1\right)$$

Chi-square

$$\text{d.f. } \bar{v} = \bar{x}$$

Erlang

$$\hat{\alpha} = \text{round } (\hat{\alpha} \text{ from Gamma})$$

$$\hat{\beta} = \frac{\hat{\alpha}}{x}$$

Exponential

$$\hat{\beta} = \frac{1}{\bar{x}} \text{ (system displays } 1/\hat{\beta}\text{)}$$

F

$$\text{num d.f.: } \hat{v} = \frac{2\hat{w}^3 - 4\hat{w}^2}{\left(s^2\left(\hat{w}-2\right)^2\left(\hat{w}-4\right)\right) - 2\hat{w}^2}$$

$$\text{den. d.f.: } \hat{w} = \frac{\max(1, 2\bar{x})}{-1+\bar{x}}$$

Gamma

$$R = \log\left(\frac{\text{arithmetic mean}}{\text{geometric mean}}\right)(0.1)$$

If $0 < R \le 0.5772$,

$$\hat{\alpha} = R^{-1}(0.5000876 + 0.1648852\,R - 0.0544274\,R)^2$$

or if $R > 0.5772$,

$$\hat{\alpha} = R^{-1}\left(17.79728 + 11.968477\,R + R^2\right)^{-1}\left(8.898919 + 9.059950\,R + 0.9775373\,R^2\right)$$

$$\hat{\beta} = \hat{\alpha}/\bar{x}$$

This is an approximation of the method of maximum likelihood solution from Johnson and Kotz.

Log Normal

$$\hat{\mu} = \frac{1}{n}\sum_{i=1}^{n}\log x_i$$

$$\hat{\alpha} = \sqrt{\frac{1}{n-1}\sum_{i=1}^{n}\left(\log x_i - \hat{\mu}\right)^2}$$

System displays:

Mean: $\exp\left(\hat{\mu} + \hat{\alpha}^2/2\right)$

Standard deviation: $\sqrt{\exp\left(2\hat{\mu} + \hat{\alpha}^2\right)\left[\exp\left(\hat{\alpha}^2\right) - 1\right]}$

Normal

$$\hat{\mu} = \bar{x}$$

$$\hat{\sigma} = s$$

22

Student's t

If $s^2 \leq 1$ or if $\hat{\upsilon} \leq 2$, then the system indicates that the data are inappropriate.

$$s^2 = \frac{\sum_{i=1}^{n} x_i^2}{n}$$

$$\hat{\upsilon} = \frac{2s^2}{-1+s^2}$$

Triangular

$$\hat{a} = \min x_i$$

$$\hat{c} = \max x_i$$

$$\hat{b} = 3\overline{x} - \hat{a} - \overline{x}$$

Uniform

$$\hat{a} = \min x_i$$

$$\hat{b} = \max x_i$$

Weibull (for solving simultaneous equations)

$$\hat{\alpha} = \frac{n}{\left[\left(\frac{1}{\hat{\beta}}\right) \sum_{i=1}^{n} x_i^{\hat{a}} \log x_i - \sum_{i=1}^{n} \log x_i\right]}$$

$$\hat{\beta} = \left(\frac{\sum_{i=1}^{n} x_i^{\hat{\alpha}}}{n}\right)^{\frac{1}{\hat{\alpha}}}$$

Chi-square for Distribution Fitting

Divide the range of data into non-overlapping classes. The classes are aggregated at each end to ensure that they have expected frequencies of at least 5.

O_i = observed frequency in class i.
E_i = expected frequency in class i from fitted distribution.
k = number of classes after aggregation.

Test statistic

$$\chi^2 = \sum_{i=1}^{k} \frac{(O_i - E_i)^2}{E_i}$$

Follow a chi-square distribution with a degrees of freedom equal to $(k - 1)$ estimated parameters.

Kolmogorov-Smirnov Test

$$D_n^+ = \max\left\{\frac{i}{n} - \hat{F}(x_i)\right\}$$

$$D_n^- = \max\left\{\hat{F}(x_i) - \frac{i-1}{n}\right\}$$

$$1 \le i \le n$$

$$D_n = \max\left\{D_n^+, D_n^-\right\}$$

where $\hat{F}(x_i)$ = estimated cumulative distribution at x_i.

ANOVA

k = number of treatments

n_t = number of observations for treatment t

$\bar{n} = n/k =$ average treatment size

$$n = \sum_{t=1}^{k} n_t$$

x_{it} = ith observation in treatment t

$$\bar{x}_t = \text{treatment mean} = \frac{\sum_{i=1}^{n_t} x_{it}}{n_t}$$

$$s_t^2 = \text{treatment variance} = \frac{\sum_{i=1}^{n_t} (x_{it} - \bar{x}_t)^2}{n_t - 1}$$

$$\text{MSE} = \text{mean square error} = \frac{\sum_{t=1}^{k} (n_t - 1) s_t^2}{\left(\sum_{t=1}^{k} n_t\right) - k}$$

$$\text{df} = \text{degrees of freedom for the error term} = \left(\sum_{t=1}^{k} n_t\right) - k$$

Standard error (internal)

$$\sqrt{\frac{s_t^2}{n_t}}$$

Standard error (pooled)

$$\sqrt{\frac{MSE}{n_t}}$$

Interval estimate

$$\bar{x}_t \pm M\sqrt{\frac{MSE}{n_t}}$$

where

Confidence interval $M = t_{n-k;\,\alpha/2}$

LSD interval $M = \frac{1}{\sqrt{2}} t_{n-k;\alpha/2}$

Tukey interval

$$M = \frac{1}{2} q_{n-k,\,k;\,\alpha}$$

where $q_{n-k,\,k;\,\alpha}$ = the value of studentized range distribution with $n-k$ degrees of freedom and k samples such that the cumulative probability equals $1-\alpha$.

Scheffe interval

$$M = \frac{\sqrt{k-1}}{\sqrt{2}} \sqrt{F_{k-1,\,n-k;\,\alpha}}$$

Cochran C-test

Follow F distribution with $\bar{n}-1$ and $(\bar{n}-1)(k-1)$ degrees of freedom.

Test statistic: $$F = \frac{(k-1)C}{1-C}$$

where $$C = \frac{\max s_t^2}{\sum_{t=1}^{k} s_t^2}$$

Bartlett Test

Test Statistic: $$B = 10^{\frac{M}{(n-k)}}$$

$$M = (n-k)\log_{10} MSE - \sum_{t=1}^{k} (n_t - 1)\log_{10} s_t^2$$

The significance test is based on

$$\frac{M(1n\,10)}{1 + \frac{1}{3(k-1)}\left[\sum_{t=1}^{k} \frac{1}{(n_t-1)} - \frac{1}{N-k}\right]} \rceil X^2{}_{k-1}$$

which follows a chi-square distribution with $k-1$ degrees of freedom.

Hartley's Test

$$H = \frac{\max\left(s_t^2\right)}{\min\left(s_t^2\right)}$$

Kruskal-Wallis Test

Average rank of treatment

$$\bar{R}_t = \frac{\sum_{i=1}^{n_t} R_{it}}{n_t}$$

If there are no ties use test statistic

$$w = \left(\frac{12}{n}\sum_{i=1}^{k} n_t \bar{R}_t^{\,2}\right) - 3(n+1)$$

To adjust for ties, let u_j = number of observations tied at any rank for $j = 1, 2, 3, \ldots, m$ where m = number of unique values in the sample.

$$W = \frac{w}{1 - \dfrac{\sum_{j=1}^{m} u_j^3 - \sum_{j=1}^{m} u_j}{n\left(n^2 - 1\right)}}$$

Significance level: W follows a chi-square distribution with $k - 1$ degrees of freedom.

Freidman Test

X_{it} = observation in the ith row, tth column

$i = 1, 2, \ldots, n \quad t = 1, 2, \ldots, k$

R_{it} = rank of X_{it} within its row.

n = common treatment size (all treatment sizes must be the same for this test).

$$R_t = \sum_{i=1}^{n} R_{it}$$

Average rank $$\bar{R}_t = \frac{\sum_{i=1}^{n_t} R_{it}}{n_t}$$

where data are ranked within each row separately.

Test statistic $$Q = \frac{12S(k-1)}{nk(k^2-1)-\left(\sum u^3 - \sum u\right)}$$

where $$S = \left(\sum_{t=1}^{k} R_t^{\,2}\right) - \frac{n^2k(k+1)^2}{4}$$

Q follows a chi-square distribution with k degrees of freedom.

Regression

Notation

$\underset{\sim}{Y}$ = vector of n observation for the dependent variable ~ $\underset{\sim}{X}$ = n by p matrix of observations for p independent variables, including constant term, if any ~ indicates a variable is a vector or matrix.

$$\bar{Y} = \frac{\sum_{i=1}^{n} Y_i}{n}$$

Statistics

Estimated coefficients: estimated by modified Gram-Schmidt orthogonal decomposition with tolerance = 1/0 E − 0.08.

$$\underset{\sim}{b} = (\underset{\approx}{X}'\underset{\approx}{X})^{-1}\underset{\sim}{X}\underset{\sim}{Y}$$

Standard error

$$S(\underset{\sim}{b}) = \sqrt{\text{diagonal elements of } (\underset{\sim}{X}'\underset{\sim}{X})^{-1}\,\text{MSE}}$$

where
$$SSE = \underset{\sim}{Y}'\underset{\sim}{Y} - \underset{\sim}{b}'\underset{\sim}{X}'\underset{\sim}{Y}$$

$$MSE = \frac{SSE}{n-p}$$

t value

$$\underset{\sim}{t} = \frac{\underset{\sim}{b}}{S(\underset{\sim}{b})}$$

Significance level: t-values follow Student's t distribution with $n - p$ degrees of freedom.

R-squared

$$R^2 = \frac{SSTO - SSE}{SSTO} \quad \text{where } SSTO = \begin{cases} \underset{\sim}{Y}' - n\bar{Y}^2 & \text{if constant} \\ \underset{\sim}{Y}\underset{\sim}{Y} & \text{if no constant} \end{cases}$$

When the no constant option is selected, the total sum of square is uncorrected for the mean. Thus, the R^2 value is of little use, since the sum of the residuals is not zero.

Adjusted R²

$$1-\left(\frac{n-1}{n-p}\right)\left(1-R^2\right)$$

Standard error of estimate

$$SE = \sqrt{MSE}$$

Predicted value

$$\hat{\underset{\sim}{Y}} = \underset{\sim}{X}\underset{\sim}{b}$$

Residual

$$\underset{\sim}{e} = \underset{\sim}{Y} - \hat{\underset{\sim}{Y}}$$

Durbin-Watson statistic

$$D = \frac{\sum_{i=1}^{n-1}(e_{i+1} - e_i)^2}{\sum_{i=1}^{n} e_i^2}$$

Mean absolute error

$$\frac{\left(\sum_{i=1}^{n} |e_i|\right)}{n}$$

Predictions

$\underset{\sim}{X}_h = m$ by p matrix of independent variables for m predictions.

Predicted value

$$\underset{\sim}{\hat{Y}}_h = \underset{\sim}{X}_h \underset{\sim}{b}$$

Standard error of predictions

$$S\left(\underset{\sim}{\hat{Y}}_{h(new)}\right) = \sqrt{\text{diagonal elements of MSE } \left(1 + \underset{\sim}{X}_h \left(\underset{\sim}{X}'\underset{\sim}{X}\right)^{-1} \underset{\sim}{X}'_h\right)}$$

Standard error of mean response

$$S\left(\underset{\sim}{\hat{Y}}_h\right) = \sqrt{\text{diagonal elements of MSE } \left(\underset{\sim}{X}_h \left(\underset{\sim}{X}'\underset{\sim}{X}\right)^{-1} \underset{\sim}{\hat{X}}_h\right)}$$

Prediction matrix results

Column 1 = index numbers of forecasts

Column 2 = $\underset{\sim}{\hat{Y}}_h$

Column 3 = $S\left(\underset{\sim}{\hat{Y}}_{h(new)}\right)$

Column 4 = $\left(\underset{\sim}{\hat{Y}}_h - t_{n-p,\,\alpha/2}\, S\left(\underset{\sim}{\hat{Y}}_{h(new)}\right)\right)$

Column 5 = $\left(\underset{\sim}{\hat{Y}}_h + t_{n-p,\alpha/2}\, S\left(\underset{\sim}{\hat{Y}}_{h(new)}\right)\right)$

Column 6 = $\underset{\sim}{\hat{Y}}_h - t_{n-p,\alpha/2}\, S\left(\underset{\sim}{\hat{Y}}_h\right)$

Column 7 = $\underset{\sim}{\hat{Y}}_h + t_{n-p,\alpha/2}\, S\left(\underset{\sim}{\hat{Y}}_h\right)$

Nonlinear Regression

$F(X,\hat{\beta})$ are values of nonlinear function using parameter estimates $\hat{\beta}$.

Estimated coefficients are obtained by minimizing the residual sum of squares using a search procedure suggested by Marquardt. This is a compromise between Gauss-Newton and steepest descent methods. The user specifies:

1. Initial estimate $\underset{\sim}{\beta}_0$
2. Initial value of Marquardt parameter λ, which is modified at each iteration. (As $\lambda \to 0$, the procedure approaches Gauss-Newton $\lambda \to \infty$, the procedure approaches steepest descent.)
3. Scaling factor used to multiply Marquardt parameter after each iteration
4. Maximum value of Marquardt parameter

Partial derivatives of F with respect to each parameter are estimated numerically.

Standard error estimated from residual sum of squares and partial derivatives

Ratio = coefficient/standard error.

R^2

$$R^2 = \frac{\text{SSTO-SSE}}{\text{SSTO}} \quad \text{where}$$

where $$\text{SSTO} = \underset{\sim}{Y}'\underset{\sim}{Y} - n\bar{Y}^2$$

and $$\text{SSE} = \text{residual sum of squares}.$$

Ridge Regression

Additional notation:

$\underset{\sim}{Z}$ = matrix of independent variables standardized so that $\underset{\sim}{Z}'\underset{\sim}{Z}$ equals correlation matrix

θ = value of ridge parameter

Parameter estimate:

$$\underset{\sim}{b}(\theta) = \left(\underset{\sim}{Z}'\underset{\sim}{Z} + \theta I_p\right)^{-1} \underset{\sim}{Z}'\underset{\sim}{Y}$$

where I_p is a $p \times p$ identity matrix.

Quality Control

k = number of subgroups

n_j = number of observations in subgroup j

$j = 1, 2, \ldots, k$

x_{ij} = ith observation in subgroup j

All formulas below for quality control assume 3 σ limits. If other limits are specified, the formulas are adjusted proportionally based on sigma for the selected limits. Also, average sample size is used unless otherwise specified.

Subgroup Statistics

Mean

$$\bar{x}_j = \frac{\sum_{i=1}^{n_j} x_{ij}}{n_j}$$

Standard deviation

$$s_j = \sqrt{\frac{\sum_{i=1}^{n_j} \left(x_{ij} - \bar{x}_j\right)^2}{\left(n_j - 1\right)}}$$

Range

$$R_j = \max\left\{x_{ij} \middle| 1 \le i \le n_j\right\} - \min\left\{x_{ij} \middle| 1 \le i \le n_j\right\}$$

X Bar Charts

$$\bar{\bar{x}} = \frac{\sum_{j=1}^{k} n_i \bar{x}_j}{\sum_{j=1}^{k} n_i}$$

$$\bar{R} = \frac{\left(\sum_{j=1}^{k} n_i R_j\right)}{\sum_{j=1}^{k} n_i}$$

$$s_p = \sqrt{\frac{\sum_{j=1}^{k} (n_j - 1) s_j^2}{\sum_{j=1}^{k} (n_j - 1)}}$$

$$\bar{n} = \frac{1}{k} \sum_{j=1}^{k} n_i$$

For a chart based on range:

$$UCL = \bar{\bar{x}} + A_2 \bar{R}$$

$$LCL = \bar{\bar{x}} - A_2 \bar{R}$$

For a chart based on sigma:

$$UCL = \bar{\bar{x}} + \frac{3 s_p}{\sqrt{\bar{n}}}$$

$$LCL = \bar{\bar{x}} - \frac{3 s_p}{\sqrt{\bar{n}}}$$

For a chart based on known sigma:

$$UCL = \bar{\bar{x}} + 3 \frac{\sigma}{\sqrt{\bar{n}}}$$

$$LCL = \bar{\bar{x}} - 3 \frac{\sigma}{\sqrt{\bar{n}}}$$

If limits other than 3 σ are used, all bounds are adjusted proportionately. If average sample size is not used, then uneven bounds are based on $1/\sqrt{n_j}$ rather than $1/\sqrt{\bar{n}}$. If the data is normalized, each observation is transformed according to

$$z_{ij} = \frac{x_{ij} - \bar{x}}{\hat{\alpha}}$$

where $\hat{\alpha}$ = estimated standard deviation.

Capability Ratio

The following indices are useful only when control limits are placed at the specification limits. To override the normal calculations, specify a subgroup size of one and select the "known standard deviation" option. Then enter the standard deviation as half of the distance between the USL and LSL. Change the position of the center line to be the midpoint of the USL and LSL and specify the upper and lower control line at one sigma.

$$C_P = \frac{USL - LSL}{6\hat{\alpha}}$$

$$C_R = \frac{1}{C_P}$$

$$C_{PK} = \min\left(\frac{USL - \bar{x}}{3\hat{\alpha}}, \frac{\bar{x} - LSL}{3\hat{\alpha}}\right)$$

R Charts

$$CL = \bar{R}$$

$$UCL = D_4\bar{R}$$

$$LCL = Max\left(0, D_3\bar{R}\right)$$

S Charts

$$CL = s_P$$

$$UCL = s_P\sqrt{\frac{\chi^2_{\bar{n}-1;\alpha}}{\bar{n}-1}}$$

$$LCL = s_P\sqrt{\frac{\chi^2_{\bar{n}-1;\alpha}}{\bar{n}-1}}$$

C Charts

$$\bar{c} = \sum u_j \qquad UCL = \bar{c} + 3\sqrt{\bar{c}}$$

$$\sum n_j \qquad LCL = \bar{c} - 3\sqrt{\bar{c}}$$

where u_j = number of defects in *jth* sample.

U Charts

$$\bar{u} = \frac{\text{number of defects in all samples}}{\text{number of units in all samples}} = \frac{\sum u_j}{\sum n_j}$$

$$UCL = \bar{u} + \frac{3\sqrt{\bar{u}}}{\sqrt{\bar{n}}}$$

$$LCL = \bar{u} - \frac{3\sqrt{\bar{u}}}{\sqrt{\bar{n}}}$$

P Charts

$$p = \frac{\text{number of defective units}}{\text{number of units inspected}}$$

$$\bar{p} = \frac{\text{number of defectives in all samples}}{\text{number of units in all samples}} = \frac{\sum p_j n_j}{\sum n_j}$$

$$UCL = \overline{p} + \frac{3\sqrt{\overline{p}\left(1-\overline{p}\right)}}{\sqrt{\overline{n}}}$$

$$LCL = \overline{p} - \frac{3\sqrt{\overline{p}\left(1-\overline{p}\right)}}{\sqrt{\overline{n}}}$$

NP Charts

$$\overline{p} = \frac{\sum d_j}{\sum n_j},$$

where d_j is the number of defectives in *jth* sample

$$UCL = \overline{n}\,\overline{p} + 3\sqrt{\overline{n}\,\overline{p}\left(1-\overline{p}\right)}$$

$$LCL = \overline{n}\,\overline{p} - 3\sqrt{\overline{n}\,\overline{p}\left(1-\overline{p}\right)}$$

CuSum Chart for Mean

Control mean = μ
Standard deviation = α
Difference to detect = Δ
Plot cumulative sums C_t versus t where

$$C_t = \sum_{i=1}^{t} (\overline{x}_i - \mu) \quad \text{for } t = 1, 2, \ldots, n$$

The V-mask is located at distance

$$d = \frac{2}{}\left[\frac{\alpha^2/\overline{n}}{} 1n \frac{1-\beta}{\alpha/2}\right]$$

in front of the last data point

$$\text{Angle of mast } = 2\tan^{-1}\frac{}{2}$$

$$\text{Slope of line } = \pm\frac{}{2}$$

Multivariate Control Charts

$\underset{\sim}{X}$ = matrix of n rows and k columns containing n observations for each k variable

S = sample covariance matrix

$\underset{\sim}{X}_t$ = observation vector at time t

$\underset{\sim}{\bar{X}}$ = vector of column average

$$T_t^2 = \left(\underset{\sim}{X}_t - \underset{\sim}{\bar{X}}\right) S^{-1} \left(\underset{\sim}{X}_t - \underset{\sim}{\bar{X}}\right)$$

$$UCL = \left(\frac{k(n-1)}{n-k}\right) F_{k,n-k;\alpha}$$

Time Series Analysis

x_t or y_t = obsrvation at time t, $t = 1, 2, \ldots, n$

n = number of observations.

Autocorrelation at Lag *k*

$$r_k = \frac{c_k}{c_0}$$

where

$$c_k = \frac{1}{n}\sum_{t=1}^{n-k}(y_t - \bar{y})(y_{t+k} - \bar{y})$$

and

$$\bar{y} = \frac{\left(\sum_{t=1}^{n} y_t\right)}{n}$$

$$\text{Standard error } = \sqrt{\frac{1}{n}\left\{1 + 2\sum_{v=1}^{k-1} r_v^{\,2}\right\}}$$

Partial Autocorrelation at Lag *k*

$\hat{\theta}_{kk}$ is obtained by solving the Yule-Walker equations:

$$r_j = \hat{\theta}_{k1} r_{j-1} + \hat{\theta}_{k2} r_{j-2} + \cdots + \hat{\theta}_{k(k-1)} r_{j-k+1} + \hat{\theta}_{kk} r_{j-k}$$

$$j = 1, 2, \ldots, k$$

$$\text{Standard error } = \sqrt{\frac{1}{n}}$$

Cross Correlation at Lag *k*

x = input time series
y = output time series

$$r_{xy}(k) = \frac{c_{xy}(k)}{s_x s_y} \quad k = 0, \pm 1, \pm 2, \ldots$$

where

$$c_{xy}(k) = \begin{cases} \dfrac{1}{n}\displaystyle\sum_{t=1}^{n-k}(x_t - \bar{x})(y_{t+k} - \bar{y}) & k = 0, 1, 2, \ldots \\ \dfrac{1}{n}\displaystyle\sum_{t=1}^{n+k}(x_t - \bar{x})(y_{t-k} - \bar{y}) & k = 0, -1, -2, \ldots \end{cases}$$

and

$$S_x = \sqrt{c_{xx}(0)}$$

$$S_y = \sqrt{c_{yy}(0)}$$

Box-Cox

$$yt = \frac{(y + \lambda_2)^{\lambda_1} - 1}{\lambda_1 g^{(\lambda_1 - 1)}} \qquad \text{if } \lambda_1 > 0$$

$$yt = g\,1n(y + \lambda_2) \qquad \text{if } \lambda_1 = 0$$

where g = sample geometric mean $(y + \lambda_2)$.

Periodogram Using Fast Fourier Transform

If n is odd:

$$\mathrm{I}(\mathrm{f}_1) = \frac{n}{2}\left(a_i^2 + b_i^2\right) \qquad i = 1, 2, \ldots, \left[\frac{n-1}{2}\right]$$

where

$$a_i = \frac{2}{n}\sum_{t=1}^{n} t_t \cos 2\pi\, \mathrm{f}_i\, t$$

$$b_i = \frac{2}{n}\sum_{t=1}^{n} y_t \sin 2\pi\, \mathrm{f}_i\, t$$

$$\mathrm{f}_i = \frac{i}{n}$$

If n is even, an additional term is added:

$$\mathrm{I}(0.5) = n\left(\frac{1}{n}\sum_{t=1}^{n}(-1)^t\, Y_t\right)^2$$

Categorical Analysis

r = number of rows in table

c = number of columns in table

f_{ij} = frequency in position (row i, column j)

x_i = distinct values of row variable arranged in ascending order; $i = 1, \ldots, r$

y_j = distinct values of column variable arranged in ascending order, $j = 1, \ldots, c$

Totals:

$$R_j = \sum_{j=1}^{c} \mathrm{f}_{ij} \qquad C_j = \sum_{i=1}^{r} \mathrm{f}_{ij}$$

$$N = \sum_{i=1}^{r}\sum_{j=1}^{c} f_{ij}$$

Any row or column that totals zero is eliminated from the table before calculations are performed.

Chi-Square

$$\chi^2 = \sum_{i=1}^{r}\sum_{j=1}^{c} \frac{\left(f_{ij} - E_{ij}\right)^2}{E_{ij}}$$

where

$$E_{ij} = \frac{R_i C_j}{N} \sim \chi^2_{(r-1)(c-1)}$$

A warning is issued if any $E_{ij} < 2$ if 20% or more of all $E_{ij} < 5$. For 2 × 2 tables, a second statistic is printed using the Yates continuity correction.

Fisher's Exact Test

Run for a 2 x 2 table when N is less than or equal to 100. For calculation details, see standard references such as *The Analysis of Contingency Tables* by B. S. Everitt.

Lambda

$$\lambda = \frac{\left(\sum_{j=1}^{c} f_{\max,j} - R_{\max}\right)}{N - R_{\max}} \quad \text{with rows dependent}$$

$$\lambda = \frac{\left(\sum_{i=1}^{r} f_{i,\max} - C_{\max}\right)}{N - C_{\max}} \quad \text{with columns dependent}$$

$$\lambda = \frac{\left(\sum_{i=1}^{r} f_{i,\max} + \sum_{j=1}^{c} f_{\max,j} - C_{\max} - R_{\max}\right)}{\left(2N - R_{\max} - C_{\max}\right)} \quad \text{when symmetric}$$

where

$f_{i\max}$ = largest value in row i

$f_{\max j}$ = largest value in column j

$R_{\max}$ = largest row total

$C_{\max}$ = largest column total.

Uncertainty Coefficient

$$U_R = \frac{U(R)+U(C)-U(RC)}{U(R)} \quad \text{with rows dependent}$$

$$U_C = \frac{U(R)+U(C)-U(RC)}{U(C)} \quad \text{with columns dependent}$$

$$U = 2\left(\frac{U(R)+U(C)-U(RC)}{U(R)+U(C)}\right) \quad \text{when symmetric}$$

where

$$U(R) = -\sum_{i=1}^{r} \frac{R_i}{N} \log \frac{R_i}{N}$$

$$U(C) = -\sum_{j=1}^{c} \frac{C_j}{N} \log \frac{C_j}{N}$$

$$U(RC) = -\sum_{i=1}^{r} \sum_{j=1}^{c} \frac{f_{ij}}{N} \log \frac{f_{ij}}{N} \qquad \text{for } f_{ij} > 0.$$

Somers' D

$$D_R = \frac{2(P_C - P_D)}{\left(N^2 - \sum_{j=1}^{c} C_j^2\right)} \quad \text{with rows dependent}$$

$$D_C = \frac{2(P_C - P_D)}{\left(N^2 - \sum_{i=1}^{r} R_i^2\right)} \quad \text{with columns dependent}$$

$$D = \frac{4(P_C - P_D)}{\left(N^2 - \sum_{i=1}^{r} R_i^2\right) + \left(N^2 - \sum_{j=1}^{c} C_j^2\right)} \quad \text{when symmetric}$$

where the number of concordant pairs is

$$P_C = \sum_{i=1}^{r} \sum_{j=1}^{c} f_{ij} \sum_{h<i} \sum_{k<j} f_{hk}$$

and the number of discordant pairs is

$$P_D = \sum_{i=1}^{r} \sum_{j=1}^{c} f_{ij} \sum_{h<i} \sum_{k>j} f_{hk} .$$

Eta

$$E_R = \sqrt{1 - \frac{SS_{RN}}{SS_R}} \quad \text{with rows dependent}$$

where the total corrected sum of squares for the rows is

$$SS_R = \sum_{i=1}^{r} \sum_{j-1}^{c} x_i^2 f_{ij} - \frac{\left(\sum_{i=1}^{r} \sum_{j-1}^{c} x_i f_{ij}\right)^2}{N}$$

and the sum of squares of rows within categories of columns is

$$SS_{RN} = \sum_{j=1}^{c} \left(\sum_{i=1}^{r} x_i^2 f_{ij} - \frac{\left(\sum_{i=1}^{r} x_i^2 f_{ij}\right)^2}{C_j} \right)$$

$$E_C = \sqrt{1 - \frac{SS_{CN}}{SS_C}} \quad \text{with columns dependent}$$

where the total corrected sum of squares for the columns is

$$SS_C = \sum_{i=1}^{r}\sum_{j=1}^{c} y_i^2 f_{ij} - \frac{\left(\sum_{i=1}^{r}\sum_{j=1}^{c} y_i f_{ij}\right)^2}{N}$$

and the sum of squares of columns within categories of rows is

$$SS_{CN} = \sum_{i=1}^{r}\left(\sum_{j=1}^{c} y_i^2 f_{ij} - \frac{\left(\sum_{j=1}^{c} y_j^2 f_{ij}\right)^2}{R_i}\right)j$$

Contingency Coefficient

$$C = \sqrt{\frac{\chi^2}{(\chi^2 + N)}}$$

Cramer's V

$$V = \sqrt{\frac{\chi^2}{N}} \text{ for } 2 \times 2 \text{ table}$$

$$V = \sqrt{\frac{\chi^2}{N(m-1)}} \text{ for all others where } m = \min(r,c)$$

Conditional Gamma

$$G = \frac{P_C - P_D}{P_C + P_D}$$

Pearson's r

$$R = \frac{\sum_{j=1}^{c}\sum_{i=1}^{r} x_i y_j f_{ij} - \frac{\left(\sum_{j=1}^{c}\sum_{i=1}^{r} x_i f_{ij}\right)\left(\sum_{j=1}^{c}\sum_{i=1}^{r} y_i f_{ij}\right)}{N}}{\sqrt{SS_R SS_C}}$$

If R = 1, no significance is printed. Otherwise, the one-sided significance is based on

$$t = R\sqrt{\frac{N-2}{1-R^2}}$$

Kendall's Tau B

$$\tau = \frac{2(P_C - P_D)}{\sqrt{\left(N^2 - \sum_{i=1}^{r} R_i^2\right)\left(N^2 - \sum_{j=1}^{c} C_j^2\right)}}$$

Tau C

$$\tau_C = \frac{2m(P_C - P_D)}{(m-1)N^2}$$

Probability Terminology

Experiment: Activity or occurrence with observable result.

Outcome: Result of experiment.

Sample point: Outcome of experiment.

Event: Set of outcomes (subset of sample space) to which a probability is assigned.

Basic Probability Principles

Consider a random sampling process in which all the outcomes solely depend on chance, i.e., each outcome is equally likely to happen. If *S* is a uniform sample space and the collection of desired outcomes is E, the probability of the desired outcomes is

$$P(E) = \frac{n(E)}{n(S)}$$

where $n(E)$ = number of favorable outcomes in E and $n(S)$ = number of possible outcomes in S. Since E is a subset of S, $0 \leq n(E) \leq n(S)$, the probability of the desired outcome is $0 \leq P(E) \leq 1$.

Random Variable

A random variable is a rule that assigns a number to each outcome of a chance experiment as shown in these two examples:

1. A coin is tossed six times. The random variable X is the number of tails noted. X can only take the values 1, 2, …, 6, so X is a discrete random variable.
2. A light bulb remains on until it burns out. The random variable Y is its lifetime in hours. Y can take any positive real value, so Y is a continuous random variable.

Mean Value $\hat{x}$ or Expected Value μ

The mean value or expected value of a random variable indicates its average or central value. It is a useful summary value of a variable's distribution.

If random variable X is a discrete mean value,

$$\hat{x} = x_1 p_1 + x_2 p_2 + \ldots + x_n p_n = \sum_{i=1}^{n} x_1 p_1$$

where pi = probability densities.

If X is a continuous random variable with probability density function $f(x)$, then the expected value of X is:

$$\mu = E(X) = \int_{-\infty}^{+\infty} x f(x) dx$$

where $f(x)$ = probability densities.

Discrete Distribution Formulas

Probability mass function, $p(x)$
Mean, μ
Variance, σ^2
Coefficient of skewness, β_1
Coefficient of kurtosis, β_2
Moment-generating function, $M(t)$
Characteristic function, $\phi(t)$
Probability-generating function, $P(t)$

Bernoulli Distribution

$$p(x) = p^x q^{x-1} \qquad x = 0,1 \qquad 0 \le p \le 1 \qquad q = 1-p$$

$$\mu = p \qquad \sigma^2 = pq \qquad \beta_1 = \frac{1-2p}{\sqrt{pq}} \qquad \beta_2 = 3 + \frac{1-6pq}{pq}$$

$$M(t) = q + pe^t \qquad \phi(t) = q + pe^{it} \qquad P(t) = q + pt$$

Beta Binomial Distribution

$$p(x) = \frac{1}{n+1} \frac{B(a+x, b+n-x)}{B(x+1, n-x+1)B(a,b)} \qquad x = 0,1,2,\ldots,n \qquad a > 0 \qquad b > 0$$

$$\mu = \frac{na}{a+b} \qquad \sigma^2 = \frac{nab(a+b+n)}{(a+b)^2(a+b+1)}$$ $B(a,b)$ is the beta function.

Beta Pascal Distribution

$$p(x) = \frac{\Gamma(x)\Gamma(\nu)\Gamma(\rho+\nu)\Gamma(\nu+x-(\rho+r))}{\Gamma(r)\Gamma(x-r+1)\Gamma(\rho)\Gamma(\nu-\rho)\Gamma(\nu+x)} \qquad x = r, r+1, \ldots \quad \nu > p > 0$$

$$\mu = r\frac{\nu-1}{\rho-1},\ \rho > 1 \qquad \sigma^2 = r(r+\rho-1)\frac{(\nu-1)(\nu-\rho)}{(\rho-1)^2(\rho-2)},\ \rho > 2$$

Binomial Distribution

$$p(x)=\binom{n}{x}p^x q^{n-x} \quad x=0,1,2,\ldots,n \quad 0\le p\le 1 \quad q=1-p$$

$$\mu=np \qquad \sigma^2=npq \qquad \beta_1=\frac{1-2p}{\sqrt{npq}} \qquad \beta_2=3+\frac{1-6pq}{npq}$$

$$M(t)=\left(q+pe^t\right)^n \qquad \phi(t)=\left(q+pe^{it}\right)^n \qquad P(t)=\left(q+pt\right)^n$$

Discrete Weibull Distribution

$$p(x)=(1-p)^{x^\beta}-(1-p)^{(x+1)^\beta} \quad x=0,1,\ldots \quad 0\le p\le 1 \quad \beta>0$$

Geometric Distribution

$$p(x)=pq^{1-x} \quad x=0,1,2,\ldots \quad 0\le p\le 1 \quad q=1-p$$

$$\mu=\frac{1}{p} \qquad \sigma^2=\frac{q}{p^2} \qquad \beta_1=\frac{2-p}{\sqrt{q}} \qquad \beta_2=\frac{p^2+6q}{q}$$

$$M(t)=\frac{p}{1-qe^t} \qquad \phi(t)=\frac{p}{1-qe^{it}} \qquad P(t)=\frac{p}{1-qt}$$

Hypergeometric Distribution

$$p(x)=\frac{\binom{M}{x}\binom{N-M}{n-x}}{\binom{N}{n}} \quad x=0,1,2,\ldots,n \quad x\le M \quad n-x\le N-M$$

$$n,M,N,\in N \qquad 1\le n\le N \qquad 1\le M\le N \qquad N=1,2,\ldots$$

$$\mu=n\frac{M}{N} \quad \sigma^2=\left(\frac{N-n}{N-1}\right)n\frac{M}{N}\left(1-\frac{M}{N}\right) \quad \beta_1=\frac{(N-2M)(N-2n)\sqrt{N-1}}{(N-2)\sqrt{nM(N-M)(N-n)}}$$

$$\beta_2 = \frac{N^2(N-1)}{(N-2)(N-3)nM(N-M)(N-n)}$$

$$\left\{N(N+1)-6n(N-n)+3\frac{M}{N^2}(N-M)\left[N^2(n-2)-Nn^2+6n(N-n)\right]\right\}$$

$$M(t) = \frac{(N-M)!(N-n)!}{N!}F(.,e^t)$$

$$\phi(t) = \frac{(N-M)!(N-n)!}{N!}F(.,e^{it})$$

$$P(t) = \left(\frac{N-M}{N}\right)^n F(.,t)$$

$F(\alpha,\beta,\gamma,x)$ is the hypergeometric function. $\alpha = -n$; $\beta = -M$; $\gamma = N - M - n + 1$.

Negative Binomial Distribution

$$p(x) = \binom{x+r-1}{r-1} p^r q^x \quad x = 0,1,2,\ldots \quad r = 1,2,\ldots \quad 0 \le p \le 1 \quad q = 1-p$$

$$\mu = \frac{rq}{p} \quad \sigma^2 = \frac{rq}{p^2} \quad \beta_1 = \frac{2-p}{\sqrt{rq}} \quad \beta_2 = 3 + \frac{p^2+6q}{rq}$$

$$M(t) = \left(\frac{p}{1-qe^t}\right)^r \quad \phi(t) = \left(\frac{p}{1-qe^{it}}\right)^r \quad P(t) = \left(\frac{p}{1-qt}\right)^r$$

Poisson Distribution

$$p(x) = \frac{e^{-\mu}\mu^x}{x!} \quad x = 0,1,2,\ldots \quad \mu > 0$$

$$\mu = \mu \quad \sigma^2 = \mu \quad \beta_1 = \frac{1}{\sqrt{\mu}} \quad \beta_2 = 3 + \frac{1}{\mu}$$

$$M(t) = \exp\left[\mu\left(e^t - 1\right)\right] \quad \sigma(t) = \exp\left[\mu\left(e^{it} - 1\right)\right] \quad P(t) = \exp\left[\mu(t-1)\right]$$

Rectangular (Discrete Uniform) Distribution

$$p(x) = 1/n \qquad x = 1, 2, \ldots, n \qquad n \in N$$

$$\mu = \frac{n+1}{2} \qquad \sigma^2 = \frac{n^2-1}{12} \qquad \beta_1 = 0 \qquad \beta_2 = \frac{3}{5}\left(3 - \frac{4}{n^2-1}\right)$$

$$M(t) = \frac{e^t\left(1-e^{nt}\right)}{n\left(1-e^t\right)} \qquad \phi(t) = \frac{e^{it}\left(1-e^{nit}\right)}{n\left(1-e^{it}\right)} \qquad P(t) = \frac{t\left(1-t^n\right)}{n(1-t)}$$

Continuous Distribution Formulas

Probability density function, $f(x)$
Mean, μ
Variance, σ^2
Coefficient of skewness, β_1
Coefficient of kurtosis, β_2
Moment-generating function, $M(t)$
Characteristic function, $\phi(t)$

Arcsin Distribution

$$f(x) = \frac{1}{\pi\sqrt{x(1-x)}} \qquad 0 < x < 1$$

$$\mu = \frac{1}{2} \qquad \sigma^2 = \frac{1}{8} \qquad \beta_1 = 0 \qquad \beta_2 \frac{3}{2}$$

Beta Distribution

$$f(x) = \frac{\Gamma(\alpha+\beta)}{\Gamma(\alpha)\Gamma(\beta)} x^{\alpha-1}(1-x)^{\beta-1} \qquad 0 < x < 1 \qquad \alpha, \beta > 0$$

$$\mu = \frac{\alpha}{\alpha+\beta} \quad \sigma^2 = \frac{\alpha\beta}{(\alpha+\beta)^2(\alpha+\beta+1)} \quad \beta_1 = \frac{2(\beta-\alpha)\sqrt{\alpha+\beta+1}}{\sqrt{\alpha\beta}\,(\alpha+\beta+2)}$$

$$\beta_2 = \frac{3(\alpha+\beta+1)\left[2(\alpha+\beta)^2 + \alpha\beta(\alpha+\beta-6)\right]}{\alpha\beta(\alpha+\beta+2)(\alpha+\beta+3)}$$

Cauchy Distribution

$$f(x) = \frac{1}{b\pi\left(1+\left(\frac{x-a}{b}\right)^2\right)} \qquad -\infty < x < \infty \qquad -\infty < a < \infty \qquad b > 0$$

$$\mu, \sigma^2, \beta_1, \beta_2, M(t) \text{ do not exist.} \qquad \phi(t) = \exp\left[ait - b|t|\right]$$

Chi Distribution

$$f(x) = \frac{x^{n-1}e^{-x^2/2}}{2^{(n/2)-1}\Gamma(n/2)} \qquad x \geq 0 \qquad n \in N$$

$$\mu = \frac{\Gamma\left(\frac{n+1}{2}\right)}{\Gamma\left(\frac{n}{2}\right)} \qquad \sigma^2 = \frac{\Gamma\left(\frac{n+2}{2}\right)}{\Gamma\left(\frac{n}{2}\right)} - \left[\frac{\Gamma\left(\frac{n+1}{2}\right)}{\Gamma\left(\frac{n}{2}\right)}\right]^2$$

Chi-Square Distribution

$$f(x) = \frac{e^{-x/2}x^{(\nu/2)-1}}{2^{\nu/2}\Gamma(\nu/2)} \qquad x \geq 0 \qquad \nu \in N$$

$$\mu = \nu \quad \sigma^2 = 2\nu \quad \beta_1 = 2\sqrt{2/\nu} \quad \beta_2 = 3 + \frac{12}{\nu} \quad M(t) = (1-2t)^{-\nu/2},\ t < \frac{1}{2}$$

$$\phi(t) = (1-2it)^{-\nu/2}$$

Erlang Distribution

$$f(x)=\frac{1}{\beta^{n(n-1)!}}x^{n-1}e^{-x/\beta} \qquad x \geq 0 \qquad \beta>0 \qquad n \in N$$

$$\mu=n\beta \qquad \sigma^2=n\beta^2 \qquad \beta_1=\frac{2}{\sqrt{n}} \qquad \beta_2=3+\frac{6}{n}$$

$$M(t)=(1-\beta t)^{-n} \qquad \phi(t)=(1-\beta it)^{-n}$$

Exponential Distribution

$$f(x)=\lambda e^{-\lambda x} \qquad x \geq 0 \qquad \lambda>0$$

$$\mu=\frac{1}{\lambda} \qquad \sigma^2=\frac{1}{\lambda^2} \qquad \beta_1=2 \qquad \beta_2=9 \qquad M(t)=\frac{\lambda}{\lambda-t}$$

$$\phi(t)=\frac{\lambda}{\lambda-it}$$

Extreme Value Distribution

$$f(x)=\exp\left[-e^{-(x-\alpha)/\beta}\right] \qquad -\infty<x<\infty \qquad -\infty<\alpha<\infty \qquad \beta>0$$

$$\mu=\alpha+\gamma\beta, \quad \gamma \doteq .5772\ldots \text{ is Euler's constant } \sigma^2=\frac{\pi^2\beta^2}{6}$$

$$\beta_1=1.29857 \qquad \beta_2=5.4$$

$$M(t)=e^{\alpha t}\Gamma(1-\beta t), \quad t<\frac{1}{\beta} \qquad \phi(t)=e^{\alpha it}\Gamma(1-\beta it)$$

F Distribution

$$f(x)\frac{\Gamma[(\nu_1+\nu_2)/2]\nu_1^{\nu_1/2}\nu_2^{\nu_2/2}}{\Gamma(\nu_1/2)\Gamma(\nu_2/2)}x^{(\nu_1/2)-1}(\nu_2+\nu_1 x)^{-(\nu_1+\nu_2)/2}$$

$$x > 0 \qquad \nu_1, \nu_2 \in N$$

$$\mu = \frac{\nu_2}{\nu_2 - 2}, \nu_2 \geq 3 \qquad \sigma^2 = \frac{2\nu_2^{\,2}(\nu_1 + \nu_2 - 2)}{\nu_1(\nu_2 - 2)^2(\nu_2 - 4)}, \quad \nu_2 \geq 5$$

$$\beta_1 = \frac{(2\nu_1 + \nu_2 - 2)\sqrt{8(\nu_2 - 4)}}{\sqrt{\nu_1}(\nu_2 - 6)\sqrt{\nu_1 + \nu_2 - 2}}, \quad \nu_2 \geq 7$$

$$\beta_2 = 3 + \frac{12\left[(\nu_2 - 2)^2(\nu_2 - 4) + \nu_1(\nu_1 + \nu_2 - 2)(5\nu_2 - 22)\right]}{\nu_1(\nu_2 - 6)(\nu_2 - 8)(\nu_1 + \nu_2 - 2)}, \quad \nu_2 \geq 9$$

$M(t)$ does not exist.

$$\phi\left(\frac{\nu_1}{\nu_2}t\right) = \frac{G(\nu_1, \nu_2, t)}{B(\nu_1/2, \nu_2/2)}$$

$B(a,b)$ is the beta function. G is defined by

$$(m+n-2)G(m,n,t) = (m-2)G(m-2,n,t) + 2itG(m,n-2,t), \; m,n > 2$$

$$mG(m,n,t) = (n-2)G(m+2,n-2,t) - 2itG(m+2,n-4,t), \; n > 4$$

$$nG(2,n,t) = 2 + 2itG(2,n-2,t), \quad n > 2$$

Gamma Distribution

$$f(x) = \frac{1}{\beta^\alpha \Gamma(\alpha)} x^{\alpha-1} e^{-x/\beta} \qquad x \geq 0 \qquad \alpha, \beta > 0$$

$$\mu = \alpha\beta \qquad \sigma^2 = \alpha\beta^2 \qquad \beta_1 = \frac{2}{\sqrt{\alpha}} \qquad \beta_2 = 3\left(1 + \frac{2}{\alpha}\right)$$

$$M(t) = (1 - \beta t)^{-\alpha} \qquad \phi(t) = (1 - \beta it)^{-\alpha}$$

Half-Normal Distribution

$$f(x) = \frac{2\theta}{\pi}\exp\left[-\left(\theta^2 x^2 / \pi\right)\right] \qquad x \geq 0 \qquad \theta > 0$$

$$\mu = \frac{1}{\theta} \qquad \sigma^2 = \left(\frac{\pi - 2}{2}\right)\frac{1}{\theta^2} \qquad \beta_1 = \frac{4 - \pi}{\theta^3} \qquad \beta_2 = \frac{3\pi^2 - 4\pi - 12}{4\theta^4}$$

Laplace (Double Exponential) Distribution

$$f(x) = \frac{1}{2\beta}\exp\left[-\frac{|x - \alpha|}{\beta}\right] \qquad -\infty < x < \infty \qquad -\infty < \alpha < \infty \qquad \beta > 0$$

$$\mu = \alpha \qquad \sigma^2 = 2\beta^2 \qquad \beta_1 = 0 \qquad \beta_2 = 6$$

$$M(t) = \frac{e^{\alpha t}}{1 - \beta^2 t^2} \qquad \phi(t) = \frac{e^{\alpha i t}}{1 + \beta^2 t^2}$$

Logistic Distribution

$$f(x) = \frac{\exp\left[(x - \alpha)/\beta\right]}{\beta\left(1 + \exp\left[(x - \alpha)/\beta\right]\right)^2}$$

$$-\infty < x < \infty \qquad -\infty < \alpha < \infty \qquad -\infty < \beta < \infty$$

$$\mu = \alpha \qquad \sigma^2 = \frac{\beta^2 \pi^2}{3} \qquad \beta_1 = 0 \qquad \beta_2 = 4.2$$

$$M(t) = e^{\alpha t}\pi\beta t \csc(\pi\beta t) \qquad \phi(t) = e^{\alpha i t}\pi\beta i t \csc(\pi\beta i t)$$

Lognormal Distribution

$$f(x) = \frac{1}{\sqrt{2\pi}\sigma x}\exp\left[-\frac{1}{2\sigma^2}(1nx - \mu)^2\right]$$

$$x > 0 \qquad -\infty < \mu < \infty \qquad \sigma > 0$$

$$\mu = e^{\mu+\sigma^2/2} \qquad \sigma^2 = e^{2\mu+\sigma^2}\left(e^{\sigma^2}-1\right)$$

$$\beta_1 = \left(e^{\sigma^2}+2\right)\left(e^{\sigma^2}-1\right)^{1/2} \qquad \beta_2 = \left(e^{\sigma^2}\right)^4 + 2\left(e^{\sigma^2}\right)^3 + 3\left(e^{\sigma^2}\right)^2 - 3$$

Noncentral Chi-Square Distribution

$$f(x) = \frac{\exp\left[-\frac{1}{2}(x+\lambda)\right]}{2^{\nu/2}} \sum_{j=0}^{\infty} \frac{x^{(\nu/2)+j-1\lambda j}}{\Gamma\left(\frac{\nu}{2}+j\right)2^{2j}j!}$$

$$x > 0 \qquad \lambda > 0 \qquad \nu \in N$$

$$\mu = \nu + \lambda \qquad \sigma^2 = 2(\nu + 2\lambda) \qquad \beta_1 = \frac{\sqrt{8}(\nu+3\lambda)}{(\nu+2\lambda)^{3/2}} \qquad \beta_2 = 3 + \frac{12(\nu+4\lambda)}{(\nu+2\lambda)^2}$$

$$M(t) = (1-2t)^{-\nu/2}\exp\left[\frac{\lambda t}{1-2t}\right] \qquad \phi(t) = (1-2it)^{-\nu/2}\exp\left[\frac{\lambda it}{1-2it}\right]$$

Noncentral F Distribution

$$f(x) = \sum_{i=0}^{\infty} \frac{\Gamma\left(\frac{2i+\nu_1+\nu_2}{2}\right)\left(\frac{\nu_1}{\nu_2}\right)^{(2i+\nu_1)/2} x^{(2i+\nu_1-2)/2} e^{-\lambda/2}\left(\frac{\lambda}{2}\right)}{\Gamma\left(\frac{\nu_2}{2}\right)\Gamma\left(\frac{2i+\nu_1}{2}\right)\nu_1!\left(1+\frac{\nu_1}{\nu_2}x\right)^{(2i+\nu_1+\nu_2)/2}}$$

$$x > 0 \qquad \nu_1, \nu_2 \in N \qquad \lambda > 0$$

$$\mu = \frac{(\nu_1+\lambda)\nu_2}{(\nu_2-2)\nu_1}, \quad \nu_2 > 2$$

$$\sigma^2 = \frac{(\nu_1+\lambda)^2 + 2(\nu_1+\lambda)\nu_2^2}{(\nu_2-2)(\nu_2-4)v_1^2} - \frac{(\nu_1+\lambda)^2\nu_2^2}{(\nu_2-2)^2\nu_1^2}, \quad \nu_2 > 4$$

Noncentral t Distribution

$$f(x) = \frac{\nu^{\nu/2}}{\Gamma\left(\frac{\nu}{2}\right)} \frac{e^{-\delta^2/2}}{\sqrt{\pi}\left(\nu + x^2\right)^{(\nu+1)/2}} \sum_{i=0}^{\infty} \Gamma\left(\frac{\nu+i+1}{2}\right)\left(\frac{\delta^i}{i!}\right)\left(\frac{2x^2}{\nu+x^2}\right)^{i/2}$$

$$-\infty < x < \infty \qquad -\infty < \delta < \infty \qquad \nu \in N$$

$$\mu_r{}' = c_r \frac{\Gamma\left(\frac{\nu-r}{2}\right)\nu^{r/2}}{2^{r/2}\Gamma\left(\frac{\nu}{2}\right)}, \quad \nu > r, \qquad c_{2r-1} = \sum_{i=1}^{r} \frac{(2r-1)!\delta^{2r-1}}{(2i-1)!(r-i)!2^{r-i}},$$

$$c_{2r} = \sum_{i=0}^{r} \frac{(2r)!\delta^{2i}}{(2i)!(r-i)!2^{r-i}}, \qquad r = 1,2,3,\ldots$$

Normal Distribution

$$f(x) = \frac{1}{\sigma\sqrt{2\pi}} \exp\left[-\frac{(x-\mu)^2}{2\sigma^2}\right]$$

$$-\infty < x < \infty \qquad -\infty < \mu < \infty \qquad \sigma > 0$$

$$\mu = \mu \qquad \sigma^2 = \sigma^2 \qquad \beta_1 = 0 \qquad \beta_2 = 3 \qquad M(t) = \exp\left[\mu t + \frac{t^2\sigma^2}{2}\right]$$

$$\phi(t) = \exp\left[\mu i t - \frac{t^2\sigma^2}{2}\right]$$

Pareto Distribution

$$f(x) = \theta a^{\theta} / x^{\theta+1} \qquad x \ge a \qquad \theta > 0 \qquad a > 0$$

$$\mu = \frac{\theta a}{\theta - 1}, \quad \theta > 1 \qquad \sigma^2 = \frac{\theta a^2}{(\theta-1)^2(\theta-2)}, \qquad \theta > 2$$

$M(t)$ does not exist.

Rayleigh Distribution

$$f(x) = \frac{x}{\sigma^2}\exp\left[-\frac{x^2}{2\sigma^2}\right] \qquad x \geq 0 \qquad \sigma = 0$$

$$\mu = \sigma\sqrt{\pi/2} \qquad \sigma^2 = 2\sigma^2\left(1-\frac{\pi}{4}\right) \qquad \beta_1 = \frac{\sqrt{\pi}}{4}\frac{(\pi-3)}{\left(1-\frac{\pi}{4}\right)^{3/2}}$$

$$\beta_2 = \frac{2-\frac{3}{16}\pi^2}{\left(1-\frac{\pi}{4}\right)^2}$$

t-Distribution

$$f(x) = \frac{1}{\sqrt{\pi\nu}}\frac{\Gamma\left(\frac{\nu+1}{2}\right)}{\Gamma\frac{\nu}{2}}\left(1+\frac{x^2}{\nu}\right)^{-(\nu+1)/2} \qquad -\infty < x < \infty \qquad \nu \in N$$

$$\mu = 0, \quad \nu \geq 2 \qquad \sigma^2 = \frac{\nu}{\nu-2}, \quad \nu \geq 3 \qquad \beta_1 = 0, \quad \nu \geq 4$$

$$\beta_2 = 3 + \frac{6}{\nu-4}, \quad \nu \geq 5$$

$M(t)$ does not exist.

$$\phi(t) = \frac{\sqrt{\pi}\Gamma\left(\frac{\nu}{2}\right)}{\Gamma\left(\frac{\nu+1}{2}\right)}\int_{-\infty}^{\infty}\frac{e^{itz\sqrt{\nu}}}{\left(1+z^2\right)^{(\nu+1)/2}}dz$$

Triangular Distribution

$$f(x) = \begin{cases} 0 & x \leq a \\ 4(x-a)/(b-a)^2 & a < x \leq (a+b)/2 \\ 4(b-x)/(b-a)^2 & (a+b)/2 < x < b \\ 0 & x \geq b \end{cases}$$

$$-\infty < a < b < \infty$$

$$\mu = \frac{a+b}{2} \qquad \sigma^2 = \frac{(b-a)^2}{24} \qquad \beta_1 = 0 \qquad \beta_2 = \frac{12}{5}$$

$$M(t) = -\frac{4\left(e^{at/2} - e^{bt/2}\right)^2}{t^2(b-a)^2} \qquad \phi(t) = \frac{4\left(e^{ait/2} - e^{bit/2}\right)^2}{t^2(b-a)^2}$$

Uniform Distribution

$$f(x) = \frac{1}{b-a} \qquad a \le x \le b \qquad -\infty < a < b < \infty$$

$$\mu = \frac{a+b}{2} \qquad \sigma^2 = \frac{(b-a)^2}{12} \qquad \beta_1 = 0 \qquad \beta_2 = \frac{9}{5}$$

$$M(t) = \frac{e^{bt} - e^{at}}{(b-a)t} \qquad \phi(t) = \frac{e^{bit} - e^{ait}}{(b-a)it}$$

Weibull Distribution

$$f(x) = \frac{\alpha}{\beta^\alpha} x^{\alpha-1} e^{-(x/\beta)^\alpha} \qquad x \ge 0 \qquad \alpha, \beta > 0$$

$$\mu = \beta\Gamma\left(1+\frac{1}{\alpha}\right) \qquad \sigma^2 = \beta^2\left[\Gamma\left(1+\frac{2}{\alpha}\right) - \Gamma^2\left(1+\frac{1}{\alpha}\right)\right]$$

$$\beta_1 = \frac{\Gamma\left(1+\frac{3}{\alpha}\right) - 3\Gamma\left(1+\frac{1}{\alpha}\right)\Gamma\left(1+\frac{2}{\alpha}\right) + 2\Gamma^3\left(1+\frac{1}{\alpha}\right)}{\left[\Gamma\left(1+\frac{2}{\alpha}\right) - \Gamma^2\left(1+\frac{1}{\alpha}\right)\right]^{3/2}}$$

$$\beta_2 = \frac{\Gamma\left(1+\frac{4}{\alpha}\right) - 4\Gamma\left(1+\frac{1}{\alpha}\right)\Gamma\left(1+\frac{3}{\alpha}\right) + 6\Gamma^2\left(1+\frac{1}{\alpha}\right)\Gamma\left(1+\frac{2}{\alpha}\right) - 3\Gamma^4\left(1+\frac{1}{\alpha}\right)}{\left[\Gamma\left(1+\frac{2}{\alpha}\right) - \Gamma^2\left(1+\frac{1}{\alpha}\right)\right]^2}$$

Variate Generation Techniques*

Let $h(t)$ and $H(t) = \int_0^t h(\tau)\, d\tau$ be the hazard and cumulative hazard functions, respectively, for a continuous nonnegative random variable T, the lifetime of the item under study. The $q \times 1$ vector z contains covariates associated with a particular item or individual. The covariates are linked to the lifetime by the function $\Psi(z)$, which satisfies $\Psi(0 = 1)$ and $\Psi(z) \geq 0$ for all z. A popular choice is $\Psi(z) = e^{\beta' z}$, where β is a $q \times 1$ vector of regression coefficients. The cumulative hazard function for T in the *accelerated life* model (Cox and Oakes, 1984) is $H(t) = H_0\,(t\Psi(z))$ where H_0 is a baseline cumulative hazard function. Note that when $z = 0$, $H_0 \equiv H$. In this model, the covariates accelerate ($\Psi(z) > 1$) or decelerate ($\Psi(z) < 1$), the rate at which the item moves through time. The *proportional* hazards model $H(t) = \Psi(z)\, H_0\,(t)$ increases ($\Psi(z) > 1$) or decreases ($\Psi(z)\ < 1$) the failure rate of the item by the factor $\Psi(z)$ for all values of t.

Generation Algorithms

The literature shows that the cumulative hazard function, $H(T)$, has a unit exponential distribution. Therefore, a random variate t corresponding to a cumulative hazard function $H(t)$ can be generated by $t = H^{-1}(-\log(u))$ where u is uniformly distributed between 0 and 1. In the accelerated life model, since time is expanded or contracted by a factor $\Psi(z)$, variates are generated by:

$$t = \frac{H_0^{-1}\left(-\log(u)\right)}{\Psi(z)}.$$

In the proportional hazards model, equating $-\log(u)$ to $H(t)$ yields the variate generation formula:

$$t = H_0^{-1}\left(\frac{-\log(u)}{\Psi(z)}\right).$$

Table A.2 shows formulas for generating event times from a renewal or nonhomogeneous Poisson process. In addition to generating individual lifetimes, these variate generation techniques may also be applied to point processes. A renewal process, for example, with time between events having

* From Leemis, L.M. (1987). Variate generation for accelerated life and proportional hazards models, *Operations Research*, Vol. 35, No. 6, Nov–Dec 1987.

TABLE A.2
Formulas for Generating Event Times from Renewal or Nonhomogeneous Poisson Process

	Renewal	NHPP
Accelerated life	$t = a + \frac{H_0^{-1}(-\log(u))}{\Psi(z)}$	$t = \frac{H_0^{-1}(H_0(a\Psi(z)) - \log(u))}{\Psi(z)}$
Proportional hazards	$t = a + H_0^{-1}\left(\frac{-\log(u)}{\Psi(z)}\right)$	$t = H_0^{-1}\left(H_0(a) - \frac{\log(u)}{\Psi(z)}\right)$

a cumulative hazard function $H(t)$, can be simulated by using the appropriate generation formula for the two cases just shown. These variate generation formulas must be modified, however, to generate variates from a nonhomogeneous Poisson process (NHPP).

In a NHPP, the hazard function $H(t)$ is equivalent to the intensity function that governs the rate at which events occur. To determine the appropriate method to generate values from an NHPP, assume that the last even in a point process occurred at time a. The cumulative hazard function for the time of the next event conditioned on survival to time a is $H_{T|T>a}(t) = H(t) - H(a)$ $t > a$. In the accelerated life model in which $H(t) = H_0(t\Psi(z))$, the time of the next event is generated by

$$t = \frac{H_0^{-1}(H_0(a\Psi(z)) - \log(u))}{\Psi(z)}.$$

If we equate the conditional cumulative hazard function to $-\log(u)$, the time of the next event in the proportional hazards case is generated by:

$$t = H_0^{-1}\left(H_0(a) - \frac{\log(u)}{\Psi(z)}\right).$$

Example

The exponential power distribution (Smith and Bain, 1975) is a flexible two-parameter distribution with cumulative hazard function

$$H(t) = e^{(t/\alpha)^{\gamma}} - 1 \quad \alpha > 0, \quad \gamma > 0, \quad t > 0$$

and inverse cumulate hazard function

$$H^{-1}(y) = \alpha\left[\log(y+1)\right]^{1/\gamma}.$$

Assume that the covariates are linked to survival by the function $\Psi(z) = e^{\beta' z}$ in the accelerated life model. If an NHPP is to be simulated, the baseline hazard function has the exponential power distribution with parameters α and γ, and the previous event occurred at time a. The next event is generated at time:

$$t = \alpha e^{-\beta' z}\left[\log\left(e^{(ae^{\beta' z}/\alpha)^{\gamma}} - \log(u)\right)\right]^{1/\gamma},$$

where u is uniformly distributed between 0 and 1.

Index

D

E

J

K

L

M

Q

Milton Keynes UK
Ingram Content Group UK Ltd.
UKHW020320111024
449327UK00040B/1448

9 780367 783761